Civil Engineering Materials

Civil Engineering Materials

Introduction and Laboratory Testing

M. Rashad Islam

Colorado State University–Pueblo

CRC Press

Taylor & Francis Group

Boca Raton London New York

CRC Press is an imprint of the

Taylor & Francis Group, an **informa** business

CRC Press
Taylor & Francis Group
6000 Broken Sound Parkway NW, Suite 300
Boca Raton, FL 33487-2742

Printed on acid-free paper

International Standard Book Number-13: 978-0-367-22482-0 (Hardback)

Library of Congress Cataloging-in-Publication Data

Names: Islam, Md. Rashadul, author.
Title: Civil engineering materials : introduction and laboratory testing /
Md Rashadul Islam, Colorado State University--Pueblo.
Description: First edition. | Boca Raton, FL : CRC Press/Taylor & Francis
Group, 2020. | Includes bibliographical references and index.
Identifiers: LCCN 2019037009 (print) | LCCN 2019037010 (ebook) | ISBN
9780367224820 (hardback ; acid-free paper) | ISBN 9780429275111 (ebook)
Subjects: LCSH: Civil engineering--Materials. | Geotechnical
engineering--Materials. | Materials--Testing.
Classification: LCC TA403 .I835 2020 (print) | LCC TA403 (ebook) | DDC
624.028/4--dc23
LC record available at https://lccn.loc.gov/2019037009
LC ebook record available at https://lccn.loc.gov/2019037010

Visit the Taylor & Francis Web site at
http://www.taylorandfrancis.com

and the CRC Press Web site at
http://www.crcpress.com

Dedication

To you, for supporting this book.

Contents

Preface

This textbook is intended for civil engineering, construction engineering, civil engineering technology, construction management engineering technology, and construction management undergraduate programs. It is better if students studying this textbook have completed a mechanics or strength of materials class. If not, a brief revision of mechanics and strength of materials is given in the first chapter of this book. Ideally this book fits well in between a strength of materials and any design-related class such as reinforced concrete design, structural steel design, pavement design, etc. The author believes in simplicity of presentation for such an introductory-level text and avoids research ambiguities or a research focus.

This textbook discusses the properties, characterization procedures, and analysis techniques of primary civil engineering materials. Without gathering too much historical literature, this book focuses on the required properties, characterization methods, design considerations, and uses of common civil engineering materials. The required theories to understand the materials and to use it at senior levels are discussed here using a sufficient number of worked-out mathematical examples. In addition, cutting-edge practice topics are included, and obsolete topics are discarded in different chapters. The important laboratory tests are described step-by-step with high-quality figures. Analysis equations and their applications have been discussed with appropriate examples and relevant practice problems. The overall organization of the textbook is such that undergraduate students can follow without any problems. Fundamentals of Engineering (FE) styled questions are also included so that this book can help to prepare students for the FE examination.

The book has been thoroughly inspected with the help of professional editors to fix typos, editorial issues, and poor sentence structures. Despite this, if there are any issues, please report them at books.mrislam@gmail.com. Any suggestions to improve this book will be considered, and any issues reported will be corrected in the next edition with proper acknowledgment. Thank you.

M. Rashad Islam
Colorado State University–Pueblo

Acknowledgments

To begin, the author would like to thank God, the Ultimate Engineer, who taught him engineering, and gave him the opportunity to serve the engineering community, and present this book to civil and construction engineering students. The author then expresses his deep and sincere gratitude to his students at the Colorado State University–Pueblo and the State University of New York at Farmingdale. Without their encouragement and continuous support, the author believes, it would not have been possible to publish this book. The author also appreciates the valuable suggestions and good wishes of his departmental colleagues: Dr. Sylvester A. Kalevela, Prof. Kevin Sparks, Prof. Michael A. Mincic, Dr. Husam Alshareef and so on. Special thanks go to Prof. Sparks for his arrangements for photo collections while laboratory testing and external visits such as to the cement plant, ready-mix concrete plant, field visits, etc. His continuous support in every phase of writing this book was invaluable.

The author would also like to thank his wife and three kids who let him think, 'day and night–the same', 'weekdays and weekend–the same'.

Permissions from the American Concrete Institute (ACI), American Wood Council (AWC), American Institute for Steel Construction (AISC), ASTM International, Portland Cement Association (PCA), Test Mark Industries, Fujian South Highway Machinery, James Instruments, Shotcrete Services Ltd., Raysky Scientific Instruments, Dr. Andrew Braham from the University of Arkansas, Concrete Reinforcing Steel Institute (CRSI), Test Resources, and so on are highly appreciated. Without their tables and/or images, it would have been impossible to complete this textbook.

My students and good friends such as Dr. Andrew Braham, Dr. Mehedi Hasan, Mr. Nick Wibb, Ms. Shelby Nesselhauf, Mr. Kyle Witzman, and so on who helped by providing photos are also thanked. The reviewers who donated their valuable time to seek out issues, point out areas for improvement, and so on are especially appreciated. The author would like to especially acknowledge the following renowned reviewers:

- Dr. Amit Bandyopadhyay, a Distinguished Professor from the State University of New York at Farmingdale. Dr. Bandyopadhyay is a licensed professional engineer in New York and New Jersey and a fellow of the American Society of Civil Engineers. He received his PhD from Penn State University.
- Dr. Fawad S Niazi, an Assistant Professor from the Purdue University Fort Wayne. Dr. Niazi obtained his PhD from Georgia Institute of Technology, USA.
- Dr. Asif Ahmed, an Assistant Professor from the State University of New York Polytechnic Institute (SUNY Poly). Dr. Ahmed earned his PhD degree from the University of Texas at Arlington.
- Dr. Harvey Abramowitz, a Professor of Mechanical Engineering from Purdue University Northwest.
- Prof. Jose Pena, an Associate Professor of Civil Engineering Technology from Purdue University Northwest.
- Dr. Zahid Hossain, an Associate Professor of Civil Engineering from Arkansas State University. He earned his PhD from the University of Oklahoma.
- Dr. Aravind Krishna Swamy, an Associate Professor of Civil Engineering from the Indian Institute of Technology Delhi. He earned his PhD from the University of New Hampshire.

Author

M. Rashad Islam is an Assistant Professor and undergraduate program coordinator at Colorado State University–Pueblo and previously at the State University of New York at Farmingdale. He is also a registered Professional Engineer (PE) in Civil Engineering and an ABET Program Evaluator for Civil Engineering, Civil Engineering Technology, and Construction Management Engineering Technology undergraduate programs.

Dr. Islam received a PhD in Civil Engineering (with the grade of 'distinction' in the research, and a GPA of 4.0 out of 4.0 in the coursework) from the University of New Mexico, USA. He has a Master's degree in Structural Engineering jointly from the University of Minho (Portugal) and the Technical University of Catalonia (Barcelona, Spain). He also obtained a Bachelor's degree in Civil Engineering from the Bangladesh University of Engineering and Technology (BUET), which is undoubtedly the best school in Bangladesh.

Dr. Islam's research focuses on sustainable and innovative civil engineering materials for a new generation and he has more than 100 publications including several reference books, several textbooks, over 60 scholarly journal articles, and several technical reports and conference papers. Dr. Islam is a member of the American Society of Civil Engineers (ASCE) and is associated with the American General Contractors (AGC), among others.

1 Introduction

Chapter 1 is a review of the prior knowledge required to study civil engineering or construction engineering materials, their properties, characterization methods, and standard laboratory methods of testing. Materials used in a structure undergo different types of stresses, such as axial stress, bending stress, shear stress, and torsional stress, due to the applied loads by users, temperature fluctuations, wind, earthquake, rain, snow, and so on. This chapter reviews the knowledge of strength of materials which is required to study this book. The strength of materials is commonly used as prerequisite knowledge for the current subject area.

1.1 BACKGROUND

Civil engineering or construction engineering is involved with, to some extent, the planning, design, construction, maintenance, and recycling of roads, bridges, tunnels, canals, dams, buildings, airports, pipelines, etc. Different structures are made up of different materials, are exposed to different environments, and must support different loads. In addition, with the increase in materials and manpower cost, satisfying a project with a limited budget has become a challenge in recent years. Therefore, a 'one-size fits all' mentality does not work with materials engineering, especially in civil and construction engineering. The knowledge of these materials, including their origin, formation, physical properties, mechanical properties, decay behavior, etc., is crucial for selecting, processing, using, maintaining, and recycling these materials. The selection of materials for any of these projects is the key factor for feasibility, stability, safety, economy, production, services, maintenance, and aesthetics. Some basic civil engineering materials are mentioned below, as well as shown in Figure 1.1:

- Aggregates
- Cement
- Cement concrete
- Asphalt binder
- Asphalt concrete
- Steel
- Wood
- Masonry
- Glass
- Composites

In recent years, there has been tremendous growth in the use of innovative, new, and recycled materials in civil engineering. This textbook is an effort to describe the basic civil engineering materials in an organized way for undergraduate students and practicing civil and construction engineers.

FIGURE 1.1 Some civil engineering materials. *Photos taken in Pueblo, Colorado.*

1.2 TYPES OF MATERIALS

There are hundreds of materials available for use in civil engineering. Most materials fall into one of four broad classes, based on the atomic bonding forces. These are:

a) Metallic
b) Ceramic
c) Polymeric
d) Composite

The first three types are distinct to each other while the fourth one, composite material, is the combination or mixture of any two or all three. Each of the first three types can be further grouped based on their composition and properties.

1.2.1 METALS

Metals are the most abundant materials in the earth, accounting for about two-thirds of all the elements and about 24% of the mass of the earth (Askeland et al. 2011). Metals have a crystalline structure and are bonded by a metallic bond with free electrons that are free to move easily from one atom to another. These free electrons make metallic materials good electrical conductors. Metals have very useful mechanical properties, such as strength, ductility, toughness, high melting points, and thermal and electrical conductivity. Metallic materials can be further classified into the following two categories, considering the presence of ferrous materials or not:

• Ferrous metals and alloys, such as irons, carbon steels, alloy steels, stainless steels, etc.
• Nonferrous metals and alloys, such as aluminum, copper, magnesium, nickel, titanium, etc.

1.2.2 CERAMICS

Ceramics have a crystalline structure most commonly bonded by an ionic bond and formed between metallic and nonmetallic elements. They have very high strengths and hardnesses. However, they are mostly brittle in nature. Depending on the elements used in the crystalline structure, its properties vary greatly from one ceramic material to another. For example, two ceramic materials, glass and diamond, have very different properties. The broad categories of ceramics can be classified as:

- Cementitious materials, usually used for the manufacturing of concrete
- Glass materials, such as window glasses, containers, optical fibers, etc.
- Structural clays, which are used for bricks, sewer pipes, wall tiles, etc.
- Whitewares (dinnerware, floor and wall tiles, electrical porcelain, etc.)
- Refractories, which are used for glasses, cements, petroleum, etc.
- Abrasives materials, such as diamond, which are used for grinding, cutting, polishing, lapping, etc.

1.2.3 POLYMERIC

The word *polymer* means many parts. Polymers are formed with a long chain of covalent-bonded atoms, held together by secondary bonds to form the polymeric material. These are primarily produced from petroleum or natural gas, raw products, and organic substances. Most polymers are created by the combination of hydrogen and carbon atoms, and the arrangement of the chains they form. The polymer plastics can be divided into two classes:

- Thermoplastics
- Thermosetting

Thermoplastic polymers melt on heating. They thus can be processed and recycled by a variety of molding and extrusion techniques. An example of thermoplastic material is asphalt binder. *Thermosetting* polymers cannot be recycled or reused since they are unable to melt. Examples of thermosetting materials are resins, epoxies, and polyesters.

1.2.4 COMPOSITES

Composite materials are a combination of two or more basic materials (namely metals, ceramics, and polymers) forming their own distinctive properties. The combined properties of the new material cannot be achieved by any of the components alone. A wide range of civil engineering materials are composite materials. For example, reinforced concrete is a composite because it is a mixture of Portland cement, aggregate, and steel. Similarly, *asphalt concrete* is a composite material consisting of asphalt binder and aggregates.

1.3 REVIEWS OF MECHANICAL BEHAVIOR OF MATERIALS

1.3.1 BASICS OF TENSION AND COMPRESSION TESTS

Tension and/or compression tests are two of the most basic tests conducted on most of the civil engineering materials. Many mechanical properties of a material can be determined using these two tests. In a tension test, a specimen is clamped into a loading frame and then pulled gradually using a hydraulic loading system, as shown in Figure 1.2. The deformation of the specimen is measured continuously with the application of loading. There are many devices used to measure the deformation, such as a gage, extensometer, or linear

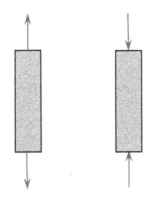

Tension Test   Compression Test

FIGURE 1.2   Tension and compression tests.

variable displacement transducer (LVDT). Stress is calculated by dividing the load by the cross-sectional area of the sample. Strain is calculated from the deformation divided by the initial dimension. Then, the stress–strain curve is plotted for the entire test until the failure occurs. The details of the stress–strain curve are discussed in the next subsection. In the compression test, the specimen is compressed uniaxially until the failure occurs. The properties of the materials can be determined following the computational methods of the tension test.

1.3.2   STRESS–STRAIN AND POISSON'S RATIO

Let us consider a compressive force of P applied on a cylindrical sample of initial diameter, D_o, initial length, L_o, and initial cross-sectional area, A_o. According to Newton's third law, every action has its own reaction of equal magnitude. Therefore, at any section of the body, the reaction force is equal to its applied force, as shown in Figure 1.3.

After applying the load, the body deforms (contracts) by ΔL ($\Delta L/2$ on both sides), and the diameter increases by ΔD ($\Delta D/2$ on both sides), as shown in Figure 1.4. The dashed area shows the initial shape and the solid area shows the final shape. *Engineering axial*

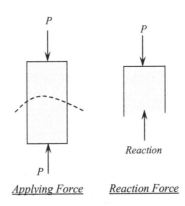

Applying Force   Reaction Force

FIGURE 1.3   Reaction force in a compression test.

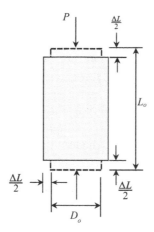

FIGURE 1.4   Deformation of a body after applying a compression load.

stress (σ_a), or simply *axial stress* (σ), is defined as the reaction force over the initial area. Thus, axial stress (σ_a) can be expressed as follows (Eq. 1.1):

$$\sigma_a = \frac{P}{A_o} \tag{1.1}$$

where,
   P = Axial force on the member
   A_o = Initial cross-sectional area of the member

Engineering axial strain, commonly known as *axial strain* (ε_a) or simply *strain* (ε), is defined as the change of length over the initial length. It can be expressed as shown in Eq. 1.2:

$$\varepsilon_a = \frac{\Delta L}{L_o} \tag{1.2}$$

where,
   ΔL = Change in length of the member
   L_o = Initial length of member

Transverse or *lateral strain* (ε_t) can be expressed as shown in Eq. 1.3:

$$\varepsilon_t = \frac{\Delta D}{D_o} \tag{1.3}$$

where,
   ΔD = Change in diameter of the member (final diameter minus the initial diameter)
   D_o = Initial diameter of the member

Poisson's Ratio (μ) is defined as the ratio of the transverse strain (ε_t) and the axial strain (ε_a), as shown in Eq. 1.4:

$$\mu = -\frac{\varepsilon_t}{\varepsilon_a} = -\frac{\dfrac{\Delta D}{D_o}}{\dfrac{\Delta L}{L_o}} = -\frac{\text{Lateral Strain}}{\text{Axial Strain}} \qquad (1.4)$$

From Figure 1.4, it is seen that if the length of the body decreases, the diameter increases (for a compressive force). If a tensile force is applied, the scenario is the opposite. This means, if the transverse strain is positive, the longitudinal strain is negative, and vice versa. Therefore, Poisson's ratio is always negative; however, the negative sign is typically omitted. Note that the transverse strain can also be written in terms of the radius or lateral dimension if the body is not cylindrical, as shown in Eq. 1.5.

$$\varepsilon_t = \frac{\Delta D}{D_o} = \frac{\Delta r}{r_o} = \frac{\Delta w}{w_o} = \frac{\Delta t}{t_o} \qquad (1.5)$$

where,

r_o = Initial radius of the member
Δr = Change in radius (final radius minus the initial radius) of the member
w_o = Initial width of the member
Δw = Change in width (final width minus the initial width) of the member
t_o = Initial thickness of the member
Δt = Change in thickness (final thickness minus the initial thickness) of the member

Note: Different books use different symbols to express the above parameters. You may choose your own symbols as well. This textbook uses the most commonly used symbols for better acceptability.

1.3.3 STRESS–STRAIN DIAGRAM OF A DUCTILE MATERIAL

Materials can be of two types, considering the amount of elongation before the failure: ductile and brittle. *Ductile* material (such as mild steel) shows significant plastic deformation before the failure. On the other hand, *brittle* material (such as glass) fails suddenly, without showing any significant plastic deformation. The schematic stress–strain (σ–ε) curve of a ductile material is shown in Figure 1.5. The diagram begins with a straight line from the origin, O, to point A. In this region, the stress is proportional to the strain. The slope at this region is constant, and is called the *modulus of elasticity*, or *Young's modulus*. The linear

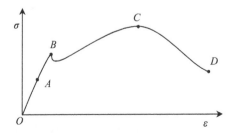

FIGURE 1.5 Stress–strain curve of a ductile material.

relationship between the stress and the strain at the initial stage of loading is expressed by Eqs. 1.6 and 1.7:

$$\sigma \propto \varepsilon \tag{1.6}$$

$$\sigma = E\varepsilon \tag{1.7}$$

where,

σ = Axial stress
ε = Axial strain
E = The modulus of elasticity or Young's modulus of the material

These equations are commonly referred to as *Hooke's law*. As the stress–strain is proportional up to point *A*, the stress at *A* is called the *proportional limit*. Hooke's law is valid up to the proportional limit.

Beyond the point *A*, the stress–strain relationship is non-linear, but still in the elastic region. Being in the elastic region means that, if a load is removed, the material will return to the initial position, i.e., the point, *O*.

However, after reaching point *B*, if the load is removed, the material will not return to its initial position. The starting point, *B*, of this behavior is called the *yield point*, or *yield stress*. After reaching the yield point, *strain hardening* occurs from points *B* to *C*, which means the stress increases with the strain due to molecular-level displacement inside the material. Eventually, the stress reaches the peak point, *C*, which is called the *ultimate stress*. After reaching the ultimate stress, a significant decrease in the cross-section occurs, stress goes down, and failure occurs by *necking*. This decrease in strain from the ultimate stress point, *C*, to the failure point, *D*, is called *strain softening*.

A few definitions can be made from Figure 1.5:

- *Modulus of Elasticity*: The slope of the stress–strain diagram up to the proportional limit (*A*).
- *Proportional Limit*: The linear portion of the stress–strain diagram (*O* to *A* region).
- *Elastic Limit*: The stress level up to which no plastic strain occurs (*B*).
- *Yield Point*: The stress level above which plastic strain occurs (*B*). Yield point is also known as the *Yield Stress* or *Yield Strength*.
- *Ultimate Stress*: The maximum stress a ductile material sustains before failing (*C*).
- *Fracture Stress*: The stress at which a material fails (*D*). Fracture stress is less than the ultimate stress for a ductile material.
- *Modulus of Resilience*: Area under the stress–strain curve up to the proportional limit.
- *Modulus of Toughness*: Area under the stress–strain curve up to the failure.
- *Strain Hardening*: For ductile material, after yielding, an increase in load can be supported until it reaches the maximum stress (ultimate stress). This rise in the stress after the yield point is called the strain hardening (region *B* to *C*).
- *Necking*: Upon applying tensile stress, the cross-sectional area of a ductile material decreases, forming a neck before the failure. This contraction of area before the failure is called the necking.
- *Strain Softening*: After the ultimate stress and before the failure point, the stress capacity keeps on decreasing with the continuous development of plastic strain due to significant molecular displacement in the material. This region is called the strain softening (region *C* to *D*).

- *Elastic* or *Recovery Strain*: The strain that is recovered fully upon the removal of the applied loading.
- *Permanent or Plastic Strain*: The strain that cannot be recovered in any amount upon removal of the applied loading.

Ductile material has the discussed four distinct behaviors: elastic, yielding, strain hardening, and necking (strain softening). Mild steel is a very common example of ductile material. Brittle material has a little or no yielding point, and fails very suddenly without showing any significant deformation. Glass is a very common example of brittle material.

Up to the proportional limit, the stain can be calculated using Eq. 1.8.

$$\varepsilon = \frac{\sigma}{E} \qquad (1.8)$$

This equation is used for a one-dimensional (1D) body, with a thin cable-type or rod-type material. In a 3D body, three-dimensional parameters are involved, as shown in Figure 1.6. The strain along the x-axis is dependent of the stress along all three perpendicular directions. Consider a rubber piece; if it is pulled, it elongates. If the rubber piece is compressed from the sides without being pulled, it also elongates.

For a 3D body, the strain can be calculated as

$$\varepsilon_x = \frac{1}{E}\Big[\sigma_x - \mu\big(\sigma_y + \sigma_z\big)\Big] \qquad (1.9)$$

$$\varepsilon_y = \frac{1}{E}\Big[\sigma_y - \mu\big(\sigma_x + \sigma_z\big)\Big] \qquad (1.10)$$

$$\varepsilon_z = \frac{1}{E}\Big[\sigma_z - \mu\big(\sigma_x + \sigma_y\big)\Big] \qquad (1.11)$$

where,

$\varepsilon_x, \varepsilon_y,$ and ε_z = The strains along the x, y, and z directions, respectively
$\sigma_x, \sigma_y,$ and σ_z = The stresses along the x, y, and z directions, respectively
μ = Poisson's ratio
E = The modulus of elasticity, commonly assumed isotropic (equal along the x, y, and z directions)

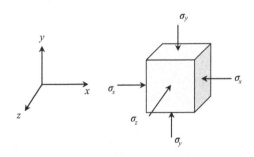

FIGURE 1.6   A three-dimensional body.

1.3.4 ELASTICITY, PLASTICITY, AND VISCOSITY

Elasticity is the property of a solid material to return to its initial shape and size immediately after the forces deforming it are removed. If the restoration to its initial shape and size takes time, then it is also a viscous property. Elasticity is not affected by the loading magnitude (small load or large load) or speed of application (fast or slow), as long as the material is within the elastic limit. Elasticity can be linear (stress–strain proportional) or non-linear. Non-linear deformation also returns to its initial shape and size upon releasing the load.

When an exterior stress is applied to a solid body, the body tends to pull itself apart. This causes the distance between atoms in the lattice to increase. Each atom tries to pull its neighbor as close as possible. This causes a force trying to resist the deformation. This resistive force over the area is known as *stress*. For equilibrium condition, resistive force equals the applied force. The total displacement of the atoms over the initial length of the sample is called the *strain*. If a graph of stress versus strain is plotted, the plot is linear for some lower values of strain. This linear area is the region where the object is deformed proportionally. The end stress of the proportional region is called the proportional limit. Hooke's law is valid up to the proportional limit. Hooke's law states that the proportional (elastic) range of the material stress is equal to the product of Young's modulus and the strain of the material (i.e., $\sigma = E\varepsilon$). Beyond the proportional limit, a material may also deform elastically non-linearly before reaching the yield point. The yield point is the point from where plastic deformation occurs. Sometimes, this non-linear region is not distinct if the very precise stress–strain data cannot be recorded. Any deformation before the yield point (linear or non-linear) returns upon releasing the load. The returning stress–strain line is parallel to the initial stress–strain curve. Any further strain which occurs after the yield point but does not return is known as the *plastic strain*. To summarize, elasticity may have two parts: linear (proportional) and non-linear. Both linear (proportional) and non-linear deformation return, following a parallel line to the initial stress–strain curve upon releasing the load. In the linear zone, stress is proportional to strain (i.e., $\sigma = E\varepsilon$). This is also known as Hooke's law. The yield point is the final point of the elastic region. Beyond the yield point, any further strain is plastic (permanent).

Plasticity is a property of a material allowing it to deform irreversibly, i.e., the deformation does not return to its initial position upon releasing the load. If a material has both the elastic and plastic properties, it deforms instantly upon applying a load. After releasing the load, the elastic part of the deformation returns to its initial position, while the plastic part does not. For a perfectly plastic material, if a load is applied, it deforms permanently and can never return to its initial state once the load is removed. If the stress–strain curve is plotted for an elasto-plastic material, after reaching the yield point (end of elastic region), plastic deformation begins. For a perfectly plastic material, there is no yield point; any deformation from the beginning is plastic deformation and does not recover upon removing the load.

Refer to Figure 1.7 to understand the elasticity and the plasticity concepts. Upon applying a compressive force, assume that the material's stress–strain (σ–ε) follows the *OAB* path. The point, *A*, is the yield point (σ_y, ε_y); the point, *B*, is the final position beyond the yield point the material reached, meaning the material experienced some plastic deformation (permanent deformation). Upon releasing the load, the material tries to go back to its initial position; however, only the *elastic strain* is recovered. Now, assume the material

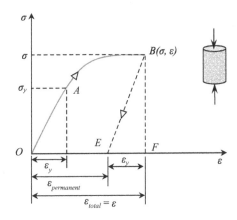

FIGURE 1.7   Elasto-plastic behavior of a material under compression.

follows the *BE* path upon releasing the load. Note that the *BE* path must be parallel to the initial part of the *OA* curve. The reason for this is that both the initial path and the returning path are elastic. At point *B*, the stress–strain is (σ, ε). When it comes to the point, *E*, the elastic strain (ε_y) is recovered. The rest of the strain that is not recovered is termed as the plastic (permanent) strain $(\varepsilon_{permanent} = \varepsilon - \varepsilon_y)$.

Another behavior of material, *viscosity*, is the time-dependent deformation behavior of material. In a viscous material, the material takes time to deform upon applying a load. Similarly, upon releasing the load, the material takes time to return to its initial position. If a material has both the elastic and viscous properties, the elastic part of the material deforms instantly upon applying a load, while the viscous part takes some time to deform. Similarly, after releasing the load, the elastic part of the deformation comes back to its initial position instantly, while the viscous part takes some time to restore.

A viscoelastic material can be thought of as a material with a response between that of a viscous material and an elastic material. A common example of a viscoelastic material is asphalt. If a viscoelastic material is held under constant strain, the level of stress decreases over a period of time. This is known as the *stress relaxation*. Recovery of strain and stress relaxation are different terms and should not be confused. A common example of stress relaxation is provided by the nylon strings in a tennis racket. We know that the level of stress, or the tension, within these things, decreases with time. If a constant stress is applied on a viscoelastic material, strain keeps on increasing over time. This property is called the *creep*. For example, if you stand on mud near a river, you continue to sink as time progresses.

The qualitative descriptions of the development of strain as a function of time for elastic, plastic, viscous, and their combinations are shown in Figure 1.8.

All these behaviors, elasticity, plasticity, and viscosity, play major roles in civil engineering materials. Ideally, no civil engineering material is perfectly elastic, plastic, or viscous. For example, mild steel, soil, aggregate layers, etc. are elasto-plastic; asphalt concrete is viscoelastic, etc. Note that the combined visco-elasto-plasto behavior of a material is still not well established. Visco-plastic behavior is also not very distinctive in civil engineering materials. Therefore, a civil engineering material can be categorized as either elasto-plastic or viscoelastic.

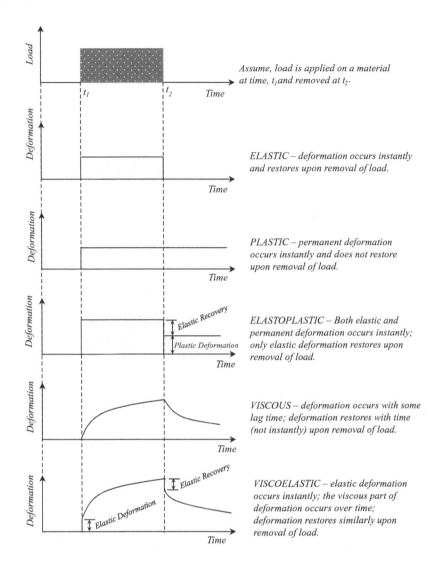

FIGURE 1.8 Stress–strain variations with time for different material behaviors.

Example 1.1 Poisson's Ratio

A tensile force of 100 kN is applied on a 0.02-m diameter and 2-m long rod. After applying the load, the diameter of the rod decreases to 0.01998 m and the length increases to 2.01 m. Determine Poisson's ratio of the material.

Solution

Given:

Applied load, $P = 100$ kN
Initial diameter, $D_o = 0.02$ m
Final diameter, $D = 0.01998$ m

Initial length, $L_o = 2$ m
Final length, $L = 2.01$ m

From Eq. 1.2: Axial strain, $\varepsilon_a = \dfrac{\Delta L}{L_o} = \dfrac{L - L_o}{L_o} = \dfrac{2.01\,\text{m} - 2.0\,\text{m}}{2.0\,\text{m}} = 0.005$

From Eq. 1.3: Lateral strain, $\varepsilon_t = \dfrac{\Delta D}{D_o} = \dfrac{D - D_o}{D_o} = \dfrac{0.01998\,\text{m} - 0.02\,\text{m}}{0.02\,\text{m}} = -0.001$

From Eq. 1.4: Poisson's Ratio, $\mu = \dfrac{\varepsilon_t}{\varepsilon_a} = \dfrac{-0.001}{0.005} = -0.2$

Answer: 0.2

Note: Poisson's Ratio of any material is always negative. However, the negative sign is commonly omitted.

Example 1.2 Modulus of Elasticity

A tensile force of 200 kN is applied on a 0.02-m diameter and 2-m long rod. After applying the load, the diameter of the rod decreases to 0.01998 m and the length increases to 2.01 m. Assuming the material is within the linear elastic region, determine the modulus of elasticity.

Solution

Given:

Applied load, $P = 200$ kN
Initial diameter, $D_o = 0.02$ m
Final diameter, $D = 0.01998$ m
Initial length, $L_o = 2$ m
Final length, $L = 2.01$ m

From Eq. 1.1: Axial stress, $\sigma_a = \dfrac{P}{A_o} = \dfrac{200\,\text{kN}}{\dfrac{\pi}{4}(0.02\,\text{m})^2} = 636{,}618\,\text{kPa}$

From Eq. 1.2: Axial strain, $\varepsilon_a = \dfrac{\Delta L}{L_o} = \dfrac{L - L_o}{L_o} = \dfrac{2.01\,\text{m} - 2.0\,\text{m}}{2.0\,\text{m}} = 0.005$

From Eq. 1.7: Modulus of elasticity, $E = \dfrac{\sigma_a}{\varepsilon_a} = \dfrac{637{,}618\,\text{kPa}}{0.005} = 127{,}523{,}600\,\text{kPa}$

Answer: 128 GPa

Note: 1 GPa $= 10^9$ Pa; 1 MPa $= 10^6$ Pa; 1 kPa $= 10^3$ Pa; 1 Pa $= 1$ N/m^2

Example 1.3 Change in Dimension

A steel rod is 0.02 m in diameter and 2.5 m in length. The modulus of elasticity of the material is 210 GPa, with a Poisson's ratio of 0.2. Calculate the change in diameter after applying a tensile force of 150 kN.

Solution

Given:

Initial diameter, $D_o = 0.02$ m
Initial length, $L_o = 2.5$ m
Modulus of elasticity, $E = 210$ GPa
Poisson's ratio, $\mu = 0.2$
Applied load, $P = 150$ kN

From Eq. 1.1: Axial stress, $\sigma_a = \dfrac{P}{A_o} = \dfrac{150\,\text{kN}}{\dfrac{\pi}{4}(0.02\,\text{m})^2} = 477,464\,\text{kPa}$

From Eq. 1.2: Axial strain, $\varepsilon_a = \dfrac{\sigma_a}{E} = \dfrac{477,464\,\text{kPa}}{210,000,000\,\text{kPa}} = 0.00227$

From Eq. 1.4: Lateral strain, $\varepsilon_t = -\mu\sigma_a = -0.2(0.00227) = -0.000454$

$$\Delta D = \varepsilon_t D_o = -0.000454(0.02\,\text{m})$$

From Eq. 1.5: Change in diameter,

$$= -0.000009\,\text{m} = -0.009\,\text{mm}$$

Answer: Contraction of 0.009 mm

Note: It is a very common practice to express the modulus of elasticity in ksi for the US customary unit and in GPa for the SI unit.

Example 1.4 Permanent Strain

For an A36 steel rod (yield stress of 36 ksi), a total of 0.011 strain occurs if a stress of 55 ksi is applied as shown in Figure 1.9. Determine the permanent strain for this stress level (55 ksi). The modulus of elasticity of steel is 29,000 ksi.

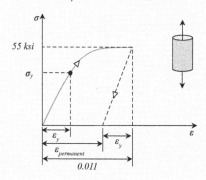

FIGURE 1.9 Elasto-plastic behavior of a material under tension for Example 1.4.

Solution

Given:

Modulus of elasticity, $E = 29 \times 10^3$ ksi

For the A36 steel, yield stress, $\sigma_y = 36$ ksi

From Eq. 1.7: Elastic/yield strain, $\varepsilon_y = \dfrac{\sigma_y}{E} = \dfrac{36 \text{ ksi}}{29 \times 10^3 \text{ ksi}} = 0.00124$

From Figure 1.9, $\varepsilon_{\text{Total}} = \varepsilon_y + \varepsilon_{\text{Permanent}}$

$$\varepsilon_{\text{Permanent}} = \varepsilon_{\text{Total}} - \varepsilon_y = 0.011 - 0.00124 = 0.0098$$

Answer: 0.0098

Note: the modulus of elasticity of steel at room temperature is always 29×10^3 ksi (210 GPa) for any kind of steel.

Example 1.5 Strain in 3D Body

On a cube element, a vertical compressive stress of 20 ksi is applied. The resulting lateral compressive stress is 12 ksi in both orthogonal lateral directions. The modulus of the material is 23,000 ksi and the Poisson's ratio is 0.25. Calculate the vertical compressive strain of the cube element.

Solution

Let us assume the compressive stress–strain is positive. The stress block is shown in Figure 1.10; the balancing stresses are not shown for simplicity.

From Eq. 1.11:

$$\varepsilon_z = \frac{1}{E}\left[\sigma_z - \mu\left(\sigma_x + \sigma_y\right)\right]$$

$$= \frac{1}{23,000 \text{ ksi}}\left[20 \text{ ksi} - 0.25\left(12 \text{ ksi} + 12 \text{ ksi}\right)\right] = 0.00061$$

Remember: $\varepsilon = \sigma/E$ is true for 1D bodies.

Answer: 0.00061

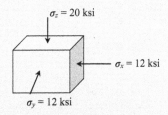

FIGURE 1.10 The three-dimensional body for Example 1.5.

Note: if you consider this example a 1-D body, then, $\varepsilon_z = \dfrac{\sigma_z}{E} = \dfrac{20\,\text{ksi}}{23{,}000\,\text{ksi}} = 0.00087$.

The error is $(0.00087 - 0.00061)/0.00061 = 0.43 = 43\%$. A rod can be considered a 1-D

body; a cube should be considered a 3-D body.

1.3.5 SHEAR TEST

Shear modulus or *modulus of rigidity* (G) is defined as the ratio of the shear stress and the shear strain. Consider the initial shape of a block is rectangular, as shown in Figure 1.11 by the dashed line. Upon applying a force, P, at the left-top corner, let the lateral displacement be Δ. Then, the change of angle is calculated as shown in Eq. 1.12.

$$\frac{\Delta}{L} = \tan\gamma \approx \gamma \tag{1.12}$$

$$\text{Shear stress, } \tau = P/A \tag{1.13}$$

$$\text{Shear strain, } \gamma = \text{Change in angle, which was initially } \pi/2. \tag{1.14}$$

$$G = \frac{\tau}{\gamma} \tag{1.15}$$

If the Poisson's ratio (μ) of the material is known, the shear modulus and the modulus of elasticity can be correlated as shown in Eq. 1.16.

$$G = \frac{E}{2(1+\mu)} \tag{1.16}$$

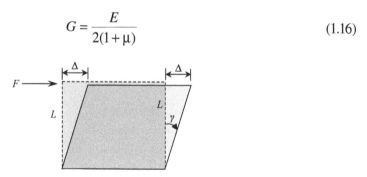

FIGURE 1.11 Shear stress–shear strain behavior under shear loading.

Example 1.6 Shear Strain

A piece of block is initially rectangular, as shown by the dashed lines in Figure 1.12. Determine the average shear strain at A, if the corners B and C are subjected to the displacements that cause the block to distort, as shown by the solid lines in Figure 1.12.

Solution

Let us assume the shear strains along the AC and AB members are θ_1 and θ_2 respectively, as shown in Figure 1.13.

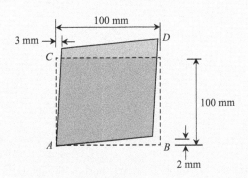

FIGURE 1.12   Shear deformation of a block for Example 1.6.

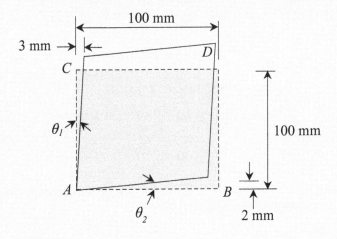

FIGURE 1.13   Analysis for Example 1.6.

From Figure 1.13, $\gamma = \theta_1 + \theta_2$

From Eq. 1.12:, $\dfrac{\Delta}{L} = \tan\gamma \approx \gamma$

Therefore, $\gamma = \theta_1 + \theta_2 = \dfrac{3\ mm}{100\ mm} + \dfrac{2\ mm}{100\ mm} = 0.03 + 0.02 = 0.05$ rad (squeezed)

Answer: 0.05 rad (squeezed)

1.3.6   FATIGUE TEST

Fatigue refers to the damage in a material due to applied cyclic/repeated loading. Cyclic loading is very common in civil engineering structures, e.g. vehicle loading on roads and bridges. Repeated cyclic loading to a stress (or strain) level less than the ultimate (or even the yield stress) can lead to failure. For example, bending a hair clip backward and forward a single time may not break it. However, the repetition of bending the clip a hundred times may break it. This is called the *fatigue failure*. The fatigue failure can be both sudden and brittle, although it occurs after many years of satisfactory service.

It is therefore potentially very dangerous for civil engineering structures. *Fatigue life* is defined as the number of cycles of loading leading to fatigue failure. It is not the time to failure, although this can of course be calculated if the frequency of loading is known. Another term commonly used in fatigue testing is called the *fatigue endurance limit*, which is the stress or the strain level at which no fatigue damage occurs. More specifically, it is the stress or the strain level at which, if a material undergoes cyclic loading, it never fails or is able to withstand a predetermined number of loadings. Figure 1.14 shows the fatigue testing setup for asphalt materials. The outer two clamps hold the asphalt beam specimen in position, and the middle two clamps force the beam down and/or up to produce repeated bending stress. The number of load repetitions to cause the failure is recorded.

a) Fatigue Test Device   b) Hardness Test Device

FIGURE 1.14   Fatigue and hardness test devices. *Photos taken at the University of New Mexico (a) and Colorado State University–Pueblo (b).*

1.3.7 HARDNESS TEST

Hardness is a surface characteristic of a material, not a fundamental physical property. It is defined as the surface resistance to indentation (penetrated by a small ball, as shown in Figure 1.14). For a certain load by a ball-type object, the smaller the indentation, the harder the material. Indentation hardness values are obtained by measuring the depth or the area of the indentation using different test methods. Chapter 9 discusses further the hardness test for steel materials.

1.3.8 IMPACT TEST

The *Charpy impact test* determines the amount of energy absorbed by a material until fracture. It is also known as *the Charpy V-notch test*. This absorbed energy is a measure of a given material's notch toughness. The test apparatus is mainly a pendulum of known mass and length. It is dropped from a known height to impact a notched specimen, as shown in Figure 1.15. The energy transferred to the material can be inferred by comparing the difference in the height of the hammer before and after the fracture (energy absorbed by the fracture event). Chapter 9 discusses further the impact test for steel material.

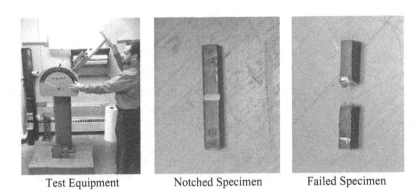

| Test Equipment | Notched Specimen | Failed Specimen |

FIGURE 1.15   Impact testing. *Photos taken at the Colorado State University–Pueblo.*

1.3.9   CREEP AND RELAXATION TESTS

Creep may be considered the opposite of fatigue. Fatigue is caused by a repeated load whereas creep is caused by a single load/stress, sustained on a body for a long time. However, creep occurs only in materials which have similar properties of viscous materials. Constant loading and/or stress is very common in civil engineering structures. For example, the stress due to the self-weight of a structure. Pure elastic or elasto-plastic materials show immediate deflection to a load. Viscoelastic materials respond to this stress by an immediate elastic strain deformation, followed by an increase in strain with time, being the creep. Typical behavior is illustrated in Figure 1.16. A stress applied at time t_1 and maintained at a constant level until removal at time t_2 results in:

- An initial elastic strain on the stress application, due to its elastic behavior
- An increase in this strain, due to creep, during the period of constant stress – rapid at first, but then at a decreasing rate
- An immediate elastic recovery occurs upon removal of stress, equal in magnitude to the initial elastic strain
- Further recovery with time (called the *creep recovery*), again at a decreasing rate and returns to its initial position with time

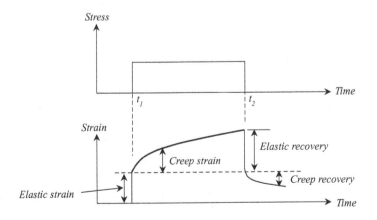

FIGURE 1.16   Creep-strain under constant stress.

If there is no plastic component in the material, then the material returns to its initial position with time. However, in reality, every viscoelastic material has some plastic component as well. Therefore, the recovery is normally less than the creep strain, meaning that the material does not return to zero position, i.e., there is some permanent deformation.

In some civil engineering structures, the applied strain is constant. For example, a cable is stretched between two fixed supports where constant strain occurs for a prolonged period. Similarly, a bolt connecting two metal plates together offers constant strain. The stress reduces with time, as shown in the Figure 1.17, if the material has some viscous property. This process is known as the *stress relaxation*. For pure viscous materials, the stress reduces to zero, i.e., the cable or bolt becomes limp.

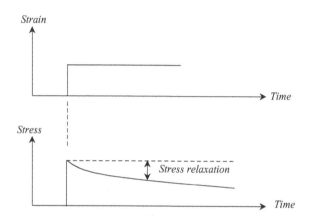

FIGURE 1.17   Stress-relaxation under a constant strain.

1.3.10   BENDING STRESS

Bending (moment) occurs if a structural member (beam) experiences a transverse load. When positive bending occurs in a member, it bends downward; the bottom fiber of the beam experiences a tensile stress, and the top fiber undergoes a compressive stress, as shown in Figure 1.18.

If the member is within the linear-elastic region, then the developed bending stress diagram along the section is linear with the maximum value at the extreme fiber, as shown in Figure 1.19.

For this case, y is the distance from the neutral axis (centroidal axis) to the point of interest. For a rectangular section, the neutral axis is located at the centerline of the beam. Thus, the neutral axis is located at an equal distance from the top or the bottom fibers. The developed bending stress is always linear and zero at the neutral axis, if the material is

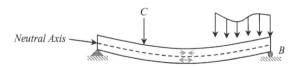

FIGURE 1.18   Arbitrary loading in a beam.

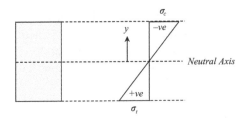

FIGURE 1.19 Bending stress in a rectangular beam section.

within the linear elastic region. The developed bending stress can be expressed as shown in Eq. 1.17.

$$\sigma_b = \frac{My}{I} \qquad (1.17)$$

where,

  σ_b = Developed bending stress
  M = Applied moment
  I = Moment of inertia of the section about the neutral axis
  y = Any distance from the neutral axis

The developed bending stress increases as the distance from the neutral axis increases; thus the maximum bending stress occurs at the extreme fiber of the beam. The maximum tensile bending stress can be calculated as shown in Eq. 1.18.

$$\sigma_b = \frac{Mc}{I} \qquad (1.18)$$

where c is the distance from the neutral axis to the most extreme fiber in the tension side, i.e., half of the depth for a rectangular section. Similarly, the maximum compressive bending stress can be calculated as shown in Eq. 1.18, where c is the distance from the neutral axis to the most extreme fiber in the compression side, i.e., half of the depth for a rectangular section.

In the above-mentioned bending stress equation, the terms I and c are constant for a section. These are geometric properties. A new term, the *section modulus* (S) is very often used as $S = \dfrac{I}{c}$. Then, the bending stress in a beam due to moment can be written as shown in Eq. 1.19.

$$\sigma_b = \frac{Mc}{I} = \frac{M}{\dfrac{I}{c}} = \frac{M}{S} \qquad (1.19)$$

where $S = \dfrac{I}{c}$ is the section modulus.

If the allowable bending stress, σ_{all}, is known, then the required section modulus (S_{req}) can be calculated as shown in Eq. 1.20.

$$S_{req} = \frac{M}{\sigma_{all}} \qquad (1.20)$$

Now, from the section catalogue of a design manual, a section can be chosen which has an S_{req} equal to or greater than that required. This is the design philosophy for bending in a beam. Sometimes, 'f_b' is used instead of 'σ_b' to express the developed bending stress.

Example 1.7 Bending Stress

For the following beam section shown in Figure 1.20, calculate the developed bending stress at the top fiber. The applied maximum moment is 100 kip.in., and the moment of inertia of the beam's cross-section about the neutral axis is 136 in.[4].

Solution

Applied maximum moment, $M = 100$ kip.in.
Moment of inertia of the section, $I = 136$ in.[4]
Distance from the neutral axis to the top fiber, $c = 3$ in.

From Eq. 1.18: The bending stress at the top fiber, $\sigma_b = \dfrac{Mc}{I} = \dfrac{100 \text{ kip.in.}(3 \text{ in.})}{136 \text{ in.}^4} = 2.2 \text{ ksi}$

Answer: 2.2 ksi

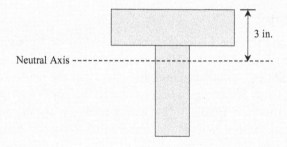

FIGURE 1.20   The cross-section of the beam for Example 1.7.

Example 1.8 Bending Stress

A beam section, as shown in Figure 1.21, has an applied maximum positive moment of 100 kip.ft. Determine the developed maximum compressive and tensile bending stresses in the section. The neutral axis is 3 in. below the top fiber and the moment of inertia about the neutral axis is 136 in.[4].

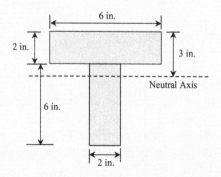

FIGURE 1.21   The cross-section of the beam for Example 1.8.

Solution

Maximum moment, $M = 100$ kip.ft $= 100$ kip.ft $\times 12$ in. per ft $= 1{,}200$ kip.in.

Distance from the neutral axis, $c = 3$ in. to the compression side (the top fiber), and
$c = 5$ in. to the tension side (the bottom fiber)

Moment of inertia, $I = 136$ in.^{4}

Positive moment causes compressive stress at the top fiber. So, the maximum compressive bending stress (from Eq. 1.18), $\sigma_b = \dfrac{Mc}{I} = \dfrac{1{,}200 \text{ kip.in.}(3 \text{ in.})}{136 \text{ in.}^4} = 26.5$ ksi

Positive moment causes tensile stress at the bottom fiber. Therefore, the maximum tensile bending stress (from Eq. 1.18), $\sigma_b = \dfrac{Mc}{I} = \dfrac{1{,}200 \text{ kip.in.}(5 \text{ in.})}{136 \text{ in.}^4} = 44.1$ ksi

Answers: 26.5 ksi (compressive), 44.1 ksi (tensile)

1.3.11 SHEAR STRESS

The above sections discuss the bending stress caused by an applied moment. Now, the developed shear stress caused by an applied shear force is to be discussed. Again the bending stress diagram is linear with the magnitude of zero at the neutral axis, with the maximum at the extreme fiber. Shear stress occurs due to the applied shear force in the beam section. In the beam shown in Figure 1.22 with an arbitrary loading, consider a very small element from the beam. At that element, the shear force V can be illustrated as shown, considering the equilibrium of the element.

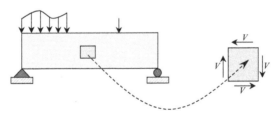

FIGURE 1.22   Concept of shear force in an arbitrary beam.

The developed shear stress (f_v or τ) due to the applied shear force can be expressed as shown in Eq. 1.21.

$$\text{Shear stress, } \tau = \frac{VQ}{It} \tag{1.21}$$

where,

V = Applied shear force at the point of interest

I = Moment of inertia of the section about the neutral axis

t = Thickness of the section at the point of interest

Q = $A\bar{y}$ (A = area of the section above or below the point of interest, $\bar{y}$ = distance to the centroid of this area (A) with respect to the neutral axis, as shown below)

To elaborate, assume that the shear stress is required to be determined at the P-line in the rectangular section shown in Figure 1.23. Then, A is the area above or below the P-line. The area above the P-line has been considered, and $\bar{y}$ is the distance to the centroid of the shaded area as shown.

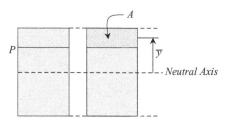

FIGURE 1.23 Rectangular cross-section of a beam.

Again, shear stress is the maximum at the neutral axis, and zero at the top and the bottom fibers. The shear stress diagram is oval-shaped, with the maximum value at the neutral axis, as shown in Figure 1.24. It shows that shear stress is zero at the upper and lower ends, and the maximum at the neutral axis or the centroid. This is true for any kind of section.

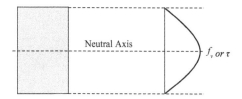

FIGURE 1.24 Shear stress diagram for a rectangular cross-section of a beam.

1.3.12 TORSIONAL STRESS

If an object is subjected to a torque, then shear stress is developed with the maximum value at the outer fiber, is zero at the center, and varies linearly along the radius of the shaft. Specifically, the shear stress distribution is considered linear and proportional to the distance from the center. Consider that a shaft of length L with a radius of r is subjected to a torque of T, as shown in Figure 1.25.

For any distance from the center, say ρ, the shear stress is τ, and at the outer fiber (the r distance from the center), the shear stress is τ_{max}. Then, it can be said that (Eq. 1.22)

$$\frac{r}{\rho} = \frac{\tau_{max}}{\tau} \tag{1.22}$$

The maximum shear stress due to the torsion in the circular-solid shafts is presented as shown in Eq. 1.23.

$$\tau_{max} = \frac{Tr}{J} \tag{1.23}$$

where,

J = Polar moment of inertia of a circular shaft about its centroidal axis
$J = \dfrac{\pi}{2}r^4$ for a solid shaft
r = Radius of solid shaft
$J = \dfrac{\pi}{2}\left(r_0^4 - r_i^4\right)$ for a hollow shaft, with r_o and r_i as the outer and the inner radii, respectively.

Chapter 9 discusses the torsion test of steel material further.

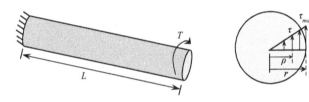

FIGURE 1.25 Torsion and shear stress due to applied torque. *Photo taken at the Colorado State University–Pueblo.*

Example 1.9 Torsional Stress

A circular shaft with an inner diameter of 12 in. and an outer diameter of 14 in. is subjected to a torque of 100 kip.ft. Determine the maximum shear stress developed at the shaft.

Solution

Given,

  Outer radius, $r_o = 14$ in./2 = 7 in.
  Inner radius, $r_i = 12$ in./2 = 6 in.
  Applied torque, $T = 100$ kip.ft = 100(12) kip.in.

  Polar moment of inertia, $J = \dfrac{\pi}{2}\left(r_o^4 - r_i^4\right) = \dfrac{\pi}{2}\left(7^4 - 6^4\right) = 1{,}736$ in.^{4}

The maximum shear stress is developed at the outmost fiber of a shaft upon applying a torque. Therefore, $r = r_o = 7$ in.

The maximum shear stress (from Eq. 1.23), $\tau_{max} = \dfrac{Tr}{J} = \dfrac{100(12)\ \text{kip.in.}(7\ \text{in.})}{1{,}736\ \text{in.}^4} = 4.84$ ksi

Answer: 4.84 ksi

1.4   REVIEWS OF NON-MECHANICAL BEHAVIOR OF MATERIALS

Non-mechanical behavior is a broad term, as it includes thermal behavior, density, corrosion, degradation, surface texture, abrasion, and wear resistance.

Thermal Behavior. *Thermal expansion* and *contraction* of material is of great interest in civil and construction engineering. It is well known that the length of a material increases or decreases with the increase or decrease in temperature, respectively. The change in length due to the change in temperature is called the *thermal deformation*. The change in length per unit length (*thermal strain*) due to the temperature change can be calculated using Eq. 1.24.

$$\varepsilon_{\text{temp}} = \frac{\delta_{\text{temp}}}{L_o} = a\Delta T \tag{1.24}$$

The total change in length of the member due to this temperature change can be calculated as shown in Eq. 1.25.

$$\delta_{\text{temp}} = aL_o\Delta T \tag{1.25}$$

where,

α = Linear coefficient of thermal expansion and contraction
ΔT = Algebraic change in temperature of the member
L_o = Initial length of the member
δ_{temp} = Algebraic change in length of the member

The α-value can be determined using Eq. 1.26:

$$a = \frac{\Delta L / L_o}{\Delta T} \tag{1.26}$$

where,

L_o is the initial length of the specimen
ΔL is the change in length due to the change in temperature by ΔT

From Hooke's law, the thermal stress can be calculated as shown in Eq. 1.27.

$$\sigma_{\text{temp}} = E\varepsilon_{\text{temp}} = Ea\Delta T \tag{1.27}$$

Density, Unit Weight, and Specific Gravity. *Density* (ρ), or more precisely, the volumetric mass density, of a substance is its mass per unit volume. *Unit weight* (γ) of a substance is its weight per unit volume. These can be presented as follows:

$$\text{Density, } \rho = \frac{\text{Mass}}{\text{Volume}} = \frac{M}{V} \tag{1.28}$$

$$\text{Unit Weight, } \gamma = \frac{\text{Weight}}{\text{Volume}} = \frac{W}{V} = \frac{Mg}{V} = \left(\frac{M}{V}\right)g = \rho g \tag{1.29}$$

where,

M = Mass of the substance
W = Weight of the substance = Mg
V = Volume of the substance
ρ = Density of the substance
γ = Unit weight of the substance = ρg
g = Gravitational acceleration (9.81 m/s² or 32.2 lb/s²)

Specific gravity (G) is the ratio of the density of a substance to the density (or unit weight) of a reference substance. Equivalently, it is the ratio of the mass (or weight) of a substance to the mass (or weight) of a reference substance of equal volume. It can be expressed as shown in Eq. 1.30.

$$\text{Specific gravity}, G = \frac{\rho}{\rho_g} = \frac{\rho g}{\rho_g g} = \frac{\gamma}{\gamma_w} \tag{1.30}$$

where,

ρ_w = Density of water (1,000 kg/m³ or 1.94 slug/ft³ but very often used as 62.4 lb/ft³)
γ_w = Unit weight of water (9.81 kN/m³ or 62.4 lb/ft³)

Corrosion. *Corrosion* is a natural process which converts a refined metal to a more chemically stable form, such as its oxide, hydroxide, or sulfide. It is the gradual destruction of a material (usually metal) by a chemical and/or electrochemical reaction with the environment. For example, if steel material is not protected from the environment by coating or cover, it corrodes readily.

Degradation. The *degradation* of material is the process of becoming damaged or poorer in quality, resulting in a diminished capacity to withstand loading and provide services. For example, wood degrades very quickly if exposed to rain, water, and fungi.

Surface Texture. *Surface texture* broadly defines the appearance of a material's exterior, such as color, smoothness, angularity, etc.

Abrasion and Wear Resistance. *Abrasion and wear resistance* describe the decay resistance of a material upon loading and environmental action. For example, the aggregate used in pavement undergoes some damage upon the impact loading of tires.

Example 1.10 Thermal Stress

A 10-m long steel rod is snugly supported by two rigid supports at the ends, as shown in Figure 1.26. The coefficient of thermal expansion and contraction of the rod is 6×10^{-6} per °F, and the elastic modulus is 210 GPa. If the temperature of the body is increased by 50°F, calculate the thermal stress developed at the supports.

Solution

Given,

Length of rod, $L = 10$ m
Modulus of elasticity, $E = 210$ GPa
Coefficient of thermal expansion and contraction, $\alpha = 6 \times 10^{-6}$ per °F

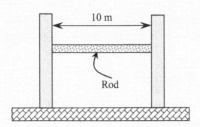

FIGURE 1.26   A steel rod supported by two rigid supports for Example 1.10.

Change in temperature, $\Delta T = 50°F$

From Eq. 1.24: Thermal strain, $\varepsilon_{temp} = \alpha(\Delta T) = 6 \times 10^{-6}$ per $°F$ $(50°F) = 3 \times 10^{-4}$

From Eq. 1.27: Thermal stress, $\sigma_{temp} = E\varepsilon_{temp} = 210\,GPa\left(3 \times 10^{-4}\right) = 630 \times 10^{-4}\,GPa$

   $= 63\,MPa$

Answer: 63 MPa

1.5   MATERIAL SELECTION

Any civil engineering structure can be constructed with different types of materials. For example, a residential building can be constructed of bricks or concrete, wood or steel, or even with mud. The material(s) to be used in a project depend on many factors, listed below.

a) Mechanical behavior
b) Non-mechanical behavior
c) Economy and availability
d) Performance
e) Aesthetic reasons

It is not always true that the costliest material is the best-suited material; sometimes, a very low-cost material can be the best choice. The material must have the desired physical and mechanical properties, be available in the desired shape, and be affordable to satisfy the needs. Sometimes, the salvage value or the recycling options after the usage are also considered.

1.5.1   MECHANICAL BEHAVIOR

The first item to be considered when selecting a material is the ability of its load carrying capacity in terms of bending, shear, and torsional capacity. The failure criterion is also a major concern. Brittle materials, such as cast iron, glass, etc., are very often not preferred, as they fail suddenly without showing any significant deformation. For example, for a 20-ft bridge where pedestrians are the only users, any material of steel, reinforced concrete, or wood would be able to withstand the pedestrian load. Then, more criteria are to be considered. Wood may be the cheapest option, but may not be appropriate if it is a high-frequency

flood-prone area. Therefore, wood may not be durable and may not be an economically via-
ble material considering the lifespan. Another case is that if the bridge is located in a family
park, then a wooden bridge may be aesthetically welcomed regardless of cost or durability.

1.5.2 NON-MECHANICAL BEHAVIOR

Non-mechanical behaviors, such as density, thermal conductivity, electrical conductivity, etc., are also important factors when considering a material. For example, when constructing a foundation, concrete is most commonly preferred due to its high density (heavy
weight), which provides stability to the structure. For long-span structures, steel is preferred
due to its high strength-to-weight ratio.

1.5.3 ECONOMY AND AVAILABILITY

Economy, the cost of the materials, is always considered when selecting materials in almost
all types of projects. This consideration also includes the following:

- Availability of skilled manpower
- Availability of time, with the least hassle
- Availability of desired shape and size
- Disposal cost
- Manufacturing and refilling cost
- Salvage value or recycling options
- Transportation, storage, placement, and maintenance

1.5.4 PERFORMANCE

The performance aspects of the section, such as serviceability, durability, 'wear and tear,'
etc., are major considerations after evaluating the above parameters. *Durability* refers to
the long-term lasting behavior; *serviceability* refers to the comfort while in use (say, high
deflection occurring even though a structure is safe); and the term 'wear and tear' means
the degradation due to the continued usage(s). For example, a wood building deflects and
vibrates greatly due to wind. On the other hand, a concrete building does not vibrate much
during windy conditions. In addition, the selected material should not pose any health or
environmental hazards.

1.5.5 AESTHETICS REASONS

Human beings, by default, are biased by beauty. In the world of competition, beauty is very
important where there is little or no monetary issue. For this reason, the usage of glass is
currently increasing, as it shines with light and attracts attention quickly. Aesthetic reasons
are often profitable in the sense that almost everybody likes to use a beautiful building.

The selection procedure of material can be presented using a flowchart, as shown in
Figure 1.27. From a group of candidate materials, a trial material is first selected. The material is then judged for every requirement of material selection. If a requirement is fulfilled,
it is then judged against the next requirement, and so on, until all the requirements are satisfied. If any of the requirements are not fulfilled, then the material is discarded and another
material is selected for the trial.

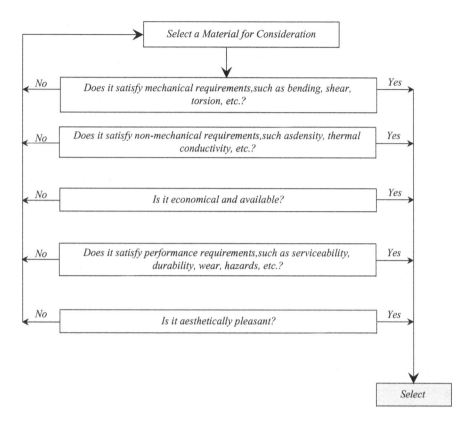

FIGURE 1.27   Materials selection procedure.

1.6   DESIGN PHILOSOPHY

It is better to understand the design philosophy of civil engineering materials at this stage to make this course more interesting. Two types of design procedures are available:

Working Stress Design (WSD): This method is nearly obsolete. It determines the working stress (or allowable stress) of material to be used in the structure by dividing the laboratory-obtained strength by a factor of safety (say, 2 or 2.5). Then, the probable stress in the material when exposed to service is calculated and compared with the allowable stress. The materials to be used must have an equal or larger capacity compared to the probable stress while in service. For example, say a cable fails at 20 ksi of stress in the laboratory. If a factor of safety of 2.0 is desired, then the allowable stress of the cable is 20 ksi ÷ 2.0 = 10 ksi. This cable can be used in a member of a structure having a probable stress of 10 ksi or less. This method is often called the allowable stress design.

Ultimate Strength Design (USD): This method is newer than the WSD method and is replacing the WSD method day by day. This method determines the design stress of material to be used in a structure by multiplying the laboratory-obtained strength by a strength reduction factor (say, 0.9 or 0.75). Then, the probable stress in the material when it is exposed to service is calculated with some factoring and is then compared to the design stress. For example, say a cable fails at 20 ksi of stress in the laboratory, which is known as the *nominal stress*. This nominal stress is decreased by a factor (say, 0.90) to obtain the design stress, i.e., 20 ksi (0.90) = 18 ksi. Then, the actual stress in the member is calculated. Say, the actual stress in the member is 10 ksi. This actual stress is increased by a factor to adopt some safety (say, 1.4), after which it is known as the *factored stress*. Here, the factored stress is 10 ksi (1.4) = 14 ksi. Finally, the design stress in the cable of 18 ksi is compared with the factored stress of 14 ksi. The design stress must be equal to or greater than the factored stress. This method is most commonly called the strength design. In steel and wood structures, this method is known as Load and Resistance Factor Design (LRFD), as the probable load (or stress) is factored to increase it, and the resistance capacity of the material is factored to decrease it. By using these two factoring techniques, the safety in the design member is ensured.

1.7 CHAPTER SUMMARY

This chapter discusses the pre-knowledge required to study the properties, behaviors, and testing procedures for different materials. Civil and construction engineers are responsible for selecting proper materials considering all factors. With the improvement in technology, new and innovative materials are being developed and used every day. Composite materials are also becoming more popular. For example, fiber-reinforced polymer (FRP) is replacing steel reinforcement, engineered wood products are replacing sawn lumber, etc. The design methodology of the new materials is different from that of the conventional materials. Thus, civil and construction engineers are required to keep up to date with innovative technology, new materials, etc.

Materials are classified into four types: metallic, ceramic, polymeric, and composite. The first three types are distinct to each other, while the composite materials are the combination or mixture of any two or all three. Metallic materials can be further classified into the following two categories, considering the presence of ferrous materials or not: ferrous metals and nonferrous metals. Ceramics have a crystalline structure and are formed between metallic and nonmetallic elements. Polymeric material is formed with a long chain of covalent-bonded atoms and secondary bonds, which hold groups of polymer chains together to form the polymeric material.

Tension and compression tests are the basic tests conducted on most of the civil engineering materials. In these tests, a specimen is clamped and a uniaxial load is applied on it until the failure occurs. Engineering stress, or simply stress, is defined as the reaction force over the initial area. Engineering strain, or simply strain, is the change in length over the initial length. Poisson's ratio is defined as the ratio of the lateral strain and the axial strain.

Considering the amount of elongation before failure, materials can be of two types: ductile and brittle. Ductile materials show remarkable amounts of deformation before the failure. Brittle materials fail without showing any noticeable deformation. Ductile materials show warning signs before failure, giving the users sufficient time to evacuate structures.

Elastic materials deform upon loading and return to their initial position immediately upon releasing the load. Plastic materials deform permanently upon loading, i.e., the deformation does not return upon releasing the load. Viscous materials take time to deform upon applying load, and the deformation takes time to return to its initial position upon releasing the load.

Fatigue is the failure of materials due to repeated loading. A small amount of load may not be able to cause a material to fail. However, repetition of this small load for a number of times may cause failure. Impact tests determine the toughness of a material, which is the amount of energy absorbed by a material until rupture.

Creep and relaxation are the two major properties of viscous materials. Creep is the phenomenon of an increase in strain upon applying a constant stress over time on the material. Relaxation is the concept of a decrease in stress upon applying a constant strain over time on the material.

Bending stress occurs in a material due to the developed moment in the member. Bending stress is the maximum at the outmost fiber and is zero at the neutral axis. Shear stress occurs due to the shear force developed in a member. Its maximum value is at the neutral axis, and it is zero at the outmost fiber. Torsion also causes shear stress in a shaft, with the maximum value occurring at the outmost fiber.

Some important properties of civil engineering materials are thermal expansion and contraction, density, corrosion, degradation, surface texture, abrasion resistance, chemical resistance, etc. All of these properties, including the mechanical properties, are considered while selecting a material for the intended use. Some other considerations in selecting a material are the availability, cost, maintenance, durability, appearance, potential hazards, etc. While designing or using a material, proper characterization is made such that the strength capacity of the material must be equal to or higher than the actual load on the member. This statement is ensured during the design phase of a structure.

REFERENCE

1. Askeland, Donald R., Fulay, Pradeep P., and Wright, Wendelin J. 2011. *The Science and Engineering of Materials*, 6th edition. Cengage Learning, Stamford, CT.

FUNDAMENTALS OF ENGINEERING (FE) EXAM STYLE QUESTIONS

FE Problem 1.1

Modulus of rigidity is defined as the

- A. Capacity of deflection resistance
- B. Initial slope of shear stress versus shear strain curve
- C. Slope of the stress–strain curve
- D. Energy required before the failure

Solution: B

Modulus of rigidity, or shear modulus, is defined as the initial slope of the shear stress versus the shear strain curve.

FE Problem 1.2

Modulus of elasticity is defined as the

- A. Capacity of deflection resistance
- B. Slope of stress versus strain curve until the elastic region
- C. Initial slope of the stress–strain curve
- D. Energy required before the failure

Solution: C

Modulus of elasticity is defined as the initial slope of the stress versus the strain curve, or any point up to the proportional limit.

FE PROBLEM 1.3

Hooke's law is valid up to the

    A. Yield point
    B. Elastic region
    C. Linear stress–strain region
    D. Any point in stress–strain curve

Solution: C

Hooke's law is valid up to the proportional limit or linear portion of the stress–strain curve.

FE PROBLEM 1.4

A 10 m long steel rod has a coefficient of thermal expansion and contraction of 6×10^{-6} per °F. If the temperature of the body is decreased by 50°F, the thermal contraction (m) is most nearly:

    A. 1.5×10^{-3}
    B. 3×10^{-3}
    C. 3×10^{-4}
    D. None of the above

Solution: B

From Eq. 1.25: Thermal contraction, $\delta_{temp} = aL_o \Delta T = 6 \times 10^{-6} (10 \text{ m}) (50°F) = 3 \times 10^{-3}$ m.

FE PROBLEM 1.5

The maximum bending stress which occurs in a rectangular beam section is most nearly:

    A. At the neutral axis
    B. At the extreme fiber
    C. At a point of minimum thickness
    D. None of the above

Solution: B

The maximum bending stress occurs at the extreme fiber in a rectangular beam section. The maximum shear stress occurs at the neutral axis in a rectangular beam section.

FE Problem 1.6

A steel pipe has a length of 6 ft, an outer diameter of 10 in., and a wall thickness of 1.0 in. The modulus of elasticity of steel is 29,000 ksi, and the Poisson's ratio is 0.30. If the pipe is subjected to an axial compression of 100 kips, a lateral strain of 0.0000366 occurs. The increase in wall thickness is most nearly:

   A. 2.38 μin.
   B. 36.6 μin.
   C. 3.66 μin.
   D. 2.38 nano in.

Solution: B

Lateral strain, $\varepsilon_t = \dfrac{\Delta t}{t_o}$ (from Eq. 1.5)

$$\Delta t = \varepsilon_t (t_o) = 0.0000366\,(1.0\ \text{in.}) = 36.6 \times 10^{-6}\ \text{in} = 36.6\ \mu\,\text{in.}$$

PRACTICE PROBLEMS

Problem 1.1

A tensile force of 50 kip is applied on a 0.75-in. diameter and 4-in. long rod. After applying the load, the length of the rod increases to 4.1 in., and the diameter decreases to 0.745 in. Assuming the material is within the proportional limit, determine the following parameters:

   a) Engineering stress
   b) Engineering strain
   c) Lateral strain
   d) Poisson's ratio
   e) Shear modulus

Problem 1.2

A steel bar has a length of 2.5 m and a rectangular cross-section of 200 mm by 100 mm. The modulus of elasticity of the material is 210 GPa and the Poisson's ratio is 0.3. Determine the change in width of the bar after applying a tensile force of 150 kN.

PROBLEM 1.3

A piece of block is initially square, as shown by the dashed lines shown in Figure 1.28. Determine the average shear strain at B, C, and D if the corners B and C are subjected to the displacements that cause the block to distort, as shown by the solid lines.

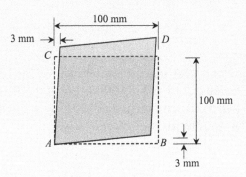

FIGURE 1.28   Schematics of the block for Problem 1.3.

PROBLEM 1.4

A cantilever square beam has a square cross-section of 8 in. by 8 in. The beam is fixed at one end and has a load of 20 kip at the other end. The span of the beam is 12 ft. Determine the absolute maximum bending stress in the beam.

PROBLEM 1.5

An 18 in. by 12 in. cantilever beam is having a triangular load, as shown in Figure 1.29. The span of the beam is 12 ft. Determine the absolute maximum bending stress in the beam.

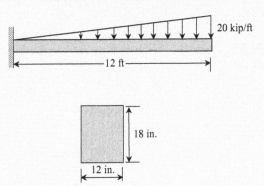

FIGURE 1.29   Schematics of beam for Problem 1.5.

PROBLEM 1.6

A 12 ft simply supported beam is having a uniformly distributed load of 3 kip/ft. The cross-section of the beam is 12 in. (width) by 18 in. (depth). Determine the maximum shear stress developed in the beam.

PROBLEM 1.7

A circular shaft of 12-in. diameter is subjected a torque of 100 kip.ft. Determine the maximum shear stress developed at the shaft.

2 Aggregates

Aggregate is the most widely used material in civil and construction engineering, especially in concrete and asphalt mixtures. Aggregate is the main load-carrying agent and fills most of the volume in concrete and asphalt mixtures. Cementitious materials such as cement, lime, or asphalt bind the aggregates together to make a solid mass and help the aggregate carry the load. Thus, the study of the sources, production, properties, selection criteria, sampling, characterization, and standard laboratory testing of aggregates is important in civil and construction engineering. This chapter discusses all these aspects of aggregates. Worked-out examples are provided where necessary for better understanding.

2.1 BACKGROUND

Aggregate (Figure 2.1) is a broad term of coarse to fine particles used in civil and construction engineering. It includes sands, gravels, crushed stones, iron-blast-furnace slags, recycled concretes, geosynthetic aggregates, and so on. Fine particles are also present in aggregate. Aggregates are the primary materials in producing *Portland cement concrete* (PCC) and *asphalt concrete* (AC). When the cementitious (binding) material, such as cement or asphalt, binds aggregate together, it forms a solid and stiff mass which can carry external loading. In addition, the aggregate-to-aggregate interlocking can also resist a fair amount of load. This material is cheaper and readily available anywhere in the world compared to cement or other similar materials. Thus, aggregate is the most widely used material in civil and construction engineering. The heavy weight adds value when it is used in foundation- and dam-type structures. Due to the relatively high hydraulic conductivity compared to soils, aggregates are widely used in drainage applications such as foundations and drains, septic drain fields, retaining wall drains, and roadside edge drains. Aggregate is also used as the base material under building foundations, roadways, and railroads to help prevent differential settling under the road or building. It is also a low-cost extender that binds with more expensive cement or asphalt to form concrete.

US domestic production and use of construction aggregates amount to 2.5 billion tons of crushed stone, sand, and gravel, valued at $25.1 billion per year, and the aggregates industry employs approximately 100,000 highly skilled employees (NSSGA 2019). Another survey by the USGS (2013) shows approximately 1.3 billion tons of crushed stone, worth about $12 billion, was produced by 1,550 companies operating 4,000 quarries, 91 underground mines, and 210 sales/distribution yards in the United States in 2012. Of the total crushed stone produced in 2012, about 69% was limestone and dolomite, 14% was granite, 7% was traprock, 5% was miscellaneous stone, and 4% was sandstone and quartzite. Of the portion of total crushed stone produced in 2012, 82% was used as a construction material, mostly for road construction and maintenance, and 10% was used for cement manufacturing. Approximately 927 million tons of construction sand and gravel worth $6.4 billion was produced in 2012. About 43% of construction sand and gravel was used as concrete aggregates; 26% for road base, coverings, and stabilization; 12% as construction fill; and 12% as asphalt concrete aggregates and used in other asphalt-aggregate products. Another survey shows about 27% of total crushed stone is used in PCC production and 9% is used in

FIGURE 2.1   Aggregates. Photo taken in Pueblo, Colorado.

AC production, while about 16% is used in road bases as unbound layers, and the remaining 48% is used in other applications. About 17% of total gravel and sand is used in PCC production and 16% is used in AC production, while about 26% is used in road bases as unbound layers, and the remaining 41% is used in other applications (Somayaji 2001).

2.2   AGGREGATE CLASSIFICATION

Aggregate can be classified in different ways:

a) Classification based on geology
b) Classification based on source
c) Classification based on unit weight
d) Classification based on size

2.2.1   CLASSIFICATION BASED ON GEOLOGY

Most aggregates are obtained from crushing stones or rocks, and the properties of aggregates are thus inherited from their parent rocks. Based on the geology of rocks, aggregates are divided into three types: *igneous*, *metamorphic* and *sedimentary* (Figure 2.2). These are briefly described here.

Igneous   Sedimentary   Metamorphic

FIGURE 2.2   Types of rocks. Photos taken in Pueblo, Colorado.

2.2.1.1 Igneous Rocks

Igneous rocks are crystalline and are developed after the cooling of magma or lava from volcanos. Much of the earth's continental crust (about 95%) and nearly all the oceanic crust are composed of igneous rocks (Somayaji 2001). As igneous rocks are solidified from a molten state, they look uniform without layers, with the grains packed together tightly. Commonly used igneous rocks are basalt and granite, which are distinctly different in compositions and textures.

2.2.1.2 Metamorphic Rocks

Metamorphic rocks are the result of the transformation of original rocks. The original rock, subjected to very high heat, water, and pressure, undergoes physical and chemical changes. These rocks are generally coarser, denser and less porous compared to their parent rocks. Examples of these rock types include marble, slate, gneiss, and schist.

2.2.1.3 Sedimentary Rocks

Igneous or metamorphic rocks are disintegrated by weathering and erosion processes over the millions of years they are formed. The disintegrated rock particles are then transported to different places and/or in water bodies with the actions of wind, rain, ice, mass movement, glaciers, and so on. The deposition and subsequent cementation of these disintegrated rock particles, at the surfaces and within bodies of water in layers of strata, become what are called sedimentary rocks. These rocks cover about 75% of the total land surface (Somayaji 2001). Calcareous rocks such as limestones, chalk, etc.; siliceous rocks such as sandstones, cherts, etc.; and argillaceous rocks such as shale are examples of common sedimentary rocks.

2.2.2 CLASSIFICATION BASED ON SOURCE

Based on the source, aggregates are divided into four types: *natural aggregates*, *crushed aggregates*, *artificial aggregates*, and *recycled aggregates* (Figure 2.3). These are briefly described here.

Natural Sand

River Rock

FIGURE 2.3 Different types of natural rocks.

2.2.2.1 Natural Aggregates

Natural aggregates consist of rock fragments that are used in their natural state, or are used after mechanical processing, such as crushing, washing, and sizing. Some natural aggregate deposits, called *pit-run gravel*, consist of gravel and sand that can be readily used in concrete after minimal processing. Natural gravel and sand are usually dug or dredged from a pit, river, lake, or seabed.

The production of aggregates from natural sources consists of

- Extraction
- Transportation to plants
- Crushing, storage
- Loading, weighing, and
- Transportation to the application site

The extraction of alluvial materials can be carried out with the help of a bulldozer, loader, or excavator from dry areas. A dredge connected to a floating conveyor is required to extract aggregate from water bodies. Drilling and blasting are required to collect aggregates from massive rocks and rocky mountains. The hole size, depth, spacing, pattern, and the amount of explosives used for each hole depends on the severity of rock break, rock soundness, and the type of explosives utilized. Small holes closely spaced together yield small rock particles, while large holes widely spaced apart yield large rock particles (Nunnally 2009).

Drilling allows inserting the explosives, such as:

- Ammonium Nitrate
- Ammonium Nitrate in Fuel Oil (ANFO)
- Slurries

Nowadays, dynamite is not used because it is costly and difficult to handle. Ammonium nitrate and ANFO are the least expensive of the explosives listed. ANFO is particularly easy to handle, as it is a liquid that can be poured into the blasthole. Common types of drilling equipment include percussion drills, rotary drills, and rotary percussion drills. Blasting breaks down the materials, which are then brought to a handling facility by means of belt conveyors, or by haul trucks or barges if a waterway is present.

2.2.2.2 Crushed Rock Aggregates

Crushed aggregate is mainly quarried or excavated stone, which is transported to a plant, crushed, and screened to the desired particle size and distribution. The particles of crushed aggregates are completely crushed in the plant. Aggregate sizes are reduced by crushing, which gives the aggregates good compaction and load-bearing capacities. Then, they are sorted using wire mesh screens. Afterwards the aggregates are often washed, as some usages require the material to be clean. The materials are stored in piles, within bins or silos, where they are sorted according to various categories. Each category of aggregate meets specific criteria based on its intended usage. All materials leaving the quarry are weighed with truck scales. Delivery is performed by trucks, barges, and trains. If concrete or asphalt is produced at the quarry site, the materials can be processed on site. Crushed stone aggregates are particularly suitable for use in the underlying layers of pavement, as well as other areas exposed to traffic.

2.2.2.3 Artificial Aggregates

Artificial aggregates are made from various waste materials. For example, steel slag (Figure 2.4) produced while manufacturing steel products, is further cut into pieces, which are used in aggregate for road construction. Artificial aggregates are sometimes produced for special purposes, such as:

- making lightweight concrete: waste from clay bricks (Figure 2.4), artificial cinders, foamed slag, expanded shales and slate, sintered fly ash, etc.
- making heavyweight concrete: steel rivet punching, steel slag, and iron ore.

Steel Mill Slag   Waste From Clay Bricks

FIGURE 2.4   Recycled aggregates.

2.2.2.4   Recycled Aggregates

Recycled aggregate is obtained from the crushing of old structures and demolition waste. Demolition of a concrete building or other structures yields primarily crushed concrete, known as *recycled concrete aggregate*. Other types of structures, such as masonry buildings, old pavements, etc., produce general recycled aggregate. While using the recycled aggregates, care should be taken to remove the inferior materials, and prevent contamination by other materials, such as steel reinforcement, asphalt, soils, glass, gypsum board, sealants, paper, plaster, wood, etc. The properties of recycled aggregates are expected to be different from the historical test data, as the aggregates were mixed up with other materials, such as cement, and might be affected by the surrounding conditions these were exposed to. In addition, the old design of the parent aggregates might be different from the current intended design and use. Recycling of concrete involves breaking, removing, transporting, and crushing existing concrete into a material with a desired size and quality. In general, applications of recycled aggregates (using the minimum processing practices) include:

- Road construction, especially for soil stabilization and work to be done beneath the surface layer
- Bulk fills, such as banks, dams, and embankments
- Bases or fills for drainage structures
- Noise barriers and landscaping

Applications of new concrete with recycled aggregates include:

- Different components of pavements, such as surface layer, underneath layers, shoulders, median barriers, sidewalks, curbs, and gutters
- Parking lots

- Bridge foundations
- Concrete bases

2.2.3 CLASSIFICATION BASED ON UNIT WEIGHT

The variability in aggregate density can be used to classify aggregates of widely different unit weights (or densities), as shown in Table 2.1. The most common classifications of aggregates, based on specific gravity (G), are as follows:

a) Lightweight aggregates
b) Normal-weight aggregates
c) Heavyweight aggregates

Lightweight concrete uses the lighter aggregates to decrease the dead load in structures. Lighter weight means this aggregate has less weight compared to the normal-weight aggregate, which has specific gravities of 2.4–2.8. Lightweight aggregate has a specific gravity less than the normal range of 2.4–2.8. Heavyweight aggregate has a specific gravity greater than the normal range of 2.4–2.8. Note that the definitions of light lightweight, normal-weight, and heavyweight are relative and not exact.

Normal-weight aggregates are common gravels, stones, ordinary sand, etc. Some of the natural lightweight aggregates are pumice, scoria, and tuff. Pumice is volcanic glass, typically whitish gray to yellow in color, and is found in volcanic areas. Scoria also comes from volcanic activities but is red or black in color. Lightweight aggregates can also be produced by expansion, using heat such as blast-furnace slag, expanded shale, expanded slate, expanded clay, expanded perlite, vermiculite, and so on. The raw materials are heated at about 2,000 °F (1,100°C) and cooled down, producing very porous, highly absorptive, and lighter aggregates. Some natural heavyweight aggregates are mineral ores and barites. Mineral ores include hematite, magnetite, limonite, etc. Manufactured steel punching, shots, bars, etc. are also heavyweight aggregates.

The primary use of lightweight concrete is in masonry construction and temperature insulation. Sometimes, lightweight concrete is made to decrease the dead load in noncritical concrete structures, such as partition walls. Heavyweight aggregates use heavy natural aggregates. The main application for heavyweight concrete is for medical or nuclear

TABLE 2.1

Classification of Aggregates by Unit Weight

Category	Specific Gravity (G)	Examples	Applications
Lightweight	$G < 2.4$	Shale, Slate, Expanded Perlite	Masonry Unit, Insulation
Normal-weight	$2.4 < G < 2.8$	Limestone, Sand, River Gravel	Regular Projects
Heavy-weight	$G > 2.8$	Hematite, Magnetite, Barites	Radiation Shielding, Counter Weights

radiation shielding, offshore structures, and ballasting of pipelines. Figure 2.5 shows photos of some lightweight and heavyweight aggregates.

Light-Weight Aggregate   Light-Weight Aggregate   Heavy-Weight Aggregates   Heavy-Weight Aggregates
(Barytes)   (Magnetite)

FIGURE 2.5 Aggregates based on unit weight.

2.2.4 CLASSIFICATION BASED ON SIZE

Aggregate may be divided based on the particle size, as shown below (Figure 2.6).

1) Boulder – greater than 12 in. (300 mm)
2) Cobbles – greater than 3 in. (75 mm) but less than 12 in. (300 mm)
3) Coarse-Grained Aggregate or Granular Aggregate
   a) Gravel – greater than 4.75 mm (#4 sieve) but less than 3 in. (75 mm)
     (i)   Coarse Gravel – 75 mm to 19 mm
     (ii) Fine Gravel – 19 mm to 4.75 mm
   b) Sand – greater than 0.075 mm (# 200 sieve) but smaller than 4.75 mm (#4 sieve)
     (i)   Coarse Sand – 4.75 mm to 2.0 mm (Sieve # 4 to #10)
     (ii) Medium Sand – 2.0 mm to 0.425 mm (Sieve # 10 to #40)
     (iii) Fine Sand – 0.425 mm to 0.075 mm (Sieve # 40 to #200)
4) Fine Aggregate – Smaller than 0.075 mm (# 200 sieve) with no minimum size
   a) Silt – cohesionless
   b) Clay – cohesive (sieving cannot separate clay and silt)

Gravel   Sand   Silt (cohesionless)   Clay (cohesive)

FIGURE 2.6 Aggregates based on size. Photos taken in Pueblo, Colorado.

Boulders and cobbles are not included in the category of aggregates. These are crushed down into other smaller sizes. Gravel is known as the particles with sizes greater than 4.75 mm (#4 sieve) and smaller than cobbles. It is further divided into coarse gravel and fine gravel. Sand is greater than 0.075 mm (# 200 sieve) but smaller than gravel. It is further divided into coarse sand, medium sand, and fine sand. Anything finer than sand is called the *fine aggregates*, or simply fines. They are further classified into silt and clay based on their cohesion properties. Silt is *cohesionless* (does not stick together) and clay has *cohesion*, i.e., the property in which a material becomes sticky upon adding water.

2.3 PROPERTIES OF AGGREGATES

Aggregates are mainly collected from local sources. Therefore, there cannot be national specifications for the materials' properties. Desired specifications are selected locally, and thus testing the properties of aggregates is crucial. Aggregate properties are defined by the characteristics of both the individual particles and the characteristics of the combined material by their

(i) Physical characteristics
(ii) Chemical characteristics, and
(iii) Mechanical characteristics

Table 2.2 lists which properties are important for PCC and which are important for AC. For example, particle shape is very important for asphalt materials, as asphalt materials are compacted by applying mechanical a load. Angular particles do not compact as readily as rounded particles because their angular surfaces tend to lock up with one another and resist compaction. However, in PCC, the degree of compaction is not as specific as AC, as PCC is compacted by rodding or vibration when it is in liquid form. The strength of PCC does not depend primarily on the compaction level as long as the minimum level of compaction is done. AC materials require a high degree of compactness, as the load is carried mainly by the aggregate to aggregate interlocking.

TABLE 2.2
Basic Aggregate Properties

	Properties	Importance PCC	AC
Physical	Particle Shape (Angularity)	Medium	High
Characteristics	Particle Shape (Flakiness, Elongation)	Medium	Medium
	Particle Surface Texture	Medium	Crucial
	Specific Gravity and Absorption	High	Medium
	Soundness	High	Medium
	Deleterious Materials	High	Medium
Chemical	Alkali–Aggregate Reactivity	High	Medium
Characteristics	Affinity for Asphalt	–	High
Mechanical	Toughness	Medium	Medium
Characteristics	Abrasion Resistance	Medium	Medium
	Compressive Strength	Medium	Unimportant
	Stiffness	Medium	High

2.3.1 PHYSICAL CHARACTERIZATIONS

Physical characterization includes, but is not limited to, the following parameters:

a) Particle shape
b) Particle surface texture
c) Specific gravity and absorption

d) Soundness
e) Deleterious materials
f) Sand equivalent test
g) Bulk density

2.3.1.1 Particle Shape

The shape of aggregate determines how the material can be packed into a dense mass and also determines the mobility of the stones within a mix. There are two considerations in the shape of the material:

a) Angularity
b) Flakiness

Angularity of aggregate particles can be characterized as either angular, subangular, subrounded, or rounded. While crushing rocks using a mechanical device, angular aggregates are produced with sharp corners and a rough texture. Due to weathering with time, the sharp corners and rough texture of the aggregates become polished, and subangular particles with a smooth texture are produced. Angular particles do not compact as easily as rounded particles because their angular surfaces tend to lock up with one another and resist compaction. However, once compacted, angular and rough-textured aggregates produce a higher stability than rounded, smooth-textured aggregates. While being transported in water, the aggregates are further polished up and become rounded. Rounded aggregates are easier to place and compact compared to angular aggregates, as their shapes make it easier for them to slide across each other. However, rounded particles carry less load compared to angular aggregates because of their tendency to slide.

Flakiness, also known as flat and elongated, describes the relationship between the dimensions of the aggregate, as shown in Figure 2.7. This is an evaluation of the coarse portion of the aggregates, but only aggregates retained on the 9.5 mm (3/8 in.) sieve are evaluated. ASTM D 4791 is used to determine the flat and elongated particles and is defined as follows:

- A flat particle is defined as one where the ratio of the middle dimension to the smallest dimension of the particle exceeds 3 to 1.
- An elongated particle is defined as one where the ratio of the largest dimension to the middle dimension of the particle exceeds 3 to 1.
- Particles are classified as flat and elongated if the ratio of the largest dimension to the smallest dimension exceeds 5 to 1.

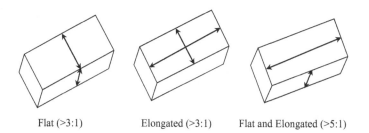

Flat (>3:1)   Elongated (>3:1)   Flat and Elongated (>5:1)

FIGURE 2.7 Concept of flakiness test for coarse aggregates.

2.3.1.2   Particle Surface Texture

Surface texture is a measure of the smoothness and roughness of aggregate by a visual examination. Aggregates are classified broadly into two groups, namely:

a) Smooth
b) Rough

Angularity is a measure of surface roughness, irregularities, or sharp angles of the aggregate particles. As mentioned earlier, angular particles do not compact as readily as rounded particles because the angular surfaces tend to lock up with one another and resist compaction, while smoother, rounded surfaces tend to pass by one another, allowing for easier compaction. If a mass of aggregate has a higher void content, it has more angular material. Angular materials are desirable in paving mixtures because they tend to lock together and resist deformation after initial compaction, whereas rounded materials may slide and produce deformation. The measured voids in an uncompacted mass are affected by the shape, angularity, and texture of the fine aggregate, the aggregate grading and specific gravity.

The angularity and texture of fine aggregates have a very strong influence on the stability of AC mixes. Fine Aggregate Angularity (FAA) is important because an excess of rounded, fine aggregate (low FAA values) can cause deformation readily upon loading. The FAA test measures the angularity of fine aggregate using the aggregate's uncompacted void content following the AASHTO T 304 or ASTM C 1252 test standards. A calibrated cylinder or container is filled with fine aggregate by letting the sample fall from a fixed height. The aggregate is then struck off and its mass is calculated by weighing. The uncompacted void content is calculated using Eq. 2.1.

$$\text{Air Void} = \frac{V - \left(\dfrac{M}{G}\right)}{V} \times 100 \tag{2.1}$$

where,

V = Volume of the cylindrical measure (cm^3, cc, or ml) (no other unit is possible)
M = Sample mass (g) (no other unit is possible)
G = Dry bulk specific gravity of the fine aggregate (discussed in the next subsection)

The term M/G in Eq. 2.1 represents the volume of the solid portion of aggregates in the filled cylinder. The details of this formulation are discussed in Section 2.3.1.3. The difference, $V - M/G$, represents the volume of void space in the cylinder. The volume of void space is expressed with respect to the total volume of the cylinder and is known as air void of fine aggregate. This air void is also known as the fine aggregate angularity (FAA). The higher the FAA, the higher the angularity and the rougher the surface.

Coarse aggregate angularity (CAA) ensures a high degree of aggregate internal friction and deformation resistance. It is defined as the percent by weight of aggregates larger than 4.75 mm with one or more fractured faces (ASTM D 5821). The CAA is estimated visually by counting fractured faces from a small amount of coarse aggregate. The percentage of particles with fractured faces by weight is taken as the CAA. A fractured face is defined as any fractured surface that occupies more than 25% of the area of the outline of the aggregate particle visible in that orientation. Another indirect method of determining CAA is similar to the method of determining FAA. A batch of aggregates is dropped into a cylinder and the spacious air void of the coarse aggregate specimen is measured (AASHTO TP 56).

Example 2.1 Air Void of Loosely filled Soil

A 100 cm³ cylindrical container is loosely filled with 230 g of fine soil with a specific gravity of 2.45. Determine the air void of the loosely filled soil.

Solution

Given
Volume of the cylindrical measure, $V = 100$ cm³
Mass, $M = 230$ g
Specific gravity, $G = 2.45$
From Eq. 2.1:

$$\text{Air Void} = \frac{V - \left(\dfrac{M}{G}\right)}{V} \times 100$$

$$= \frac{100 \text{ cm}^3 - \left(\dfrac{230 \text{ g}}{2.45}\right)}{100 \text{ cm}^3} \times 100$$

$$= 6.12\%$$

Answer: 6%

Note: Voids in loose aggregate are reported to the nearest %.

2.3.1.3 Specific Gravity and Absorption

Specific gravity (G) is a measure of a material's unit weight (weight per unit volume) as compared to the unit weight of water. Therefore, it can be expressed as shown in Eq. 2.2.

$$G = \frac{\dfrac{\text{Weight of Substance}}{\text{Volume}}}{\dfrac{\text{Weight of Water}}{\text{Volume}}} = \frac{\dfrac{\text{Weight of Substance}}{\text{Volume}}}{\text{Unit Weight of Water}}$$

$$= \frac{\text{Weight of Substance}}{\text{Volume}\left(\text{Unit Weight of Water}\right)} = \frac{W}{V\gamma_w} \tag{2.2}$$

$$= \frac{Mg}{V\rho_w g} = \frac{M}{V\rho_w}$$

where,
W = Weight of the substance
M = Mass of the substance
V = Volume of the substance
ρ_w = Density of water (1.94 slug/ft³, 62.4 lb/ft³ or 1,000 kg/m³)
$\gamma_w = (\rho_w)g$ = Unit weight of water (62.4 lb/ft³ or 9,810 N/m³)
g = Gravitational acceleration (9.81 m/s² or 32.2 ft/s²)

Water has a density of 1 g/cm^3. Therefore, if the mass of the substance is expressed in g (grams), and the volume of the substance is expressed in cm^3, then the specific gravity equation from Eq. 2.2 can be written as shown in Eq. 2.3.

$$G = \frac{M}{V \rho_w} = \frac{M \text{ gram}}{V \text{ cm}^3 \left(1.0 \frac{\text{gram}}{\text{cm}^3}\right)} = \frac{M \text{ gram}}{V \text{ cm}^3} \tag{2.3}$$

Specific gravity (G) can also be written as shown in Eq. 2.4, where equal volumes of the substance and water are considered.

$$G = \frac{\dfrac{\text{Weight of Substance}}{\text{Volume}}}{\dfrac{\text{Weight of Water}}{\text{Volume}}} = \frac{\text{Weight of Substance}}{\text{Weight of Water}}$$

$$= \frac{\text{Mass of Substance}}{\text{Mass of Water}} \tag{2.4}$$

Eq. 2.3 dictates that water has a specific gravity of 1.0. Aggregate specific gravity is needed to determine the weight-to-volume relationships and various volume-related quantities of asphalt mixtures. The details of the weight-to-volume relationship of asphalt mixes are discussed in the asphalt mix design chapter. Three types of specific gravities are commonly used, based on how the volume of water and permeable voids, or pores, within the aggregate are addressed. These are mentioned below:

a) *Bulk Specific Gravity* (Bulk Dry Specific Gravity or simply Specific Gravity), G_{sb} (G=specific gravity; s=stone; b=bulk): The volume measurement which includes the overall volume of the aggregate particle, as well as the volume of the water accessible/permeable voids, as shown by the *dashed line* boundary in Figure 2.8. Since it includes the water-permeable void volume, bulk specific gravity is less than apparent specific gravity. The weight of air is considered the dry weight.

b) *Bulk Saturated-Surface Dry* (SSD) *Specific Gravity* (Bulk SSD Specific Gravity), G_{sb} (G=specific gravity; s=stone; b=bulk): The volume measurement which includes the overall volume of the aggregate particle, as well as the volume of the water accessible/permeable voids. Since it includes the water-permeable void volume, bulk specific gravity is less than apparent specific gravity. The weight of air is considered the Saturated-Surface Dry (SSD) weight. In SSD condition, pores are saturated with water, but the surface is dry. This is discussed in detail later in this chapter.

c) *Apparent Specific Gravity*, G_{sa} (G=specific gravity; s=stone; a=apparent): The volume measurement which only includes the net volume of the aggregate particles (the *solid line* boundary in Figure 2.8). It does not include the volume of any water permeable voids.

To determine the bulk-specific gravity (G_{sb}) and the apparent specific gravity (G_{sa}) in the laboratory, a batch of aggregate is submerged in water for about 24 hours, and the weight of the aggregate is measured in the submerged condition. This is called the *submerged weight of aggregate*, C. At this condition, the permeable pores of aggregates are filled with

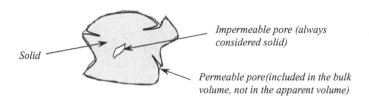

FIGURE 2.8 A porous aggregate piece.

water. The submerged weight of aggregate is the weight of dry aggregate minus the water displaced by the aggregate, according to the Archimedes Principle. Then, the aggregate is removed, and the surface is dried using a damp towel. This condition is called the SSD condition (pores are saturated with water, but the surface is dry). It is then weighed in the SSD condition and the weight is denoted by B. The SSD aggregate is then dried, using an oven, for 24 hours and then is weighed in air. This dried weight is denoted as A.

Then, the bulk dry specific gravity (or bulk specific gravity or specific gravity), G_{sb} is calculated as shown in Eq. 2.5.

$$G_{sb} = \frac{A}{B-C} \tag{2.5}$$

where,

A = Dry weight of aggregate (or the substance)
B = SSD weight of aggregate (or the substance)
C = Submerged weight of aggregate (or the substance)

Simply, specific gravity means the bulk-dry specific gravity. Now, let us see how the ratio, $\dfrac{A}{B-C}$, is the representation of the bulk-specific gravity (G_{sb}).

$$
\begin{aligned}
\frac{A}{B-C} &= \frac{A}{\left(A + \text{Weight of Pore Water}\right) - \left(A - \text{Weight of Water Displaced}\right)} \\[2mm]
&= \frac{A}{\left(\text{Weight of Pore Water} + \text{Weight of Water Displaced}\right)} \\[2mm]
&= \frac{\text{Weight of Substance}}{\text{Total Weight of Water}} = G_{sb}
\end{aligned}
\tag{2.6}
$$

A volume measurement includes the overall volume of the aggregate particle as well as the volume of the water permeable voids. The mass measurement includes the aggregate particle as well as the water within the water permeable voids. ASTM C 127 and AASHTO T 85 are used to determine the bulk-specific gravity of aggregates. The basic procedure is to (Figure 2.9):

- Soak some aggregates in water
- Weigh the aggregates at the saturated-surface dried condition (B)
- Weigh the aggregates inside water (C)
- Finally, weigh the aggregates in air after oven drying (A)

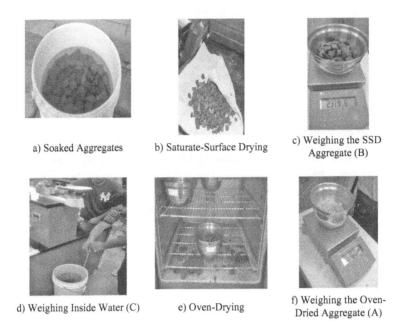

a) Soaked Aggregates b) Saturate-Surface Drying c) Weighing the SSD Aggregate (B)

d) Weighing Inside Water (C) e) Oven-Drying f) Weighing the Oven-Dried Aggregate (A)

FIGURE 2.9 Determining specific gravity and absorption. Photos taken at the Farmingdale State College of the State University of New York.

Similarly, the bulk SSD specific gravity is calculated as shown in Eq. 2.7.

$$G_{sb}(\text{SSD}) = \frac{B}{B - C} \tag{2.7}$$

Apparent dry specific gravity only includes the net volume of the aggregate particle; it does not include the volume of any water permeable voids and is determined as shown in Eq. 2.8.

$$\text{Apparent dry specific gravity, } G_{sa} = \frac{A}{A - C} \tag{2.8}$$

Note: The term 'dry' is mostly omitted. Thus, if nothing is mentioned (dry or wet) while presenting specific gravity, it is supposed to be the dry one.

Now, let us see how the ratio, $\dfrac{A}{A - C}$, is the representation of the bulk-specific gravity (G_{sa}).

$$\frac{A}{A - C} = \frac{A}{(A) - (A - \text{Weight of Water Displaced})}$$

$$= \frac{A}{\text{Weight of Water Displaced}} = G_{sa} \tag{2.9}$$

The volume of water displaced is the net volume of aggregate.

Absorption is the amount of water a batch of aggregate can absorb if it is submerged under water and is expressed as a percent of the dry weight of the aggregate. To determine the absorption, a batch of oven-dry aggregate is submerged under water for 24 hours to determine the amount of water it can absorb. The water absorbed is the increase in the mass of the aggregate due to the water that entered into the pores of the aggregate particles. Absorption is calculated as the amount of water absorbed with respect to the dry weight and is expressed as a percentage as shown in Eq. 2.10.

$$\% \text{ Absorption} = \frac{B-A}{A} \times 100 \qquad (2.10)$$

Aggregate absorption is a very useful quality because:

- High absorption values indicate the aggregate has the capacity to absorb more water when in service, thus becoming non-durable aggregate.
- Absorption can indicate the amount of asphalt binder the aggregate may absorb if the aggregate is used for an asphalt mix.

Very often, the *moisture content* of aggregate is required, especially for concrete mix design. Moisture content is defined as the amount of moisture present in the aggregate and is expressed as a percent with respect to its dry mass. Using an equation, it can be written as:

$$\text{Moisture Content} = \frac{\text{Mass of Moisture}}{\text{Mass of Dry Aggregate}} = \frac{D-A}{A} \times 100 \qquad (2.11)$$

where,
D = The received mass (or moist mass) of aggregate
A = Dry mass of aggregate

Highly absorptive aggregates in AC are generally not desirable. This is because asphalt binder that is absorbed by the aggregate is not available to coat the aggregate particle surface and is therefore not conducive for bonding. Therefore, the highly absorptive aggregates require more asphalt binder to develop the same film thickness as the less absorptive aggregates. Finally, the highly absorptive aggregates make the resulting AC more expensive. Aggregate absorption is also used in the PCC mix design. Highly absorptive aggregates require more water, as some amount of the water is to be absorbed.

Example 2.2 Specific Gravity

A batch of aggregates weighing 2.465 lb was received from a client. After oven drying the weight was found to be 2.415 lb. The sample was then submerged under normal water, weighing in at 1.775 lb. The sample was taken out and the surfaces were dried using a damp towel. It was then weighed again, and found to be 2.995 lb. Determine the following properties of the batch of aggregates, using these test data:

   a) The moisture content of the received sample
   b) Absorption of the sample in percent of the dry weight
   c) Bulk dry specific gravity

d) Bulk SSD specific gravity
e) Apparent dry specific gravity

Solution

Given

As received weight, D = 2.465 lb
Dry weight, A = 2.415 lb
Submerged weight, C = 1.775 lb
SSD weight, B = 2.995 lb

a) (From Eq. 2.11)

$$\text{Moisture Content} = \frac{\text{Mass of Moisture}}{\text{Mass of Dry Aggregate}} = \frac{D-A}{A} \times 100$$

$$= \frac{2.465 - 2.415}{2.415} \times 100$$

$$= 2.07\%$$

b) (From Eq. 2.10)

$$\% \text{ Absorption} = \frac{B-A}{A} \times 100$$

$$= \frac{2.995 - 2.415}{2.415} \times 100$$

$$= 24.02\%$$

c) (From Eq. 2.5) Bulk dry specific gravity,

$$G_{sb} = \frac{A}{B-C}$$

$$= \frac{2.415}{2.995 - 1.775}$$

$$= 1.980$$

[Note: The terms 'bulk' and/or 'dry' are mostly omitted]

d) (From Eq. 2.7) Bulk SSD specific gravity,

$$G_{sb}(\text{SSD}) = \frac{B}{B-C}$$

$$= \frac{2.995}{2.995 - 1.775}$$

$$= 2.455$$

e) (From Eq. 2.8) Apparent dry specific gravity,

$$G_{sa} = \frac{A}{A-C}$$

$$= \frac{2.415}{2.415 - 1.775}$$

$$= 3.773$$

[**Note again:** The terms 'bulk' and/or 'dry' are mostly omitted. Thus, specific gravity means the bulk dry specific gravity]

Answers:

a) *Moisture Content = 2.1%*
b) *Absorption = 24.0%*
c) *Bulk dry specific gravity = 1.980*
d) *Bulk SSD specific gravity = 2.455*
e) *Apparent dry specific gravity = 3.773*

Note: Specific gravity is expressed in three significant figures after the decimals and moisture content is expressed to the nearest 0.1%.

2.3.1.4 Soundness

Aggregate abrasion resistance is important, especially in AC, as aggregate in AC is intended to resist crushing, degradation, and disintegration due to the impact loading by traffic. The Los Angeles (LA) abrasion test (Figure 2.10) is used to determine aggregate toughness

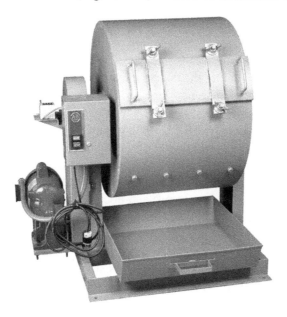

FIGURE 2.10 Los Angeles Abrasion Test apparatus. Courtesy of Test Mark Industries, 995 North Market St., East Palestine, Ohio. Used with Permission.

and abrasion characteristics. In the LA abrasion test by ASTM C 131 and C 535, coarse aggregate samples and steel spheres are placed in a drum and subjected to abrasion, impact, and grinding by rotating the drum. Weaker aggregates break down with the rotation of the drum. After rotating for 500 revolutions at a constant speed of 30–33 rpm, the percentage of the aggregate broken down is measured, which is known as the LA abrasion value.

2.3.1.5   Cleanliness and Deleterious Materials

Aggregates are desired to be fairly clean when used in concrete and asphalt mixtures. The presence of some deleterious pollutants such as vegetation, weaker particles, clay lumps, excess dust, and vegetable matter are not desirable, as these adversely affect the aggregates' performance. Vegetation degrades and loosens the bonding and creates hollow space once degraded. Weaker particles, clays, dust, etc. have less strength and make the aggregate softer. Some of the adverse effects of several deleterious substances are listed below:

- Organic impurities delay the setting and hardening of concrete and reduce strength gain.
- Coal, lignite, clay lumps, and friable particles increase popouts and reduce durability.
- Fines passing a No. 200 sieve (smaller than 0.075 mm) weaken the bond between asphalt and aggregates.

To test for clay lumps or friable particles, a batch of specimen is washed and dried to remove the fines. The remaining specimen is separated into different sizes. Each size is then weighed and soaked in water for 24 hours. Particles that can be broken down into fines by hand are classified as the clay lumps or friable material. The amount of this material is calculated by percentage of the total specimen's weight.

2.3.1.6   Sand Equivalent (SE) Test

Every batch of aggregate (especially fine aggregate) has a mixture of desirable coarse–fine particles, such as sand and undesirable clay, plastic fines, and dust. Some amount of fines is required to fill the voids among the coarse particles. On the other hand, too much clay, plastic fines, or dust coat aggregate particles and prevent proper asphalt binder–aggregate bonding. In addition, too many fines means too much asphalt binder is required to produce a certain binder film thickness. Too much asphalt binder is a waste of money and produces a highly deformable mix. Thus, the amount of fines must be optimized.

The sand equivalent (SE) test is a rapid field test used to determine the proportion of fine dust or clay-like materials in fine aggregate. According to the AASHTO T 176, a sample of aggregate passing the No. 4 (4.75 mm) sieve and flocculating agent are poured into a cylinder, as shown in Figure 2.11. The cylinder is agitated to loosen the clay-like coatings from the sand particles. The clay-like material tends to float above the sand. After a period, the height of suspended clay and the height of sand are determined. The ratio of the height of sand over the height of clay is considered the SE, as shown in Eq. 2.12. The larger the sand equivalent value, the cleaner (less fine dust or clay-like material) the aggregate.

$$SE = \frac{\text{Sand Height}}{\text{Clay Height}} \times 100 \qquad (2.12)$$

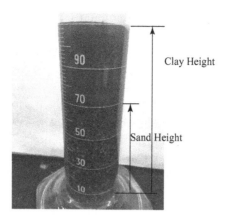

FIGURE 2.11 Sand equivalent testing.

Example 2.3 Sand Equivalency

In a sand equivalent test, a clay layer of 12-mm height is deposited on a 22-mm height of sand particles. Calculate the SE of the specimen.

Solution

Sand Height = 22 mm
Clay Height = 22 + 12 mm = 34 mm
From Eq. 2.12: Sand equivalent,

$$SE = \frac{\text{Sand Height}}{\text{Clay Height}} \times 100$$

$$= \frac{22}{34} \times 100$$

$$= 64.7\%$$

Answer: 65

Note: As per the AASHTO T 176, if the calculated sand equivalent is not a whole number, report it as the next higher whole number.

2.3.1.7 Bulk Density

Bulk density, or the rodded density, is the mass of aggregate per unit bulk volume. *Bulk volume* is the volume with respect to the whole volume of a batch of aggregate, considering the solid aggregates and the voids among the aggregates. A *void* is the open space among aggregate particles in a batch of aggregates (not occupied by solid particles). The coarser the aggregate particles, the larger the void spaces between the particles. Sometimes, the term 'unit weight' is used instead of bulk density. *Unit weight* is the weight (mass × gravitational acceleration) of aggregate per unit bulk volume. Bulk density or unit weight is required to estimate the weight of the material required to fill a certain volume. For example, a space of 10 ft by 20 ft by 5 ft is to be filled with natural sand having a unit weight of 110 lb/ft³. Let us calculate the required amount of sand to fill this space. The volume of the space is $10 \times 20 \times 5 = 1,000$ ft³. The amount of sand required is 110 lb/ft³ $\times$ 1,000 ft³ = 110,000 lb.

According to the ASTM C 29 test protocol, one-third of a specimen is placed in the mold first. The surface is then leveled using fingers and rodded 25 times evenly. Another third is filled up and rodded 25 times in a similar manner. Finally, the mold is filled up, overflowing the mold, and rodded 25 times again. The surface is leveled with fingers and then smoothed using a straightedge in such a way that any slight projections of the larger pieces of coarse aggregate approximately balance the larger voids in the surface below the top of the mold. This can be difficult for coarse aggregate. The mass of the filled material is measured by subtracting the mold mass from the mass of mold filled with the material. The final steps of determining the bulk density and the voids are shown in Figure 2.12. The bulk density and voids are then calculated using Eqs. 2.13 and 2.14.

$$\text{Bulk Density}, \rho_{\text{Rod}} = \frac{\text{Mass of Aggregate}}{\text{Volume of Mold}} \qquad (2.13)$$

$$\text{Percent Void}, \% \text{ Void} = \left[\frac{(G_{sb}\, \rho_w) - \rho_{Rod}}{G_{sb}\, \rho_w} \right] \times 100 \qquad (2.14)$$

| Equipment Required | Filling the Last Layer |
| Rodding the Last layer | Leveling the Surface |

FIGURE 2.12 Different steps of determining bulk density and air voids of fine aggregates. Photos taken at the Colorado State University–Pueblo.

where,

G_{sb} = Bulk dry specific gravity determined earlier

ρ_w = Density of water (62.4 pcf or 1,000 kg/m³)

Example 2.4 Air Void

A batch of aggregates weighing 22.4 lb was received from a client. After oven drying, the weight was found to be 20.2 lb. The sample was then submerged under normal water, weighing in at 17.7 lb. The sample was taken out and the surfaces were dried using a damp towel. It was then weighed again, and found to be 24.9 lb. Some dry aggregates were removed and placed in a 0.3 ft³ mold from this batch of aggregates to conduct the rodded density test. The rodded density was found to be 150 pcf (lb/ft³).

Determine the air void of the aggregate while conducting the rodded density test.

Determine the total void space volume inside the mold when filled with aggregate.

Solution

Given

As received weight, D = 22.4 lb

Dry-weight, A = 20.2 lb

Submerged-weight, C = 17.7 lb

SSD weight, B = 24.9 lb

Rodded density, ρ_{Rod} = 150 pcf

From Eq. 2.14: $\%\text{Void} = \left[\dfrac{(G_{sb}\,\rho_w) - \rho_{Rod}}{G_{sb}\,\rho_w} \right] \times 100$

This equation requires G_{sb} to determine the air void.

From Eq. 2.5: Bulk dry specific gravity,

$$G_{sb} = \frac{A}{B-C} = \frac{20.2}{24.9-17.7} = 2.806$$

$$\%\text{Void} = \left[\frac{(G_{sb}\,\rho_w) - \rho_{Rod}}{G_{sb}\,\rho_w} \right] \times 100$$

$$= \left[\frac{(2.806 \times 62.4\,\text{pcf}) - 150\,\text{pcf}}{2.806 \times 62.4\,\text{pcf}} \right] \times 100$$

$$= 14.3\%$$

Volume of void = 0.3 ft³ × 14.3% = 0.0429 ft³ = 0.0429 × 12³ = 74.13 in.³

Answers:

Air void = 14%

Void space = 74.13 in.³

Note: ASTM C 29 recommends reporting the void content to the nearest 1%. There is no recommendation for the void space.

2.3.2 CHEMICAL CHARACTERIZATIONS

Two major chemical characterizations are performed on aggregates. These are:

a) Alkali-aggregate reactivity
b) Affinity for asphalt

2.3.2.1 Alkali-Aggregate Reactivity

Alkali-aggregate reaction is the expansive reaction that occurs in PCC between alkali (available in cement) and silica (available in aggregates). In the presence of moisture, the alkalis found in cement break down the silica in the aggregate, producing an expansive gel. This expansion causes tensile forces in PCC, leading to loss of strength and resulting in map or pattern cracking, shown in Figure 2.13. This reaction can be controlled by:

- Avoiding susceptible aggregates, such as siliceous limestone, chert, shale, volcanic glass, synthetic glass, sandstone, opaline rocks, and quartzite. River rocks are sometimes susceptible.
- By using the pozzolanic admixture, as the silica contained in a pozzolan may react with the alkali in the cement, leaving less alkali available for the silica within the aggregate.
- Using low-alkali cement
- Lowering the water–cement ratio, which limits the supply of water to the alkali–silica gel formation.

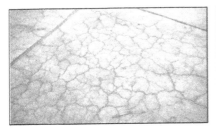

FIGURE 2.13   Pattern cracking due to the effect of alkali–silica reactivity. Photos taken in Pueblo, Colorado.

Several laboratory tests are available to check the potential alkali–aggregate reactivity, such as ASTM C 227, ASTM C 289, and ASTM C 586.

2.3.2.2 Affinity for Asphalt

The *affinity for asphalt* is an important property of aggregate when it is used in the preparation of an asphalt mixture. If the aggregates have a greater affinity for water than asphalt (hydrophilic), they attract water while in service. After doing so, the bond between the aggregate and asphalt breaks up. Two conditions may arise:

- If the breaking of the bond begins at the bottom of the asphalt layer and propagates upward, it is called *stripping*. This happens if there is a source of water at the bottom of the asphalt layer, such as capillary rise, infiltrated water that could not be drained, a high ground water table, etc.

- When the breaking of the bond begins at the surface of the asphalt layer and propagates downward, it is called *raveling*. This happens if there is a source of water at the surface of the asphalt layer, such as rainfall, snow melting, etc.

Some aggregates have a greater affinity for asphalt than water (hydrophobic), which is very good for an asphalt mixture. These aggregates (or mixes prepared using these aggregates) do not suffer from stripping and raveling problems. Thus, aggregate chemistry is an important factor for stripping and raveling. There are many standard tests to measure the stripping and raveling susceptibility of aggregates. In the ASTM D 1075, a batch of compacted asphalt specimens are subjected to freeze–thaw conditioning. The decrease in strength is measured and compared with the values of unconditioned specimens. In the ASTM D 3625, a batch of compacted asphalt specimens are boiled in water, and the percentage of particles stripped is measured. Every agency sets its threshold values for its own region considering local conditions.

2.3.3 MECHANICAL CHARACTERIZATIONS

Mechanical characterization relates to the response of a material under a mechanical load. The most common mechanical properties of aggregates include, but not limited to, the following:

a) Resilient modulus
b) California bearing ratio
c) *R*-value

These tests are commonly used in pavement design and thus, are covered in that course.

2.4 GRADATION OF AGGREGATES

The particle size distribution, or the gradation of aggregates, is a very influential aggregate characteristic in determining the aggregate's performance as a pavement material. In AC, the gradation helps determine almost every important property, including stiffness, stability, durability, permeability, workability, fatigue resistance, frictional resistance, and moisture susceptibility (Brown et al. 2009). In PCC, the gradation helps determine durability, porosity, workability, cement and water requirements, strength, and shrinkage. Because of this, the gradation is a primary concern in AC and PCC mix designs, so most agencies specify the allowable aggregate gradations for both.

2.4.1 SIEVE ANALYSIS

A gradation test (Figure 2.14) is performed on a sample of aggregate in a laboratory. A typical sieve analysis involves a nested column of sieves with a wire mesh cloth (screen). A representative weighed sample is poured into the top sieve, which has the largest screen openings. The column is typically placed in a mechanical shaker. The shaker shakes the column for a fixed amount of time. After the shaking is complete, the material on each sieve is weighed. The weight of the sample on each sieve is then divided by the total weight of the batch to calculate the percentage retained on each sieve. Then, the 'Cumulative % Retained' is calculated by summing up the '% Retained' in the corresponding sieve plus

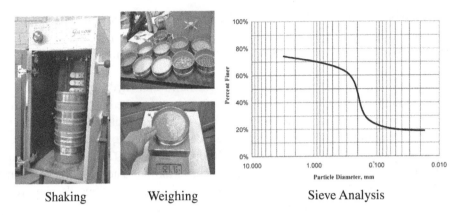

Shaking	Weighing	Sieve Analysis

FIGURE 2.14 Sieve analysis procedure. Photos taken at the Farmingdale State College of the State University of New York.

that on the larger sieves. The 'Percent Finer' for a certain sieve is calculated as 100 minus the 'Cumulative % Retained' for that sieve. The results of sieve analysis are provided in a graphical form to identify the type of gradation of the aggregate. A common practice is to present a 'Percent Finer versus Particle Diameter' curve.

2.4.2 MAXIMUM AGGREGATE SIZES

Maximum aggregate size can affect AC and PCC in several ways. In AC, instability may result from excessively small maximum sizes, and poor workability and/or segregation may result from excessively large maximum sizes (Brown et al. 2009). In PCC, large maximum sizes may not fit between reinforcing bar openings, but they generally cause an increase in the PCC strength because the water–cement ratio can be lowered. ASTM C 125 defines the maximum aggregate size in one of two ways:

Maximum Size. The traditional definition of the *maximum size* is the smallest sieve through which 100% of the aggregate sample particles pass. *Superpave* defines the maximum aggregate size as one sieve larger than the nominal maximum size.

Nominal Maximum Aggregate Size (NMAS). The traditional definition of the *NMAS* is the largest sieve that retains some of the aggregate particles, but generally not more than 10% by weight. Superpave defines NMAS as one sieve size larger than the first sieve, to retain more than 10% of the material. Each stock of aggregate is represented by its NMAS.

Example 2.5 Aggregate Size

After the sieve analysis, the mass retained on each sieve is shown in Table 2.3:

a) Draw the sieve analysis curve (% Finer vs. Sieve Size)
b) Determine the NMAS (Superpave Criteria)
c) Determine the maximum aggregate size (Superpave Criteria)

TABLE 2.3
Sieve Analysis Data for Example 2.5

Sieve No.	Sieve Size (mm)	Mass Retained (g)
1.0 in.	25	0
0.75 in.	19	0
0.375 in.	9.5	31
No. 4	4.75	62
No. 8	2.4	178
No. 16	1.19	210
No. 30	0.6	298
No. 50	0.297	278
No. 100	0.149	118
No. 200	0.075	33
Pan	Pan	9

d) Determine the NMAS (Traditional Criteria)
e) Determine the maximum aggregate size (Traditional Criteria)

Solution

a) Draw the sieve analysis curve (% Finer vs. Sieve Size)
  - First, calculate the % retained in each sieve, as follows (Eq. 2.15):

$$\% \text{ Retained} = \frac{\text{Amount Retained}}{\text{Total Amount}} \times 100 \qquad (2.15)$$

  - Then, the 'Cumulative % Retained' is calculated by summing up the '% Retained' on the corresponding sieve plus that on the larger sieves. For example, 'Cumulative % Retained' in No. 16 = 17.3 + 14.6 + 5.1 + 2.5 + 0 + 0 = 39.5, shown in Table 2.4.
  - '% Finer' is calculated as 100 – 'Cumulative % Retained'.
  - Plot the '% Finer' versus 'Sieve Size' curve, as shown in Figure 2.15.
b) Determine the NMAS (Superpave Criteria)
  1. Superpave defines NMAS as one sieve size larger than the first sieve to retain more than 10% of the material. Here, the first sieve to retain more than 10% of the material is No. 8 (2.4 mm). Therefore, the NMAS of the aggregate is 4.75 mm, by the Superpave criteria.
c) Determine the maximum aggregate size (Superpave Criteria)
  2. Superpave defines the maximum aggregate size as one sieve larger than the nominal maximum size. As the NMAS of the aggregate is 4.75 mm, then the maximum size is 9.5 mm by the Superpave criteria.
d) Determine the NMAS (Traditional Criteria)
  3. The traditional definition of NMAS is the largest sieve that retains some aggregate, but generally not more than 10% by weight. Therefore, the NMAS of the aggregate is 9.5 mm by the traditional criteria.

TABLE 2.4
Sieve Analysis for Example 2.5

Sieve No	Sieve Size (mm)	Mass Retained (g)	% Retained (from Eq. 2.15)	Cumulative % Retained	% Finer
1.0 in.	25	0	0	0	100
0.75 in.	19	0	0	0.0	100
0.375 in.	9.5	31	2.5	2.5	97
No. 4	4.75	62	5.1	7.6	92
No. 8	2.4	178	14.6	22.3	78
No. 16	1.19	210	17.3	39.5	60
No. 30	0.6	298	24.5	64.0	36
No. 50	0.297	278	22.8	86.9	13
No. 100	0.149	118	9.7	96.5	3
No. 200	0.075	33	2.7	99.3	1
Pan	Pan	9	0.7	100.0	0
		Total = 1,217			

Note: MS Excel has been used to perform the calculations. Some rounding off error may occur if a calculator is used. This type of error is always acceptable in engineering.

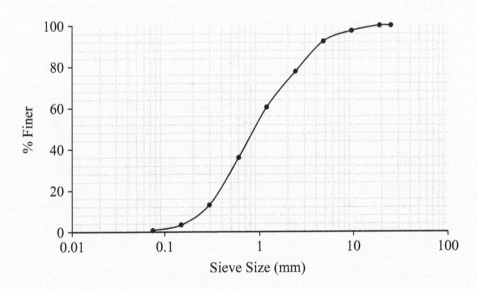

FIGURE 2.15 Sieve analysis curve for Example 2.5.

e) Determine the maximum aggregate size (Traditional Criteria)

4. The traditional definition of the maximum size is the smallest sieve through which 100% of aggregate sample particles pass. Therefore, the maximum size of the aggregate is 19 mm by the traditional criteria.

2.4.3 DESIRED GRADATION

Gradation has a solid effect on material performance. The optimum gradation is a complicated topic. The desired characteristics, loading, environmental, material, structural, and mix property inputs vary depending on the material (AC or PCC). This section presents some important guidelines applicable to common, dense-graded mixes. The optimum gradation produces the maximum density. This involves a particle arrangement in which smaller particles are inserted among the larger particles, reducing the void space between particles. This creates more particle-to-particle contact, increases stability and reduces water infiltration in AC. In PCC, this reduced void space reduces the amount of cement paste required. However, having some minimum amount of void space present is necessary to:

- Provide adequate volume for the binder
- Promote rapid drainage and resistance to frost action for base and subbases

Therefore, although it may not be the optimum aggregate gradation, a maximum density gradation does provide a common reference. A widely used equation to describe the maximum density gradation is shown in Eq. 2.16.

$$P = \left(\frac{d}{D}\right)^{n} \times 100 \tag{2.16}$$

where,
   P = % finer than the sieve
   d = Aggregate size being considered
   D = Maximum aggregate size to be used
   n = Parameter which adjusts the fineness or coarseness, = 0.45 by the Federal Highway Administration (FHWA) (Brown et al. 2009)

Table 2.5 shows an example of calculating the densest gradation curve for a 19-mm NMAS aggregate. The 0.45 power curve is drawn on plain axes, with Column 2 in the x-axis and Column 4 in the y-axis. If this gradation could be achieved for a 19-mm aggregate batch, then that would be the densest gradation. It is important to reiterate that the densest gradation is not commonly desired; some amount of void space is essential. Thus, the optimum gradation is sought out. The procedure to find out the optimum gradation is discussed in the concrete mix design and asphalt mix design chapters, as each of these mix designs has its own requirement.

2.4.4 GRADATION TYPES

Several common terms are used to classify gradation, as shown in Figure 2.16. These are not precise technical terms, but rather terms that refer to gradations which share common characteristics:

Dense or Well Graded. Typical dense gradations are near the 0.45 power curve but may not be directly on it.

TABLE 2.5
Calculations for a 0.45 Power Curve Using a 19-mm
Aggregate

Particle Size (mm)	(Size, mm)$^{0.45}$	%Passing Calculation	%Passing
Column 1	Column 2	Column 3 (from Eq. 2.16)	Column 4
19.0	3.762	$P = \left(\dfrac{19.0}{19.0}\right)^{0.45} \times 100$	100
12.5	3.116	$P = \left(\dfrac{12.5}{19.0}\right)^{0.45} \times 100$	82.8
9.5	2.754	$P = \left(\dfrac{9.5}{19.0}\right)^{0.45} \times 100$	73.2
2.00	1.366	$P = \left(\dfrac{2.00}{19.0}\right)^{0.45} \times 100$	36.3
0.300	0.582	$P = \left(\dfrac{0.30}{19.0}\right)^{0.45} \times 100$	15.5
0.075	0.312	$P = \left(\dfrac{0.075}{19.0}\right)^{0.45} \times 100$	8.3

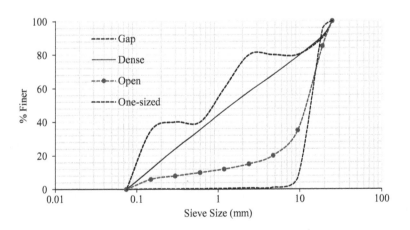

FIGURE 2.16   Different types of gradations.

Gap Graded. Gap graded refers to the gradation containing only a small percentage of aggregate particles in the mid-size range. The curve is flat in the mid-size range. Some PCC mix designs use gap graded aggregate to provide a more economical mix, since less sand may be used for a given workability. AC gap graded mixes can be prone to segregation during the placement.

Open Graded. Open graded refers to a gradation that contains only a small percentage of aggregate particles in the small range. This results in a greater amount of air voids because there are not enough small particles to fill the voids between the larger particles. The curve is nearly vertical in the mid-size range, and flat and near zero in the small-size range.

One-Sized or Uniformly Graded. Uniformly graded refers to the gradation containing most of the particles in a very narrow size range. All the particles are very similar in size. The curve is steep and only occupies the narrow size range specified.

2.4.5 FINENESS MODULUS

The *Fineness Modulus* (FM) is an index value obtained by summing the cumulative percentage of the sample of an aggregate retained on a specified series of sieves, and then dividing the sum by 100 (shown in Eq. 2.17). It is described in the ASTM C 125, and is a single number used to describe a gradation curve. The larger the FM, the coarser the aggregate. A typical fineness modulus for fine aggregate used in PCC is between 2.70 and 3.00. The standard sieves used are No. 100, No. 50, No. 30, No. 16, No. 8, as well as No. 4, 3/8-in., ¾-in., 1½-in., 3-in., and 6.0-in. No other sieves are used in the calculation.

$$FM = \frac{\sum \text{"Cumulative \% Retained" on standard sieves}}{100} \qquad (2.17)$$

The FM commonly ranges from 2.2 to 3.2 with the following classification:

- Fine Sand: FM of 2.2–2.6
- Medium Sand: FM of 2.6–2.9
- Coarse Sand: FM of 2.9–3.2

Example 2.6 Fineness Modulus

The sieve analysis results of a batch of aggregates are shown in Table 2.6. Calculate the fineness modulus of the aggregate batch.

TABLE 2.6
Sieving Output for Example 2.6

Sieve	Sieve Size (mm)	Mass Retained (g)
2.00 in.	50	0
1.50 in.	37.5	8
1.00 in.	25	19
0.75 in.	19	44
0.375 in.	9.5	55
No. 4	4.75	62
No. 8	2.4	178
No. 16	1.19	210
No. 30	0.6	298
No. 50	0.297	278
No. 100	0.149	118
No. 200	0.075	33
Pan	Pan	9

Solution

Step 1. Calculate the cumulative % retained in each sieve (Table 2.7).

TABLE 2.7

Sieve Analysis for Example 2.6

Sieve	Sieve Size (mm)	Mass Retained (g)	% Retained	Cumulative % Retained
2.00 in.	50	0	0.0	0.0
1.50 in.	37.5	8	0.6	0.6
1.00 in.	25	19	1.4	2.1
0.75 in.	19	44	3.4	5.4
0.375 in.	9.5	55	4.2	9.6
No. 4	4.75	62	4.7	14.3
No. 8	2.4	178	13.6	27.9
No. 16	1.19	210	16.0	43.9
No. 30	0.6	298	22.7	66.6
No. 50	0.297	278	21.2	87.8
No. 100	0.149	118	9.0	96.8
No. 200	0.075	33	2.5	99.3
Pan	Pan	9	–	–

Note: MS Excel has been used to perform the calculations. Some rounding off error may occur if a calculator is used. This type of error is acceptable.

Step 2. Identify the standard sieves. Here, all sieves except 2.0 in., 1.00 in., and No. 200 are standard sieves.

Step 3. From Eq. 2.17:

$$FM = \frac{\sum \text{"Cumulative \% Retained" on standard sieves}}{100}$$

$$= \frac{0.6 + 5.4 + 9.6 + 14.3 + 27.9 + 43.9 + 66.6 + 87.8 + 96.8}{100}$$

$$= 3.53$$

Answer: 3.53

2.5 BLENDING OF AGGREGATES

AC requires the combining of two or more aggregates, having different gradations, to produce an aggregate blend that meets the gradation specifications for a particular asphalt mix. Generally, a single aggregate source is unlikely to meet the gradation requirements for PCC or AC mixes. Thus, the blending of aggregates from two or more sources is required to satisfy the specifications.

2.5.1 BLENDING PROCEDURES

Depending on the number of sources, aggregate blending can be performed by trial and error or by a graphical method. These are described below:

2.5.1.1 Two Aggregate Sources

Although trial and error is the most common method of blending aggregates, a graphical method can give more latitude to the materials engineer in selecting the proportion of aggregate piles. However, the graphical method is suitable only for two aggregates stokes. An example is shown here to describe the graphical method step by step. Assume the gradation of two aggregates and the target specification of the combined aggregate are as listed in Table 2.8. Now, let us see how the proportions of each aggregate can be found using the graphical method.

TABLE 2.8

Example Data for Aggregate Blending from Two Sources

Size	Aggregate A % Finer	Aggregate B % Finer	Specification % Finer
19.0 mm	100	100	100
12.5 mm	70	90	80–85
9.5 mm	65	80	70–75
No. 4 (4.75 mm)	45	65	50–60
No. 8 (2.36 mm)	35	60	40–50
No. 30 (0.6 mm)	20	40	30–40
No. 100 (0.15 mm)	10	30	20–30
No. 200 (0.075 mm)	5	10	7–10

The following steps are followed to determine a satisfactory aggregate blend using the graphical method:

1) Plot the percentages passing (or % Finer) through each sieve on the right vertical axis for aggregate A and on the left vertical axis for aggregate B, shown in Figure 2.17.
2) The lower horizontal axis from the left to right is the '% of Aggregate A' to be used in the blend. The upper horizontal axis from the right to left is the '% if Aggregate B' to be used in the blend.
3) For each sieve size (19 mm, 12.5 mm, etc.), connect the left and right axes.
4) Plot the specification limits of each sieve on the corresponding sieve lines; that is, place a mark on the 9.5-mm sieve line corresponding to 70% and 75% on the vertical axis.
5) Connect the upper- and lower-limit points on each sieve line, as shown by the dashed lines.
6) Draw vertical lines through the rightmost point of the upper-limit line and the leftmost point of the lower-limit line, as shown by the hashed area. If the upper- and lower-limit lines overlap, then no combination of the aggregates meets the specifications.
7) Any vertical line drawn between these two vertical lines identifies an aggregate blend that does meet the specification. The intersection with the upper axis defines the percentage of aggregate B required for the blend. The projection to the lower

axis defines the percentage of aggregate A required. It can be seen that 40–48% of A, and 52–60% of B, meets the specification. Specifically, 45% of A and 55% of B would be a good proportion.

8) Projecting intersections of the blend line and the sieve lines horizontally gives an estimate of the gradation of the blended aggregate.

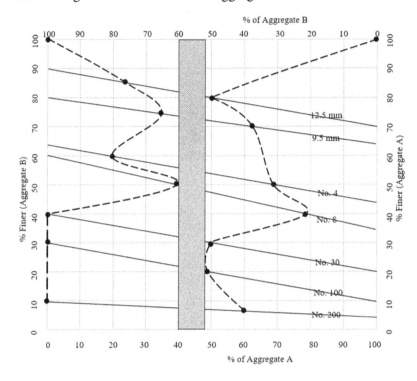

FIGURE 2.17   Graphical method for determining two-source aggregates blend.

2.5.1.2   More than Two Aggregate Sources

A trial and error procedure is the most common method of blending aggregates when more than two sources of aggregates are sought out. The steps can be summarized as follows:

Step 1 – Obtain the required data
  a.   The gradation of each material must be determined.
  b.   The design limits for the type of mix must be obtained.
Step 2 – Select a target value for trial blend (optional)
      The target value for the combined gradation must be within the design limits of the specification – the value desired by the design agency. This value now becomes the target for the combined gradation. This is an optional step; for learning purposes, the average value of the specification range can be used.
Step 3 – Estimate the proportions
      Estimate the correct percentage of each aggregate needed to obtain a combined gradation near the target value. The formula expressing the combination is shown in Eq. 2.18.

$$p = Aa + Bb + Cc + \cdots \qquad (2.18)$$

where,

p = Percent of materials passing a given sieve for the combined aggregates

$A, B, C, \ldots$ = Percent of materials passing a given sieve for each aggregate, A, B, C....

$a, b, c,\ldots$ = Proportions of aggregate A, B, C, ... to be used in the blend $[a + b + c + \ldots = 1.0]$

Step 4 – Calculate the combined gradation

This calculation shows the results of the estimate from Step 3. The method of calculating the combined gradation is shown in Example 2.7.

Step 5 – Compare the result with the target value.

If the calculated gradation is within the specification (or close to the target value), no further adjustments need to be made. If this is not the case, an adjustment in the proportions must be made and the calculations repeated. The second trial should be closer due to the knowledge received from the first trial. The trials are continued until the proportions of each aggregate which meet the specification are determined. If the aggregates do not combine within the desired range, it may be necessary to use or add different materials.

Example 2.7 Blending of Three Batches

Determine the proportions of Aggregate A, Aggregate B, and Aggregate C, shown in Table 2.9 to prepare a blend to satisfy the job specification.

TABLE 2.9
Sieving Output for Example 2.7

Size	Aggregate 1 % Finer	Aggregate 2 % Finer	Aggregate 3 % Finer	Specification % Finer
25.0 mm	100	100	100	100
19.0 mm	100	100	92	90–100
12.5 mm	80	99	47	90 maximum
No. 8	50	66	36	28–49
No. 200	15	10	0	2–8

Solution

Step 1 – Obtain the required data

Given in the problem

Step 2 – Select a target value for trial blend

For learning purposes, any value within the range of the specification will work.

Step 3 – Estimate the proportions

The first estimate used for this trial blend is 15% of Aggregate 1, 40% of Aggregate 2, and 45% of Aggregate 3. Remember, the sum of the proportions (15 + 40 + 45) must always equal 100.

Steps 4 and 5 – Calculate the combined gradation and compare with the target value, shown in Table 2.10.

TABLE 2.10

Comparison of the Combined Gradation with the Specification for Example 2.7

Size	Agg. 1 % Finer	Agg. 2 % Finer	Agg. 3 % Finer	Total Blend	Specification % Finer	Comment
25.0 mm	100×0.15	100×0.40	100×0.45	100	100	Okay
19.0 mm	100×0.15	100×0.40	92×0.45	96	90–100	Okay
12.5 mm	80×0.15	99×0.40	47×0.45	73	90 maximum	Okay
No. 8	50×0.15	66×0.40	36×0.45	50	28–49	Okay
No. 200	15×0.15	10×0.40	0×0.45	6.25	2–8	Okay

The blended values meet the specification. Therefore, 15% of Aggregate 1, 40% of Aggregate 2, and 45% of Aggregate 3 can be recommended. In real life, the design agency may recommend a target value from the specification limits. Several iterations may be required to satisfy those targets.

2.5.2 Properties of Blended Aggregates

Other than specific gravity and density, the properties of blended aggregates such as angularity, absorption, strength, and modulus can be determined using Eq. 2.19:

$$X = \sum P_i X_i = P_1 X_1 + P_2 X_2 + P_3 X_3 + \cdots \tag{2.19}$$

where,

X = Blended property
$P_i\ (P_1, P_2, ..)$ = Proportions of different stockpiles
$X_i\ (X_1, X_2, ..)$ = Properties of different stockpiles

There are some properties of aggregates which do not apply to the whole stockpiles, but rather are specific to the coarse aggregates or the fine aggregates. For example, the coarse aggregate angularity applies to only the coarse fraction of the stockpiles; the fine aggregate angularity applies to only the fine fraction of the stockpiles. The properties of blended aggregates for these properties can be determined using Eq. 2.20:

$$X = \frac{\sum P_i x_i p_i}{\sum P_i p_i} = \frac{P_1 x_1 p_1 + P_2 x_2 p_2 + P_3 x_3 p_3 + \cdots}{P_1 p_1 + P_2 p_2 + P_3 p_3 + \cdots} \tag{2.20}$$

where,

X = Blended property
$P_i\ (P_1, P_2, ..)$ = Proportions of different stockpiles
$x_i\ (x_1, x_2, ..)$ = Properties of different stockpiles
$p_i\ (p_1, p_2, ..)$ = Fraction of the aggregate that possesses the property, x_i

The specific gravity of the blended aggregate is obtained by Eq. 2.21:

$$G = \frac{1}{\dfrac{P_1}{G_1} + \dfrac{P_2}{G_2} + \dfrac{P_3}{G_3} + \cdots} \tag{2.21}$$

where,

P_1, P_2, .. = Proportions of different stockpiles
G_1, G_2, .. = Specific gravity of different stockpiles

To derive this equation, let us consider that a specimen from Aggregate Stockpile 1 has a mass of M_1 in grams and a volume of V_1 in cm^3.

Therefore, the specimen from Aggregate Stockpile 1's relative density (specific gravity), $G_1 = M_1/V_1$

Therefore, the specimen from Aggregate Stockpile 2's relative density (specific gravity), $G_2 = M_2/V_2$

where M_2 is the mass in grams and V_2 is the volume in cm^3 of Aggregate Particle 2.

If these aggregates are considered together, the relative density (specific gravity) of the combination is the total mass divided by the total volume, as shown in Eq. 2.22:

$$\text{Combined Specific Gravity, } G = \frac{M_1 + M_2 + \cdots}{V_1 + V_2 + \cdots} \tag{2.22}$$

$$G = \frac{1}{\dfrac{V_1 + V_2 + \cdots}{M_1 + M_1 + \cdots}}$$

$$G = \frac{1}{\dfrac{V_1}{M_1 + M_1 + \cdots} + \dfrac{V_2}{M_1 + M_1 + \cdots} + \cdots}$$

$$G = \frac{1}{\dfrac{V_1}{M_1 + M_1 + \cdots}\left(\dfrac{M_1}{M_1}\right) + \dfrac{V_2}{M_1 + M_1 + \cdots}\left(\dfrac{M_2}{M_2}\right) + \cdots}$$

$$G = \frac{1}{\dfrac{M_1}{M_1 + M_1 + \cdots}\left(\dfrac{V_1}{M_1}\right) + \dfrac{M_2}{M_1 + M_1 + \cdots}\left(\dfrac{V_2}{M_2}\right) + \cdots}$$

Now, $\dfrac{M_1}{M_1 + M_1 + \cdots} = P_1$ (Fraction of specimen from Aggregate Stockpile 1)

$\dfrac{M_2}{M_2 + M_2 + \cdots} = P_2$ (Fraction of specimen from Aggregate Stockpile 2)

$$\frac{1}{G_1} = \frac{V_1}{M_1}$$

$$\frac{1}{G_2} = \frac{V_2}{M_2}$$

Therefore,

$$G = \frac{1}{\dfrac{P_1}{G_1} + \dfrac{P_2}{G_2} + \cdots} \tag{2.23}$$

If there are n number of stockpiles of aggregates, then the combined specific gravity can be written as shown in Eq. 2.22.

$$G = \frac{1}{\dfrac{P_1}{G_1} + \dfrac{P_2}{G_2} + \cdots + \dfrac{P_n}{G_n}} \tag{2.24}$$

where,
P_n = Fraction of the specimen in the combined aggregate from Aggregate Stockpile n
G_n = Specific gravity of the specimen from Aggregate Stockpile n

Example 2.8 Modulus of Blended Aggregates

Two stockpiles, A and B, are blended together at the ratio of 3:7. Stockpile A has an average modulus value of 27 ksi, and Stockpile B has an average modulus value of 36 ksi. Calculate the modulus value of the blended aggregates.

Solution

Let Stockpile A be 1 and Stockpile B be 2. The blending fractions are 30% and 70%, i.e., 0.30 and 0.70 respectively, for Stockpile A and Stockpile B.
   Thus, given data:

Proportion of stockpile A, $P_1 = 0.30$
Proportion of stockpile B, $P_2 = 0.70$
Modulus value of stockpile A, $X_1 = 27$ ksi
Modulus value of stockpile B, $X_2 = 36$ ksi
From Eq. 2.19: $X = \sum P_i X_i = P_1 X_1 + P_2 X_2 + P_3 X_3 + \cdots$

Therefore, $X = 0.30 \ (27 \text{ ksi}) + 0.70 \ (36 \text{ ksi}) = 33.3$ ksi

Answer: 33.3 ksi

Example 2.9 Fine Aggregate Angularity of Blended Aggregates

Two stockpiles, 1 and 2, are blended together at the ratio of 2:8. Stockpile 1 has 45 coarse aggregates and Stockpile 2 has 76 coarse aggregates. If the fine aggregate angularities of Stockpile 1 and Stockpile 2 are 48 and 37, respectively, then calculate the fine aggregate angularity of the blended aggregates.

Solution

The blending fractions are 20% and 80%, i.e., 0.2 and 0.8 respectively, for Stockpile 1 and Stockpile 2. Stockpile 1 has 55 fine aggregates and Stockpile 2 has 24 fine aggregates.
Thus, given data:

Proportion of stockpile 1, $P_1 = 0.2$
Proportion of stockpile 2, $P_2 = 0.8$
Fraction of fine aggregates in stockpile 1, $p_1 = 1 - 0.45 = 0.55$
Fraction of fine aggregates in stockpile 2, $p_2 = 1 - 0.76 = 0.24$
Fine-aggregate angularities Stockpile 1, $x_1 = 48$
Fine-aggregate angularities Stockpile 2, $x_2 = 37$
From Eq. 2.20:

$$X = \frac{\sum P_i x_i p_i}{\sum P_i p_i} = \frac{P_1 x_1 p_1 + P_2 x_2 p_2 + P_3 x_3 p_3 + \cdots}{P_1 p_1 + P_2 p_2 + P_3 p_3 + \cdots}$$

Therefore,

$$X = \frac{(0.2)(48)(0.55) + (0.8)(37)(0.24)}{(0.2)(0.55) + (0.8)(0.24)} = 41$$

Answer: 41

Example 2.10 Specific Gravity of Blended Aggregates

Two stockpiles, 1 and 2, have bulk-specific gravities of 2.45 and 2.67, respectively. If 25% of the Aggregate 1 and 75% of the Aggregate 2 are blended together, calculate the specific gravity of the blended aggregates.

Solution

Given:

Proportion of Stockpile 1, $P_1 = 0.25$
Proportion of Stockpile 2, $P_2 = 0.75$
Specific gravity of Stockpile 1, $G_1 = 2.45$
Specific gravity of Stockpile 2, $G_2 = 2.67$
From Eq. 2.21:

$$G = \frac{1}{\dfrac{P_1}{G_1} + \dfrac{P_2}{G_2} + \dfrac{P_3}{G_3} + \cdots} = \frac{1}{\dfrac{0.25}{2.45} + \dfrac{0.75}{2.67}} = 2.611$$

Answer: 2.611

2.6   SAMPLING OF AGGREGATES

Sampling is a regular activity when dealing with materials. It is performed in almost all steps of aggregate production for quality control and assurance and provides material data to the concerned engineers and contractors. Samplers must exercise caution to obtain samples that are true in nature and representative of the materials. Simply, the collected samples should carry the behavior and condition of the actual aggregate piles. ASTM D 75 covers the sampling of coarse and fine aggregates for the preliminary investigation of the source, as well as acceptance or rejection of the materials. This standard describes the sample collection strategy, procedure, size, etc. for the following cases:

a) Sampling from a flowing aggregate stream (bins or belt discharge)
b) Sampling from a conveyor belt
c) Sampling from stockpiles
d) Sampling from transportation units
e) Sampling from a roadway (bases and subbases)
f) Sampling stone from quarries or ledges
g) Sampling roadside or bank run sand and gravel deposits

Samples for the initial investigations of the aggregate sources are obtained by the company responsible for collecting aggregate from the sources, such as mountains, water bodies, rock quarries, etc. Aggregate samples from the processing plants are obtained by the manufacturer, contractor, or other responsible parties for performing the work. Samples for the acceptance or rejection decisions by the purchaser are obtained by the purchaser.

Samples brought to the laboratory must be reduced by sample splitters or by quartering. ASTM C 702 covers three methods (mechanical splitter, quartering, and miniature stockpile sampling) for the reduction of large samples of aggregate to the appropriate size for testing. It employs techniques to minimize variations in tested properties, between the test specimen selected and the larger sample. Figure 2.18 shows a mechanical splitter and quartering device.

Mechanical Splitter   Quartering

FIGURE 2.18   Aggregate splitting devices. Photos taken at the Colorado State University–Pueblo.

2.7   CHAPTER SUMMARY

Aggregate is a very versatile, inhomogeneous material used in the daily lives of civil and construction engineering professionals. It is composed of very fine material to very coarse material, river deposits to steel slag, Styrofoam to rocks, and so on. It can be classified based on geology, source, density, size, etc. Some physical properties used in civil and construction engineering are particle angularity, shape (flat, elongated), surface texture (smooth or rough), soundness (abrasion resistance), specific gravity, cleanliness, and the presence of fine dust. Chemical properties include alkali-aggregate reactivity and affinity for asphalt. Mechanical properties used in civil and construction engineering are the resilient modulus, California bearing ratio, and R-value.

Gradation is the very basic test conducted to determine the aggregate size distribution in a batch of aggregate. Then, the present gradation is very often (especially for highway aggregate) compared with the maximum density gradation possible. Blending is often required to combine aggregates of different sizes from different stockpiles. Blending of two different aggregate sources can be performed graphically, but the blending of more than two sources requires a trial and error analysis.

The properties of blended aggregates, such as angularity, absorption, strength, and modulus can be determined using the weighted average value, except for the specific gravity and density. Sampling of aggregate should be conducted following the standard specification, as the whole stockpile is evaluated based on the properties of the sampled aggregate. Thus, representative samples must be collected to avoid bias and error.

ORGANIZATIONS DEALING WITH AGGREGATES

National Stone, Sand and Gravel Association (NSSGA). The National Stone, Sand and Gravel Association (NSSGA) is the leading voice and advocate for the aggregates industry. Its members are responsible for the essential raw materials found in every home, building, road, bridge, and public works project, and represent more than 90% of the crushed stone and 70% of the sand and gravel produced annually in the United States.

Location: Alexandria, VA.
Website: https://www.nssga.org

ASTM International. ASTM International, formerly known as American Society for Testing and Materials, is an international standards organization that develops and publishes voluntary consensus technical standards for a wide range of materials, products, systems, and services. Founded in 1898 as the American Section of the International Association for Testing Materials, about 12,575 ASTM voluntary consensus standards operate globally.

Location: West Conshohocken, Pennsylvania
Website: https://www.astm.org/

AASHTO – The American Association of State Highway Transportation Officials. AASHTO is a nonprofit, nonpartisan association representing highway and transportation departments in the 50 states, the District of Columbia, and Puerto Rico. It represents all transportation modes, including air, highways, public transportation, active transportation, rail, and water. Its primary goal is to foster the development, operation, and maintenance of an integrated national transportation system.
Location: Washington, DC.
Website: https://www.transportation.org/

REFERENCES

Brown, E., Kandhal, P., Roberts, F., Kim, Y., Lee, D., and Kennedy, T. 2009. *Hot-Mix Asphalt Materials, Mixture Design, and Construction*, 3rd Edition. NAPA Research and Education Foundation, Lanham, MD.

NSSGA. 2019. National Stone, Sand and Gravel Association (NSSGA), https://www.nssga.org (last accessed 2/22/2019).

Nunnally, S. W. 2009. *Construction Methods and Management*, 7th Edition. Pearson, Upper Saddle River, NJ.

Somayaji, S. 2001. *Civil Engineering Materials*, 2nd Edition. Pearson, Upper Saddle River, NJ.

US Geological Survey (USGS). 2013. *Mineral Commodity Summaries 2013*. US Department of the Interior, US Geological Survey, Reston, VA.

FUNDAMENTAL OF ENGINEERING (FE) EXAM STYLE QUESTIONS

FE PROBLEM 2.1

Particles, smaller than 0.075 mm, are called fines. The recommended minimum size of fines is closest to:

A. 0.00075 µm
B. 0.00425 µm
C. 10^{-12} m
D. No minimum size

Solution: D

FE PROBLEM 2.2

What kinds of particles do not compact readily?

A. Angular
B. Rounded
C. Flat and elongated
D. Smooth

Solution: A

Angular particles do not compact as readily as rounded particles because their angular surfaces tend to lock up with one another and resist compaction.

FE PROBLEM 2.3

Which of the following statements is false?

A. Gravel retains on #200 sieves

B. Sand retains on #200 sieves
C. Fine aggregate passes #200 sieves
D. Silt is cohesionless

Solution: A

Gravel retains on #4 sieves.

PRACTICE PROBLEMS

PROBLEM 2.1

Determine the percentage of flat and elongated particles in the following coarse aggregate:

Weight of oven-dry aggregates: 498.2 g
Weight of flat aggregates: 4.2 g
Weight of elongated aggregates: 11.3 g
Weight of flat and elongated aggregates: 52.1 g
Flat and elongation limit set by the design agency: 20%

PROBLEM 2.2

Samples of coarse aggregate from a stockpile are brought to the laboratory, where the following weights are found:

Mass of aggregate submerged in water = 3,750 g
Mass of oven-dry aggregate = 4,750 g
Mass of SSD aggregate = 5,750 g

Determine the following parameters:

a) Bulk dry specific gravity
b) Bulk SSD specific gravity
c) Apparent dry specific gravity
d) Absorption

PROBLEM 2.3

Samples of coarse aggregate from a stockpile are brought to the laboratory, where the following weights are found:

Mass of aggregate submerged in water = 1,515 g
Mass of oven-dry aggregate = 2,165 g

Mass of SSD aggregate = 2,354 g

Determine the following parameters:

a) Bulk dry specific gravity
b) Bulk SSD specific gravity
c) Apparent dry specific gravity
d) Absorption

PROBLEM 2.4

A wet aggregate specimen weighs 23.2 N. After oven drying, the weight decreases to 21.87 N. If the absorption of the aggregate is 1.87%, calculate the free water, in grams, of the wet aggregate.

PROBLEM 2.5

An oven-dry aggregate specimen weighs 2,135 g. The absorption of the aggregate is 1.87%. Determine the amount of free water if 100 g of water is added to the oven-dry aggregate.

PROBLEM 2.6

Aggregate A has an oven-dry weight of 853 g and moisture content of 1.32%. Aggregate B has an oven-dry weight of 912 g and moisture content of 0.87%. If both of the aggregates have an absorption capacity of 2.9%, calculate how much water is to be added to saturate the mixture of these two aggregates.

PROBLEM 2.7

Aggregate A has an oven-dry weight of 3,853 g, moisture content of 5.32%, and absorption capacity of 2.9%. Aggregate B has an oven-dry weight of 4,333 g, moisture content of 0.67%, and absorption capacity of 4.12%. If both aggregates are mixed together, calculate how much water will need to be added to saturate the mixture of these two aggregates.

PROBLEM 2.8

The sieve analysis results of a batch of aggregate are shown in Table 2.11.

a) What are the NMAS and maximum aggregate size based on the Superpave definition?

b) What are the NMAS and maximum aggregate size based on the traditional definition?
c) What is the fineness modulus?
d) Draw the sieve analysis curve on semi-log graph paper.
e) Compare the gradation with the maximum density line.

TABLE 2.11

Sieving Results of a Batch of Aggregate for Problem 2.8

Sieve	Sieve Size (mm)	Mass Retained (g)
2.00 in.	50	0
1.50 in.	37.5	8
1.00 in.	25	19
0.75 in.	19	44
0.375 in.	9.5	55
No. 4	4.75	78
No. 8	2.4	256
No. 16	1.19	189
No. 30	0.6	254
No. 50	0.297	278
No. 100	0.149	118
No. 200	0.075	33
Pan	Pan	9

PROBLEM 2.9

The sieve analysis results of a batch of aggregate are shown in Table 2.12.

a) What are the NMAS and maximum aggregate size based on the Superpave definition?
b) What are the NMAS and maximum aggregate size based on the traditional definition?
c) What is the fineness modulus?
d) Draw the sieve analysis curve on semi-log graph paper.
e) Compare the gradation with the maximum density line.

TABLE 2.12

**Sieving Results of a Batch of Aggregate
for Problem 2.9**

Sieve	Sieve Size (mm)	Mass Retained (g)
2.00 in.	50	0
1.50 in.	37.5	0
1.00 in.	25	0
0.75 in.	19	12
0.375 in.	9.5	145
No. 4	4.75	111
No. 8	2.4	118
No. 16	1.19	180
No. 30	0.6	311
No. 50	0.297	216
No. 100	0.149	11
No. 200	0.075	3
Pan	Pan	0

PROBLEM 2.10

The sieve analysis results of a batch of aggregate are shown in Table 2.13.

TABLE 2.13

**Sieving Results of a Batch of Aggregate
for Problem 2.10**

Sieve	Sieve Size (mm)	Mass Retained (g)
2.00 in.	50	0
1.50 in.	37.5	36
1.00 in.	25	188
0.75 in.	19	180
0.375 in.	9.5	178
No. 4	4.75	322
No. 10	2.4	412
No. 16	1.19	123
No. 40	0.6	211
No. 50	0.297	500
No. 100	0.149	66
No. 200	0.075	500
Pan	Pan	39

a) What are the NMAS and maximum aggregate size based on the Superpave definition?

b) What are the NMAS and maximum aggregate size based on the traditional definition?

c) What is the fineness modulus?

d) Draw the sieve analysis curve on semi-log graph paper.

e) Compare the gradation with the maximum density line.

PROBLEM 2.11

Determine the proportions of Aggregate A, Aggregate B, and Aggregate C, shown in Table 2.14, to prepare a blend to satisfy the job specification.

TABLE 2.14

Gradation of Three Aggregate Sources for Problem 2.11

Size	Aggregate 1 % Finer	Aggregate 2 % Finer	Aggregate 3 % Finer	Specification % Finer
25.0 mm	100	100	100	100
19.0 mm	90	100	85	90–100
12.5 mm	60	92	42	70 max
No. 8	40	52	23	32–40
No. 200	15	0	13	5–8

PROBLEM 2.12

Determine the proportions of Aggregate A, Aggregate B, and Aggregate C, shown in Table 2.15, to prepare a blend to satisfy the job specification.

TABLE 2.15

Gradation of Three Aggregate Sources for Problem 2.12

Size	Aggregate 1 % Finer	Aggregate 2 % Finer	Aggregate 3 % Finer	Specification % Finer
25.0 mm	100	100	100	100
19.0 mm	90	100	85	90–100
12.5 mm	60	92	42	70 min
No. 4	50	76	30	50–60
No. 50	45	62	28	45–55
No. 100	40	52	23	40–45
No. 200	15	0	13	2–8

PROBLEM 2.13

Two stockpiles, A and B, are blended together at the ratio of 4:7. Stockpile A has an absorption capacity of 3.12% and Stockpile B has an absorption capacity of 3.92%. Calculate the combined absorption capacity of the blended aggregates.

PROBLEM 2.14

Two stockpiles, 1 and 2, are blended together at the ratio of 1:2. Stockpile 1 has 45 coarse aggregates and Stockpile 2 has 76 coarse aggregates. If the coarse aggregate angularities of Stockpile 1 and Stockpile 2 are 48 and 37, respectively, then calculate the coarse aggregate angularity of the blended aggregates.

PROBLEM 2.15

Two stockpiles, 1 and 2, have bulk-specific gravities of 2.620 and 2.670, respectively. If 35% of Aggregate 1 and 65% of Aggregate 2 are blended together, calculate the specific gravity of the blended aggregates.

PROBLEM 2.16

A 200-cm^3 container is loosely filled with 470 g of fine soil with a specific gravity of 2.65. Determine the air void of the loosely filled soil.

PROBLEM 2.17

A batch of aggregates weighing 4.85 kg was received from a client. After oven drying, the weight was found to be 4.56 kg. The sample was then submerged under normal water, weighing in at 3.85 kg. The sample was taken out and the surfaces were dried using a damp towel. It was then weighed again and found to be 5.15 kg. Determine the following properties of the batch of aggregates, using the mentioned test data:

a) The moisture content of the received sample
b) Absorption of the sample in percent of the dry weight
c) Bulk dry specific gravity
d) Bulk SSD specific gravity
e) Apparent dry specific gravity

PROBLEM 2.18

In a sand equivalent test, a clay layer of 2.35 in. is deposited on the 7.65 in. of sand particles. Calculate the SE of the specimen.

PROBLEM 2.19

A batch of aggregates weighing 10 kg was received from a client. After oven drying, the weight was found to be 8.55 kg. The sample was then submerged under normal water, weighing in at 7.20 kg. The sample was taken out and the surfaces were dried using a damp towel. It was then weighed again and found to be 10.35 kg. Some dry aggregates were placed in a mold from this batch of aggregates to determine the rodded density test. The rodded density was found to be 2,500 kg/m^3. Determine the air void of the aggregate while conducting the rodded density test. Also, calculate the total void space volume inside the mold when filled with aggregate.

3 Portland Cements

Cement is the binding material primarily used in concrete mixtures. Sometimes, cement is used in asphalt concrete, as well as for soil stabilization, aggregate stabilization, etc. This chapter discusses the production, chemical components, types, properties, and standard laboratory testing methods of cement.

3.1 BACKGROUND

The word *cement* means a material with the binding property capable of binding loose materials into a solid mass. This definition indicates a large number of cementing materials, such as asphalt binder, fly ash, glue, lime, Portland cement, etc. In civil and construction engineering, the term *cement* (Figure 3.1) is restricted to the binding materials used with stones, sand, bricks, building blocks, etc. A survey conducted by the USGS (2013) shows that the United States used approximately 79 million tons of hydraulic cement in 2011, worth about $6.5 billion. In 2005, approximately 111 million tons of cement was consumed. In the United States in 2009, approximately 36% of cement utilized was used for streets and highways; about 26% was used for residential building; about 4% for public building; and about 4% for commercial building (PCA 2013).

Broadly, two types of cement are used in civil engineering: *hydraulic* and *non-hydraulic*. Both are used to construct buildings, homes, sidewalks, and bridges, as well as for repairs on properties and structures. There are several differences between hydraulic cement and non-hydraulic cement:

- Hydraulic cement hardens due to hydration, i.e., chemical reactions with water. Non-hydraulic cement hardens due to carbonation, i.e., exposure to carbon dioxide in the air. Therefore, hydraulic cements can be used underwater, while non-hydraulic versions cannot.
- Hydraulic cement is made with limestone, gypsum, and clay, which is burned at high temperatures. Non-hydraulic cement is made with lime, gypsum, and oxychloride.
- Hydraulic cement dries and hardens in a few minutes. Non-hydraulic cement can take up to a month or more in order to reach usable conditions.

Examples of hydraulic cements consist of:

- Portland cements
- Natural cements
- High-alumina cements
- Blended cement
- Slag cement
- White cement
- Masonry cement
- Plastic cement

FIGURE 3.1   Portland cement. *Photo taken at the Colorado State University–Pueblo.*

Portland cement is the most widely used cement in civil engineering. This chapter deals with the manufacture of Portland cement, and its structure and properties, both when not hydrated and in a hardened state. The name, *Portland cement*, was given originally due to the resemblance in color to the hardened cement to Portland stone. Portland stone is a limestone quarried on the Isle of Portland, Dorset, England. The term *Portland cement* is used all over the world to describe a cement obtained by intimately mixing together calcareous and argillaceous (or other alumina, silica, and iron oxide-bearing) materials, burning them at a clinkering temperature, and grinding the resulting clinker. With the improvement of technology, other materials may also be added or blended to attain special properties of the cement.

Examples of non-hydraulic cements consist of:

- Lime
- Gypsum

Lime is produced from natural limestone by burning the stone in a kiln until only quicklime (calcium oxide) is left behind. The quicklime is then mixed with a small amount of water to create *hydrated lime*. Hydrated lime may be included in the cement or mixed with water for use as a mortar. Lime hardens by slowly absorbing carbon dioxide over time, eventually returning to limestone.

Gypsum is a naturally occurring mineral, mined from deposits formed by ancient seabeds as a raw material. Composed of calcium sulfate and water, gypsum is used for a variety of manufacturing, industrial, and agricultural purposes. Gypsum is often added to other cement and concrete mixtures to slow down the setting of cement. If not added, the cement sets immediately after the mixing of water, leaving no time for placing and finishing. The concept of setting of cement is discussed later in this chapter.

3.2   MANUFACTURE OF PORTLAND CEMENT

Portland cement is made primarily from a calcareous material, such as limestone or chalk, alumina, iron oxide, and silica (found as clay or shale). A mixture of calcareous and argillaceous materials, called *Marl*, is also used. Raw materials for the manufacture of Portland cement are found in nearly all countries, thus Portland cement plants operate all over the world. The process of manufacturing cement consists essentially of grinding the raw materials, mixing them intimately in certain proportions, and burning them in a large rotary kiln at a temperature of up to 1,480°C, until the material sinters and partially fuses into balls known as *clinker*. The clinker is cooled down and ground down to a fine powder. Then, some amount of gypsum is added, and the resulting product is the commercial Portland cement so widely used throughout the world. Some steps of the manufacturing of cement are shown in Figure 3.2.

To begin the production of Portland cement, raw materials are collected from quarries by excavation. The raw materials are then transported to crushing plants, as the collected raw materials are mostly large masses. Several stages of crushing may be required to attain the desired gradation. The crushed sand and clay-type materials are then proportioned as appropriate and ground together. Then, the blended materials are passed through a kiln with a length commonly greater than 690 ft (210 m), diameter of about 16 ft (5 m), and temperature near 1,480°C. After burning, ball-type materials are produced which are known as the clinker. Then, the clinker is cooled down, and a quality check is performed. If it passes the quality check, it is finally ground, otherwise the material is returned to the kiln. Some gypsum or other materials may be added prior to the packaging. Gypsum is added to Portland cement in order to delay the initial setting of cement with the addition of water. Otherwise, cement starts hardening immediately after adding water. Air-entraining agents are added as well, for air-entraining cements. Air-entraining agents provide space for the expansion of

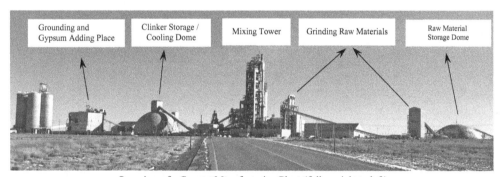

Overview of a Cement Manufacturing Plant (follow right to left)

| Mixing Tower | Large Rotary Kiln (1,450 °C) | Hot Clinker Just Out of Clinker | Gypsum Adding Tank |

FIGURE 3.2 Manufacture of Portland cement. *Photos taken in Pueblo, Colorado.*

minute droplets of waters in the concrete, due to freezing and thawing in cold areas. The effects of air-entraining in Portland cement or concrete are discussed in the next chapter. After the production of Portland cement, it stored in silos until it is transported to customers. About 97% of US cement is transported to customers by trucks, while the remaining quantities travel by barge or rail (PCA 2013).

3.3   CHEMICAL COMPOSITIONS OF PORTLAND CEMENT

The raw materials used in the manufacture of Portland cement consist mainly of lime, silica, alumina, and iron oxide. These compounds interact with one another in the kiln to form a series of more complex products. Four compounds are usually regarded as the major constituents of cement. They are listed in Table 3.1, along with their abbreviated symbols. These shortened notations, used by cement chemists, describe each oxide by one letter, such as:

TABLE 3.1
Main Compounds of Portland Cement

Chemical Name	Chemical Formula	Abbreviation	% by Weight
Tricalcium Silicate	$3CaO.SiO_2$	C_3S	45–60
Dicalcium Silicate	$2CaO.SiO_2$	C_2S	15–30
Tricalcium Aluminate	$3CaO.Al_2O_3$	C_3A	6–12
Tetracalcium Aluminoferrite	$4CaO.Al_2O_3.Fe_2O_3$	C_4AF	6–8

Calcium oxide, $CaO = C$
Silicon dioxide, $SiO_2 = S$
Aluminum oxide, $Al_2O_3 = A$
Iron oxide, $Fe_2O_3 = F$

The silicates in cement are not pure compounds, as they contain minor oxides in a solid solution. These oxides have significant effects on the atomic arrangements, crystal forms, and hydraulic properties of the silicates. As stated earlier, gypsum is often added during the cement manufacturing process, upon the cooling of clinker.

3.4   HYDRATION OF PORTLAND CEMENTS

The chemical compounds of Portland cement undergo a series of chemical reactions when it is mixed with water and begins to harden (or set). These chemical reactions require the addition of water, which reacts with the basic chemical compounds listed in Table 3.1. This chemical reaction with water is known as the *hydration*. Each one of these reactions occurs at a different time and rate. Together, the results of these reactions determine how Portland cement hardens and gains its strength.

- *Tricalcium silicate* (C_3S) hydrates, hardens rapidly, and is largely responsible for the initial setting and early strength. Portland cement with higher percentages of C_3S exhibits high early strength.
- *Dicalcium silicate* (C_2S) hydrates, hardens slowly, and is largely responsible for the strength increase which occurs after 7 days.
- *Tricalcium aluminate* (C_3A) hydrates and hardens the fastest. It liberates a large amount of heat almost immediately and contributes somewhat to the early strength. Gypsum is added to Portland cement to retard C_3A hydration. Without gypsum, C_3A hydration causes Portland cement to set almost immediately after adding water.
- *Tetracalcium aluminoferrite* (C_4AF) hydrates rapidly but contributes very little to strength. It allows for lower kiln temperatures during the manufacturing of Portland cement. The color of Portland cement is affected by the C_4AF.

The equation for the hydration of tricalcium silicate can be shown as:

$$\text{Tricalcium Silicate} + \text{Water} \rightarrow \text{Calcium Silicate Hydrate}$$

$$+ \text{Calcium Hydroxide} + \text{Heat}$$

Upon the addition of water, tricalcium silicate reacts rapidly to release calcium ions, hydroxide ions, and a large amount of heat. The generation of heat is caused by the breaking and making of chemical bonds during hydration. The reaction slowly continues producing calcium and hydroxide ions until the system becomes saturated. The hydration continues as long as water is present and there are still compounds which are not yet hydrated in the cement paste.

Dicalcium silicate also affects the strength of concrete through its hydration. Dicalcium silicate reacts with water in a manner similar to that of tricalcium silicate but much more slowly. The heat released is less than that by the hydration of tricalcium silicate because the dicalcium silicate is much less reactive. The products from the hydration of dicalcium silicate are the same as those for tricalcium silicate:

$$\text{Dicalcium Silicate} + \text{Water} \rightarrow \text{Calcium Silicate Hydrate}$$

$$+ \text{Calcium Hydroxide} + \text{Heat}$$

The other major components of Portland cement, tricalcium aluminate and tetracalcium aluminoferrite, also react with water in a complex manner. However, as their reactions do not contribute significantly to strength gain, these are not discussed any further here.

3.5 TYPES OF PORTLAND CEMENT

In the United States, AASHTO M 85 and ASTM C 150 recognize eight basic types of Portland cement concrete (Table 3.2). There are also many other types of blended and proprietary cements not mentioned here. The modification is added during the final stage of cement production to incorporate the desired property.

TABLE 3.2
ASTM Types of Portland Cement

Type	Name	Purpose
I	Normal	General-purpose cement suitable for most purposes.
IA	Normal-Air Entraining	An air-entraining modification of Type I.
II	Moderate Sulfate Resistance	Used as a precaution against moderate sulfate attack. It usually generates less heat at a slower rate than Type I cement.
IIA	Moderate Sulfate Resistance-Air-Entraining	An air-entraining modification of Type II.
III	High Early Strength	Used when high early strength is needed. It has more C_3S than Type I and has been ground finer to provide a higher surface-to-volume ratio. Strength gain is double that of Type I cement in the first 24 hours.
IIIA	High Early Strength-Air Entraining	An air-entraining modification of Type III.
IV	Low Heat of Hydration	Used when hydration heat must be minimized in large volume applications, such as gravity dams. Contains about half the C_3S and C_3A, and double the C_2S of Type I cement.
V	High Sulfate Resistance	Used as a precaution against severe sulfate action – principally where soils or groundwater have high sulfate contents. It gains strength at a slower rate than Type I cement. High sulfate resistance is attributable to low C_3A content.

3.6 PHYSICAL PROPERTIES

Portland cements are commonly characterized by their physical properties for quality control purposes. Their physical properties can be used to classify and compare Portland cements. Specification values of cement can be found in the ASTM C 150. Some of these properties are listed below and described further in this section.

a) Fineness
b) Soundness
c) Setting time
d) Specific gravity
e) Heat of hydration
f) Loss on ignition
g) Strength of cement

3.6.1 FINENESS

The *fineness* of cement describes how fine (tiny) the cement particle is. After adding water, the hydration of cement starts at the surfaces of the cement particles. It is the total surface area of the cement particles that represents the material available for the hydration. Thus, the rate of the hydration depends on the fineness (surface area) of the cement particles. The smaller the particle size, the greater the surface area, meaning that there is more area available for the water–cement interaction. For the rapid development of strength, a high fineness is desired. The effects of having greater fineness on the strength gain can be observed during the first 7 days.

Several standard test methods are available to characterize the fineness of cement; AASHTO T 98, ASTM C 115, AASHTO T 128, ASTM C 184, AASHTO T 153, ASTM C 204, AASHTO T 192, and ASTM C 430. ASTM C 204 are very popular for determining the fineness of cements. This method uses the *Blaine air permeability apparatus* for determining the fineness of Portland cement. The components of the apparatus are a stainless-steel test cell, plunger, perforated disk, calibrated U-tube manometer, rubber aspirator, and bulb, as shown in Figure 3.3. This apparatus draws air through a bed of known porosity. The rate of air flow is determined by the pore volume in the bed. The flow is dependent on the particle size. The Blaine number is the specific surface area of fine materials, in square centimeters per gram (cm^2/g), in the test specimen. The larger the Blaine number, the larger the surface area (the finer the particles).

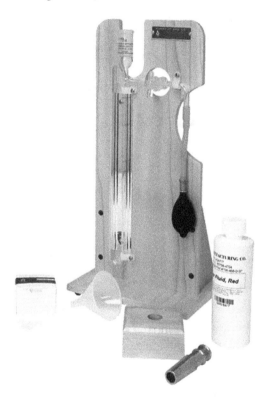

FIGURE 3.3 Blaine air permeability apparatus. *Courtesy of Test Mark Industries, 995 North Market St., East Palestine, Ohio. Used with permission.*

3.6.2 Soundness

Soundness of cement is the ability of a hardened cement paste to retain its volume after setting, without the delayed destructive expansion. This destructive expansion is caused by excessive amounts of free lime (CaO) or magnesia (MgO). Most Portland cement specifications limit the magnesia content, and thus the expansion. The typical expansion test places a small specimen of cement paste into an autoclave (a high-pressure steam vessel, shown in Figure 3.4). The autoclave is slowly brought to 2.03 MPa (295 psi) and is then kept at that

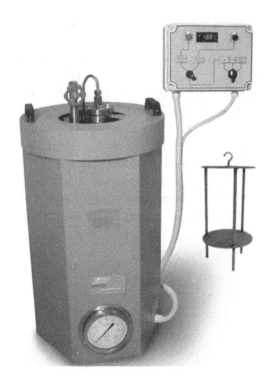

FIGURE 3.4 Autoclave equipment. *Courtesy of Impact Test Equipment, Stevenston Ayrshire, Scotland, UK.*

pressure for 3 hours. The change in the specimen length due to its time in the autoclave is measured and reported as a percentage. ASTM C 150 specifies a maximum autoclave expansion of 0.80% for all types of Portland cement. Other standard test methods used to determine soundness are AASHTO T 107 and ASTM C 151.

3.6.3 SETTING TIME

When cement is mixed with water, a cement paste is formed which remains plastic (deformable) for only a short time; the paste sets and stiffens quickly. Normally, two setting times are defined:

a) *Initial set* – occurs when the paste begins to stiffen considerably.
b) *Final set* – occurs when the paste has hardened to the point of sustaining a load.

Before the time of initial setting, the paste can be deformed or disturbed without any injury. Once the initial set has occurred, the paste can be disturbed but loses its plasticity (deformability) and may contain some internal discontinuous fractures. After the final setting, the concrete becomes rigid, and continuous fracture/cracking may occur if heavy stress is applied. Setting and hardening are often confused. Setting is the stiffening process of the paste, and hardening is the strength-gaining process. These properties can be thought of as two sides of a coin.

The setting time of cement paste is affected by many factors, such as the fineness of cement, water–cement ratio, mix temperature, and the presence of chemicals (especially gypsum) and admixtures. Setting times are shorter for finer cement, as more cement surface is available for the hydration reaction with water. High water–cement ratios, gypsum, and cold temperatures cause an increase in setting time. Admixtures (discussed in the next chapter) may decrease or increase the time of setting, depending on the type.

Setting tests are used to characterize how a cement paste sets. For construction purposes, the initial set must not occur too soon, and the final set must not occur too late. Additionally, setting times can give some indication of whether or not a cement is undergoing the normal hydration. These times are just arbitrary points used to characterize the cement; they do not have any fundamental chemical significance. They essentially describe the setting of the cement and are not tied to the setting time of concrete.

Some standard test methods used to determine the setting times are AASHTO T 131, ASTM C 191, AASHTO T 154, and ASTM C 266. In the ASTM C 191, penetration tests are performed on the paste by allowing a 1-mm Vicat needle to settle into this paste. The *Vicat initial time of setting* is calculated as the time elapsed between the initial contact of cement and water and the time when the penetration reaches 25 mm. The *Vicat final time of setting* is calculated as the time when the needle does not sink visibly into the paste. In the ASTM C 266, the *initial time of setting* is the time required for the test specimen to bear the initial Gillmore needle without appreciable indentation, while the time required for the test specimen to bear the final Gillmore needle without appreciable indentation is the *final time of setting*. The initial setting-time needle has a weight of 113.4 g and a tip diameter of 2.12 mm. The final setting time needle has a weight of 453.6 g and a tip diameter of 1.06 mm. Both the Vicat needle and the Gillmore needle are shown in Figure 3.5. The Vicat needle test is more common and tends to give shorter times than the Gillmore needle test. Specific setting specifications are listed in Table 3.3.

False set and *flash set* are two other forms of undesirable setting in cement paste. False set is the rapid (within 1–5 minutes) development of rigidity in a freshly mixed paste or concrete, without any heat generation. The mix can be remixed to attain its properties. Flash set is the rapid development of the rigidity with some heat generation. A proper amount of gypsum and continuous remixing can avoid these problems.

Vicat Needle Apparatus   Gillmore Apparatus

FIGURE 3.5 Setting time test equipment. *Courtesy of Test Mark Industries, 995 North Market St., East Palestine, Ohio. Used with permission.*

TABLE 3.3
ASTM C 150 Specified Set Times

Test Method	Set Type	Time Specification (min)
Vicat	Initial	≥ 45
	Final	≤ 375
Gillmore	Initial	≥ 60
	Final	≤ 600

3.6.4 SPECIFIC GRAVITY

Specific gravity is normally used in the mixture proportioning calculations. The specific gravity of Portland cement is generally around 3.15, while the specific gravity of Portland-blast-furnace-slag and Portland-pozzolan cements may have specific gravities near 2.90. Standard test methods for determining the specific gravity are AASHTO T 133 and ASTM C 188.

3.6.5 HEAT OF HYDRATION

The *heat of hydration* is the heat generated when water and Portland cement react. The heat of hydration is mainly influenced by the proportions of C_3S and C_3A in the cement, but it is also influenced by the water–cement ratio, fineness, and curing temperature. As each one of these factors is increased, the heat of hydration increases. In large concrete structures, such as gravity dams, hydration heat is produced faster than it can be dissipated (especially in the center of large concrete masses). This creates high temperatures in the center of these large concrete masses that, in turn, may cause undesirable stresses as the concrete cools down to the ambient temperature. Conversely, the heat of hydration can help maintain favorable curing temperatures during winter. The standard test method to determine the heat of hydration is ASTM C 186.

3.6.6 LOSS ON IGNITION

The *loss on ignition* is calculated by heating up a cement specimen to 900–1,000°C (1,650–1,830°F), until a constant weight is obtained. The weight loss of the specimen due to the heating is then determined. A high loss on ignition can indicate the pre-hydration and the carbonation, which may be caused by improper and prolonged storage, as well as corruption during transport or transfer. Some popular standard test methods to determine the loss on ignition are AASHTO T 105 and ASTM C 114.

3.6.7 STRENGTH OF CEMENT

The *strength of cement* is the compressive strength of a cement mortar, and *not* the strength of the cement paste nor the strength of the concrete. The compressive strength of the mortar is affected by the water–cement ratio, cement–sand ratio, age of specimen, types of sand, grading of sand, preparation procedure, curing conditions, shape of the specimen, loading rate, etc. The compressive strength of the concrete is roughly proportional to the compressive strength of the mortar. However, the exact relationship between the strength of mortar and the strength of concrete is unknown, as the compressive strength of concrete depends on the aggregate characteristics, water–cement ratio, construction procedures, etc.

The mortar is produced using 1 part cement and 2.75 parts sand, proportioned by mass. The compressive strength of mortar is determined by testing a 2-in. (50 mm) cube specimen under a compressive loading until failure occurs in 20–80 seconds. After the curing, the cube is compressed, and the compressive force over the cross-sectional area gives the compressive strength of the cement mortar. Table 3.4 shows the ASTM C 150 compressive strength specifications.

TABLE 3.4
Portland Cement Mortar Compressive Strength in psi (MPa)

Curing Time	Portland Cement Type							
	I	IA	II	IIA	III	IIIA	IV	V
1 day	–	–	–	–	1,800 (12.4)	1,450 (10.0)	–	–
3 days	1,800 (12.4)	1,450 (10.0)	1,500 (10.3)	1,200 (8.3)	3,500 (24.1)	2,800 (19.3)	–	1,200 (8.3)
7 days	2,800 (19.3)	2,250 (15.5)	2,500 (17.2)	2,000 (13.8)	–	–	1,000 (6.9)	2,200 (15.2)
28 days	–	–	–	–	–	–	2,500 (17.2)	3,000 (20.7)

Note: Type II and IIA requirements can be lowered if either an optional heat of hydration or chemical limit on the sum of C_3S and C_3A is specified.

3.7 CHAPTER SUMMARY

The word *cement* describes a material which can bind two materials together. In civil and construction engineering, cement specifically means Portland cement, which can bind aggregate or stone pieces together and is used in binding civil engineering materials. Portland cement gets its name from the resemblance of the color of the hardened cement to Portland stone found in Dorset, England. The production of Portland cement broadly consists of heating crushed raw materials in a kiln at 1,480°C. The burned material, called the clinker, is cooled down, ground again, and different materials (such as gypsum) are added to obtain special properties. If water is mixed with cement, each component of cement reacts with water and hardens slowly. There are different types of cement, such as normal cement, normal-air entraining, moderate sulfate resistant, etc. These properties are gained by adding special materials with the clinker during production.

Among the different physical properties of cement, the fineness of Portland cement is one of the major properties. The smaller the particle size, the larger the surface area, meaning a stronger chemical reaction with water and thus a greater gain in strength. Soundness of cement is the ability of a hardened cement paste to retain its volume after setting, without the delayed destructive expansion. Two types of setting times are important: initial set and final set. Initial set occurs when the paste begins to stiffen considerably. Any cement mixed with water and other materials must be used at the job site before the initial setting time. Thus, a higher initial set time is sometimes good, as more time is available to use it. Final set occurs when the cement has hardened to the point where it can carry some load. Thus, a smaller final set time is desirable, as load can be applied quickly.

The heat of hydration is the heat generated when water and Portland cement react. In large concrete structures, the hydration heat is produced faster than it can be dissipated. This creates high thermal stress inside the hardened concrete, potentially causing the concrete to crack.

ORGANIZATIONS DEALING WITH CEMENT

The Portland Cement Association (PCA), founded in 1916, is the premier policy, research, education, and market intelligence organization serving America's cement manufacturers. PCA members represent 93% of US cement production capacity, with facilities in all 50 states.

Location: Skokie, IL.
Website: www.cement.org
ASTM International is also a resource for cements.

REFERENCES

PCA. 2013. *Design and Control of Concrete Mixtures, EB001*, 15th edition. Portland Cement Association, Skokie, IL.
United States Geological Survey (USGS). 2013. *2011 Minerals Yearbook*, *Cement. Advance Release*. US Department of the Interior, United States Geological Survey, Reston, VA.

FUNDAMENTALS OF ENGINEERING (FE) EXAM STYLE QUESTIONS

FE PROBLEM 3.1

By weight, which of the following compounds occupies the largest weight of Portland cement?

A. Gypsum
B. Tricalcium silicate
C. Dicalcium silicate
D. Tricalcium aluminate

Solution: B

Tricalcium silicate is about 50% of the weight in Portland cement.

FE PROBLEM 3.2

Which of the following types is the general-purpose Portland cement?

A. Type I
B. Type II

C. Type III
D. Type IV

Solution: A

FE PROBLEM 3.3

Which of the following statements is false for Portland cement?

A. Initial set occurs when the paste begins to stiffen considerably
B. Final set occurs when Portland cement has hardened to the point at which it can sustain some load
C. The specific gravity of Portland cement is commonly 3.15
D. None of the above

Solution: D

All three statements are true for Portland cement.

PRACTICE PROBLEMS

PROBLEM 3.1

Why is the fineness of Portland cement important?

PROBLEM 3.2

Classify the types of Portland cement.

PROBLEM 3.3

Discuss the initial and final set times of Portland cement.

PROBLEM 3.4

Discuss the effects that the fineness of cement, water–cement ratio, mix temperature, and presence of gypsum have on the setting of Portland cement.

PROBLEM 3.5

Discuss the false set and the flash set of Portland cement.

4 Concrete Mix Design

The strength and performance of concrete are primarily dependent on the mix design of concrete, i.e., the proportions of coarse and fine aggregates, type of cement, volume of water, presence of admixture, etc. Some other factors, such as workability, curing, and placement environment are also important in achieving the desired performance. Beyond the mix design, proper handling, curing, and controlled early-age cracking must also be considered in order to attain the desired strength. Ready-mix concrete is being used nowadays to ensure the quality in mix design. This chapter discusses the procedure of proportioning the ingredients to achieve the desired strength, performance, workability, and durability. Proper handling, curing, and early-age cracking control methods are also discussed. Finally, the production and delivery methods of ready-mix are also addressed.

4.1 BACKGROUND

The previous chapter discusses the basics of Portland cement, its production, types, properties, etc. This chapter discusses how the proportions of cement, aggregates, and water are determined based on the strength requirement, service environment, and other conditions. The properties and characterizations of hardened concrete will be discussed in the next chapter.

The terms, 'mixture design' and 'mixture proportioning' are often confused. *Mix design* leads to the development of a concrete specification. The process of determining the required and specifiable performance characteristics of a concrete mixture is called the mix design. The characteristics of the concrete mixture design can include:

- Fresh concrete properties, such as air content and workability
- Required mechanical properties of hardened concrete, such as strength and durability requirements
- The inclusion, exclusion, or limits on specific ingredients

The output from the mixture design process becomes a part of the input for mixture proportioning. In other words, mix design is the selection of the proportions of coarse aggregate, fine aggregates, cement, water, and likely admixtures. *Mixture proportioning* refers to the process of determining the quantities of concrete ingredients using local materials to achieve the specified characteristics of the concrete. The following units are commonly used for mixture proportioning:

- Coarse aggregate (CA), fine aggregate (FA), water, and cement: lb/yd^3 or kg/m^3
- Air content: percent (%)
- Admixtures: ml or (fl oz)

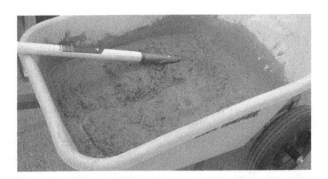

FIGURE 4.1   Concrete mix. Photo taken in Pueblo, Colorado.

A properly proportioned concrete mix possesses the following qualities:

- Acceptable workability of the freshly mixed concrete
- Durability, strength, and uniform appearance of the hardened concrete
- Economy: although economy is not directly calculated, rather local materials are used in the mix

A small portion of a freshly mixed concrete is shown in Figure 4.1.

4.2   CONCRETE MIX BEHAVIOR

Some knowledge of concrete mix design factors, such as water–cement ratio, slump test, workability, air void requirements, etc. are required. Hence, these topics are discussed in this section before discussing the mix design procedure.

4.2.1   WATER–CEMENT RATIO

The water–cement (W/C) ratio is the ratio of the weight of water to the weight of cement used in a concrete mix. A lower water–cement ratio leads to higher strength and durability. However, a lower water–cement ratio makes the mix difficult to handle and place on the job site. Plasticizers or superplasticizers can be used to increase the ease of working with concrete without adding more water. If such is used, the ratio of water to cement plus pozzolan, W/(C+P) is used. Pozzolan is a fly ash or blast furnace slag. It includes many other materials, such as silica fume, rice husk, ash, or natural pozzolans. Pozzolans also strengthen concrete.

Concrete hardens because of the chemical reaction between cement and water, known as the hydration. The hydration produces heat, which is called the heat of hydration. An amount of water (about 35% of cement) is required to complete the hydration reactions. However, a mix with a W/C ratio of 0.35 may not mix thoroughly, and may not flow well enough to be placed with ease. More water is therefore used than is technically necessary to complete the hydration. A water–cement ratio of 0.45 to 0.55 is commonly used. For a higher-strength concrete, lower W/C ratios are used, along with a plasticizer, to increase the workability. Disadvantages of excess water include:

- Causes segregation of the aggregates in the mix
- Excess water not used by the hydration evaporates, resulting in microscopic pores (bleeding) and reduced the strength

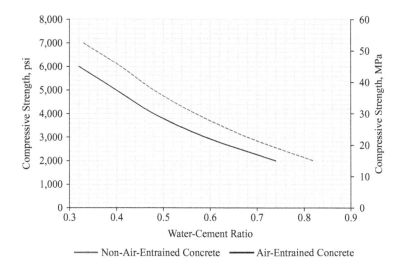

FIGURE 4.2   Variations of compressive strength with the water–cement ratio. Based on concrete using 3⁄4-in. to 1-in. (19-mm to 25-mm) aggregate cured 28 days per ASTM C 31. Courtesy of American Concrete Institute, Standard Practice for Selecting Proportions for Normal, Heavyweight and Mass Concrete, ACI 211.1-91, American Concrete Institute, Farmington Hills, Michigan, Reapproved 2002. Table 6.3.4a. Used with permission.

- Experiences more shrinkage as excess water evaporates, resulting in internal cracks and visible fractures (around inside corners), which again reduces the strength

Figure 4.2 shows how the compressive strength decreases with the increase in the W/C ratio for both the air-entrained and non-air-entrained concretes. In *air-entrained* concrete, air-admixtures are added to produce microscopic air bubbles inside the concrete, which are required when concrete is exposed to deicing, aggressive chemicals, moisture, or free water prior to freezing. For example, in Texas, air-entrained concrete is not practiced as there is no expansion issue for concrete, whereas in New York, air-entrained concrete is required to avoid cracking due to the expansion of ice during the freezing weather. Air entrainment also increases the workability of concrete. The details of air-entrainment requirements are discussed further in this chapter.

4.2.2 Workability and Slump

Workability is the ease with which the concrete components can be mixed, transported, placed, compacted, and finished, with respect to a homogeneous condition. Sufficient workability in concrete is required for proper compaction, and prevents segregation of concrete, resulting in a strong and durable concrete. Increasing water content in concrete increases the workability of concrete. However, an increase in the water ratio causes a decrease in the strength of the concrete, and enhances the segregation. There are several ways of improving the workability without increasing the W/C ratio, such as:

- Use of smooth textured and round aggregates improves workability, as these types of aggregates can slide over one another very easily. Angular aggregates tend to interlock and are difficult to work with.
- Use of well-graded, fine aggregates provides better workability.

- Use of larger aggregates causes an increase in the workability.
- Decrease in the ambient temperature can increase the workability of concrete.
- Chemical admixtures, such as water-reducers, air-entraining superplasticizers, and workability agents, are often used to improve the workability of concrete.

Slump is the indirect measurement of workability. The choice of slump is a choice of mix workability. Workability can be described as a combination of several different, but related, PCC properties related to its rheology, such as:

- Ease of mixing
- Ease of placing
- Ease of compaction
- Ease of finishing

Workability should not be confused with consistency. *Consistency* is the ability of the freshly mixed concrete to flow, which includes the entire range of fluidity from the driest condition to the wettest condition. *Plastic consistency* is the condition in which the mix undergoes continuous deformation, without rupture upon applying some stress. Plastic mixtures have cohesion among the particles, do not crumble, and flow sluggishly without segregation. Generally, mixes of the stiffest consistency that can still be placed adequately should be used.

The *concrete slump test* measures the consistency of fresh concrete before it sets. It is performed to check the workability of freshly mixed concrete, and therefore, the ease with which the concrete flows. It can also be used as an indicator of an improperly mixed batch. The test is popular due to the simplicity of the involved apparatus and procedure. The slump test is used to ensure uniformity for different loads of concrete under field conditions. The test is carried out using a metal conical frustum mold, known as the *slump cone* or the Abrams cone. It is open at both ends, and has handles as shown in Figure 4.3. The cone has an internal diameter of 4.0 in. (100 mm) at the top and of 8.0 in. (200 mm) at the bottom, with a height of 12 in. (300 mm). The cone is placed on a hard, flat surface. This cone is filled with fresh concrete in three layers. Each layer is tamped 25 times with a 2 ft (600 mm)-long metal rod, measuring 5/8 in. (16 mm) in diameter. At the end of the third stage, the concrete is struck off at the top of the mold. The mold is carefully lifted vertically upward, so as not to disturb the freshly compacted concrete cone.

Slump Cone   Tamping   Removing Cone   Measurement

FIGURE 4.3 Slump test procedure. Photos taken at the Farmingdale State College of the State University of New York.

Collapse   Shear-Type Slump   True Slump

FIGURE 4.4   Slump test result interpretation. Photos taken at the Colorado State University–Pueblo.

The freshly compacted concrete then subsides (slumps). The slump of the concrete is measured by measuring the distance from the top of the slumped concrete to the level of the top of the slump cone. The slumped concrete takes various shapes. According to the profile of slumped concrete, slump is termed as *collapse slump*, *shear slump*, or *true slump*, as shown in Figure 4.4.

- Collapse – the concrete collapses completely
- Shear – the top portion of the concrete shears off and slips sideways
- True – the concrete simply subsides, keeping more or less its shape

For a shear or collapse slump, the test is repeated using a fresh sample. A collapse slump indicates that the mix is too wet or has a very high workability. Recommendations for the selection of the degree of slump are listed later in this chapter.

4.2.3   DURABILITY AND FREEZE–THAW CYCLES

One of the forces that can damage PCC is the *freeze–thaw* cycles (freezing of water within the concrete and subsequent thawing). Since water expands when it freezes, concrete must have some amount of air voids so that any water in concrete which expands due to freezing does not create pressure. Entrained air bubbles of greater than 3 μm in diameter provide adequate resistance to damage caused by freezing and thawing. Bubbles should be spaced no further apart than 0.2 mm inside the mix, which can be achieved by the addition of air-entraining admixtures or air-entraining cement. If the freezing water creates pressure, the concrete could be fractured. If a PCC has adequate air voids and has cured to a sufficient strength, it should be able to resist environmental damage due to the freeze–thaw cycles. Water expands by about 9% during freezing. Therefore, an air content of about 9%, by volume of the mortar fraction, should be sufficient to protect the concrete.

Some air may be trapped in concrete during the mixing process, and additional voids can be created when excess water in the concrete mix evaporates. However, the resulting system of air voids is haphazard and may not be adequate to resist the freeze–thaw pressure. In most cases, it is desirable to deliberately increase the air content in the mix, as well as to ensure that the air is uniformly distributed as tiny bubbles throughout the entire mix. This process is called the *air entrainment*, and can be achieved mainly by incorporating chemical admixtures and adjustments to the mix design.

| Compacting | Leveling | Injecting Water | Pumping |

FIGURE 4.5 Measuring the air content of a concrete mix using the pressure method. Photos taken at the Farmingdale State College of the State University of New York.

There are different methods of determining the air content in freshly mixed concrete, such as:

- Pressure Method (ASTM C 231)
- Volumetric Method (ASTM C 173)
- Gravimetric Method (ASTM C 138)
- Chase Air Content (AASHTO 199)

The determination of air voids using the pressure method is shown in Figure 4.5. To begin, the measuring bowl is filled with freshly mixed concrete in three layers with 25 blows to each. Then, the bowl is capped properly. Some water is injected until it drains from the other side, in order to remove any entrapped air by the capping. Then, pressure is applied and allowed to stabilize for few seconds. Afterward, the air valve is opened between the air chamber and the measuring bowl. The percentage of air on the dial of the pressure gage can be read.

The volumetric method to determine the air void in freshly mixed concrete is discussed in the laboratory testing section of this text.

4.3   SELECTING MIX PROPORTIONS

The proportions of mixture ingredients (aggregates, cement, water, and any admixtures) are selected based on the desired strength, intended use, exposure conditions, size and shape of member, and the physical properties of the concrete. The characteristics should reflect the needs and economic-related aspects of the usage. For example, the mixture should have resistance to freeze–thaw if it is expected to be exposed to freezing conditions with the presence of salts, moisture, or any harmful chemicals. Once the mix characteristics are selected, the mixture can be proportioned, fulfilling all the requirements step by step, as listed below.

Step 1. Strength requirements
Step 2. Determination of the water–cement (W/C) ratio
Step 3. Estimation of the coarse aggregate mass
Step 4. Air entrainment requirements
Step 5. Workability needs

Step 6. Estimation of the water content
Step 7. Determination of the cement content requirements
Step 8. Evaluation of the admixture needs
Step 9. Estimation of the fine aggregate mass
Step 10. Determination of the moisture corrections
Step 11. Trial mix procedures

Step 1. Strength Requirements
At the beginning of the mix design, the required strength (f_{cr}') is determined to be slightly greater than the specified compressive strength (f_c'). The specified compressive strength, f_c', at 28 days is the strength that is expected to be equal to or exceeded by the average of any set of three consecutive strength tests. The average strength should equal the specified strength plus an allowance to account for any variations in materials, mixing methods, transportation, and placement of concrete, as well as curing, testing, concrete cylinder specimens, and so on. The required strength (f_{cr}') is determined from the specified compressive strength (f_c') in three different ways:

a) If the standard deviation (s) of the tested results of the compressive strength is calculated using at least 30 specimens, the required average strength (f_{cr}') is the larger of the following values, obtained from the equations listed in Table 4.1:

TABLE 4.1

Required Average Compression Strength When Data Is Available for More than 30 specimens

Specified Compressive Strength (f_c'), psi (MPa)	Required Average Compression Strength (f_{cr}'), psi	Equations
≤5,000 (34.5)	$f_{cr}' = f_c' + 1.34s$	(4.1)
	$f_{cr}' = f_c' + 2.33s - 500$	(4.2)
	Use the larger value	
>5,000 (34.5)	$f_{cr}' = f_c' + 1.34s$	(4.1)
	$f_{cr}' = 0.90 f_c' + 2.33s$	(4.3)
	Use the larger value	

Courtesy of American Concrete Institute, ACI 318-11, table 5.3.2.1. Used with permission.

In the case of metric units, Eq. 4.2 is to be replaced by Eq. 4.4. Eqs. 4.1 and 4.3 remain the same.

$$f_{cr}' = f_c' + 2.33s - 3.45 \text{ in MPa} \tag{4.4}$$

b) If the standard deviation (s) is calculated using 15–29 samples, then the modification factors listed in Table 4.2 are multiplied with the obtained standard deviation (s). More specifically, the calculated standard deviation from the

TABLE 4.2

Modification Factor for the Standard Deviation with Number of Test

Number of Test	Modification Factor for the Standard Deviation, F
15	1.16
20	1.08
25	1.03
30≤	1.00

Courtesy of American Concrete Institute, ACI 318-11, table 5.3.1.2. Used with permission.

15–29 specimens is increased by multiplying this modification factor. After that, Eqs. 4.1 to 4.4 can be used as appropriate to calculate the required average strength (f_{cr}').

c) If fewer than 15 samples are tested, or no standard deviation (s) data is available, then the modification equations (Eqs. 4.5 to 4.7) listed in Table 4.3 are to be used as appropriate to calculate the required average strength (f_{cr}').

In metric units, Eqs. 4.5, 4.6, and 4.7 are replaced by Eqs. 4.8, 4.9, and 1.10 respectively, as listed in Table 4.4.

TABLE 4.3

Required Compressive Strength for Less Than 15 Specimens

Specified Compressive Strength (f_c'), psi	Required Average Compression Strength (f_{cr}'), psi	Equations
<3,000	$f_{cr}' = f_c' + 1,000$	(4.5)
3,000–5,000	$f_{cr}' = f_c' + 1,200$	(4.6)
>5,000	$f_{cr}' = 1.10 f_c' + 700$	(4.7)

Courtesy of American Concrete Institute, ACI 318-11, table 5.3.2.2. Used with permission.

TABLE 4.4

Required Compressive Strength for Less Than 15 Specimens in Metric Units

Specified Compressive Strength (f_c'), MPa	Required Average Compression Strength (f_{cr}'), MPa	Equations
<21	$f_{cr}' = f_c' + 7.0$	(4.8)
21–35	$f_{cr}' = f_c' + 8.5$	(4.9)
>35	$f_{cr}' = 1.10 f_c' + 4.8$	(4.10)

Note: 1 psi = 6.89 kPa (0.00689 MPa). To convert unit from psi to MPa or vice-versa, a gross rounding off is very often made by the standard specifications. For example, 3,000 psi = 20.67 MPa (20,670 kPa). However, a rounded value of 21 MPa is assumed.

Step 2. Determination of the Water–Cement (W/C) Ratio

In the next step, the W/C ratio is determined using the required average strength (f_{cr}') determined in Step 1, as follows:

a) Use Table 4.5 for estimating the water–cement ratios for the trial mixes when no other data is available.
b) Use Table 4.6 to check the determined W/C ratio for exposure conditions.
c) Use Table 4.7 to check the determined W/C ratio for exposure to sulfates in soil or water.
d) Check Tables 4.5 to 4.7, select the best-suited W/C ratio.

TABLE 4.5

Relationship between Water–Cement Ratio and Compressive Strength of Concrete

Compressive Strength at 28 days, f_c' psi (MPa)	Water–Cement Ratio by Weight	
	Non-Air-Entrained Concrete	Air-Entrained Concrete
7,000 (48)	0.33	–
6,000 (41)	0.41	0.32
5,000 (35)	0.48	0.40
4,000 (28)	0.57	0.48
3,000 (21)	0.68	0.59
2,000 (14)	0.82	0.74

Courtesy of American Concrete Institute, Standard Practice for Selecting Proportions for Normal, Heavyweight and Mass Concrete, ACI 211.1-91, American Concrete Institute, Farmington Hills, Michigan, Reapproved 2002, table 6.3.4a. Used with permission.

TABLE 4.6
Maximum Water–Cementitious Material Ratios and Minimum Design Strengths for Various Exposure Conditions

Exposure Condition	Maximum Water–Cement Ratio by Mass for Concrete	Minimum Design Compressive Strength, f_c' psi (MPa)
Concrete protected from exposure to freezing and thawing, application of deicing chemicals, or aggressive substances	Select water–cement ratio on the basis of strength, workability, and finishing needs	Select strength based on structural requirements
Concrete expected to have low permeability when exposed to water	0.50	4,000 (28)
Concrete exposed to freezing and thawing in moist conditions or deicers	0.45	4,500 (31)
Reinforced concrete exposed to chlorides from deicing salts, salt water, brackish water, seawater, or spray from these sources	0.40	5,000 (35)

Courtesy of American Concrete Institute, Building Code Requirements for Structural Concrete, ACI 318-02, American Concrete Institute, Farmington Hills, Michigan, 2002, table 4.2.2. Used with permission.

TABLE 4.7
Requirements for Concrete Exposed to Sulfates in Soil or Water

Sulfate Exposure	Water-Soluble Sulfate (SO₄) in Soil, % by weight	Sulfate (SO₄) in Water, ppm	Cement Type	Maximum W/C Ratio by Weight
Negligible	<0.10	<150	No special type required	–
Moderate	0.10–0.20	150–1,500	II, II(MH), IP(MS), IS(<70)(MS), IT(P≥S)(MS), IT(P<S<70)(MS), MS	0.50
Severe	0.20–2.00	1,500–10,000	V, IP(HS), IS(<70)(HS), IT(P≥S) (HS), IT(P<S<70)(HS), HS	0.45
Very Severe	>2.00	>10,000	V, IP(HS), IS(<70)(HS), IT(P≥S) (HS), IT(P<S<70)(HS), HS	0.40

Courtesy of American Concrete Institute, Building Code Requirements for Structural Concrete, ACI 318-02, American Concrete Institute, Farmington Hills, Michigan, 2002, table 4.3.1. Used with permission.

Step 3. Estimation of the Coarse Aggregate Mass
The next step is to determine the suitable aggregate characteristics for the trial mixes. The course aggregate proportion is determined based on the possible nominal maximum aggregate size (NMAS) and the fineness modulus (FM) of the fine aggregate.

a) First, the possible NMAS is determined as the smallest one from Table 4.8.
b) Once NMAS is determined, the coarse aggregate proportion is determined from Table 4.9, using the NMAS and FM of fine aggregate.
     The NMAS requirements listed in Table 4.8 can be waived if the mixture possesses sufficient workability such that the concrete can be properly placed without honey-comb or large voids.

TABLE 4.8
Determining the NMAS of Aggregates

Situation	NMAS
Form dimensions	1/5 of minimum clear distance
Clear space between reinforcement or prestressing tendons	¾ of minimum clear space
Clear space between reinforcement and form	¾ of minimum clear space
Unreinforced slab	1/3 of thickness or depth

Data tabulated from the recommendations by the ACI 318-11 (pg. 44).

TABLE 4.9
Bulk Volume of Coarse Aggregate (CA) per Unit Volume of Concrete

NMAS of Aggregate, in. (mm)	Bulk Volume of Dry-Rodded Coarse Aggregate per Unit Volume of Concrete for Different Fineness Moduli of Fine Aggregate			
	Fineness Modulus (FM)			
	2.40	2.60	2.80	3.00
3/8 (9.5)	0.50	0.48	0.46	0.44
1/2 (12.5)	0.59	0.57	0.55	0.53
3/4 (19)	0.66	0.64	0.62	0.60
1.0 (25)	0.71	0.69	0.67	0.65
1.5 (37.5)	0.75	0.73	0.71	0.69
2.0 (50)	0.78	0.76	0.74	0.72
3.0 (75)	0.82	0.80	0.78	0.76
6.0 (150)	0.87	0.85	0.83	0.81

Courtesy of American Concrete Institute, Standard Practice for Selecting Proportions for Normal, Heavyweight and Mass Concrete, ACI 211.1-91, American Concrete Institute, Farmington Hills, Michigan, Reapproved 2002, table 6.3.6. Used with permission.

Step 4. Air-Entrainment Requirements

After selecting the size and proportion of the coarse aggregate, the target air content is determined. First, the need for air entrainment is evaluated. Air entrainment is required if the proposed concrete is expected to be exposed to freeze–thaw conditions and deicing salts, as well as sometimes for workability improvement. Whether air entrainment is to be used or not, the target percent of air content requirements for different nominal maximum sizes of aggregates is determined from Table 4.10. The air content of the job specifications should be specified to be delivered within –1% to +2% points of the table target value, for moderate and severe exposures.

TABLE 4.10

Approximate Target Percent Air Content Requirements for Different Nominal Maximum Sizes of Aggregates

	NMAS, in. (mm)							
	3/8 (9.5)	1/2 (12.5)	3/4 (19)	1.0 (25)	1.5 (37.5)	2.0 (50)	3.0 (75)	6.0 (150)
Non-Air-Entrained Concrete	3	2.5	2	1.5	1	0.5	0.3	0.2
Air-Entrained Concrete[a]								
Mild Exposure	4.5	4.0	3.5	3.0	2.5	2.0	1.5	1.0
Moderate Exposure	6.0	5.5	5.0	4.5	4.5	4.0	3.5	3.0
Severe Exposure	7.5	7.0	6.0	6.0	5.5	5.0	4.5	4.0

[a] The air content in job specifications should be specified to be delivered within −1 to +2 percentage points of the table target value for moderate and severe exposures.

Courtesy of American Concrete Institute, Standard Practice for Selecting Proportions for Normal, Heavyweight and Mass Concrete, ACI 211.1-91, American Concrete Institute, Farmington Hills, Michigan, Reapproved 2002, table A1.5.3.3. Used with permission.

Table 4.10 requires the input of exposures which are described below:

Mild Exposure. This exposure includes indoor or outdoor service in a climate where concrete may not be exposed to freezing or deicing agents. Air entrainment can be used for workability and strength.

Moderate Exposure. Service in a climate where freezing is expected, but where the concrete may not be continually exposed to moisture or free water for long periods prior to freezing and may not be exposed to deicing or other aggressive chemicals. Examples include exterior beams, columns, walls, girders, or slabs that are not in contact with wet soil, and are so located that they may not receive direct applications of deicing chemicals.

Severe Exposure. Concrete that is exposed to deicing or other aggressive chemicals, or moisture or free water prior to freezing. Examples include pavements, bridge decks, curbs, gutters, sidewalks, canal linings, exterior water tanks, or sumps.

Step 5. Workability Needs

Concrete must always be made with a workability, consistency, and plasticity suitable for job conditions. Workability is a measure of how easy or difficult it is to place, consolidate, and finish concrete. The slump test or slump value is used as an indication of workability. as shown in Table 4.11.

Step 6. Estimation of the Water Content

Then, the amount of water required to produce the desired slump for angular aggregate is determined from Table 4.12, using the NMAS of coarse aggregate and the air entrainment requirements.

To remind you, Table 4.12 lists the water requirements for angular-coarse aggregate. If coarse aggregate is not angular, the amount of water may be decreased using Table 4.13.

The water content of concrete is influenced by a number of factors, including the presence of admixtures, air content, content and type of cementing materials,

TABLE 4.11
Recommended Slumps for Several Types of Construction

Concrete Construction	Slump, in. (mm) Maximum[a]	Minimum
Reinforced foundation walls and footings	3 (75)	1 (25)
Plain footings, caissons, and substructure walls	3 (75)	1 (25)
Beams and reinforced walls	4 (100)	1 (25)
Building columns	4 (100)	1 (25)
Pavements and slabs	3 (75)	1 (25)
Mass concrete	2 (50)	1 (25)

[a] May be increased by 25 mm (1.0 in.) for consolidation by hand methods, such as rodding and spading. Plasticizers can safely provide higher slumps.
Courtesy of American Concrete Institute, Standard Practice for Selecting Proportions for Normal, Heavyweight and Mass Concrete, ACI 211.1-91, American Concrete Institute, Farmington Hills, Michigan, Reapproved 2002, table 6.3.1. Used with permission.

TABLE 4.12
Approximate Mixing Water in lb/yd^3 (kg/m^3) for Different Slumps and NMASs

Slump, in. (mm)	Nominal Maximum Aggregate Size (NMAS), in. (mm)							
	3/8 (9.5)	1/2 (12.5)	3/4 (19)	1.0 (25)	1.5 (37.5)	2.0 (50)	3.0 (75)	6.0 (150)
Non-Air-Entrained Concrete								
1–2 (25–50)	350 (207)	335 (199)	315 (190)	300 (179)	275 (166)	260 (154)	220 (130)	190 (113)
3–4 (75–100)	385 (228)	365 (216)	340 (205)	325 (193)	300 (181)	285 (169)	245 (145)	210 (124)
6–7 (150–175)	410 (243)	385 (228)	360 (216)	340 (202)	315 (190)	300 (178)	270 (160)	–
Air-Entrained Concrete								
1–2 (25–50)	305 (181)	295 (175)	280 (168)	270 (160)	250 (150)	240 (142)	205 (122)	180 (107)
3–4 (75–100)	340 (202)	325 (193)	305 (184)	295 (175)	275 (165)	265 (157)	225 (133)	200 (119)
6–7 (150–175)	365 (216)	345 (205)	325 (197)	310 (184)	290 (174)	280 (166)	260 (154)	–

Courtesy of American Concrete Institute, Standard Practice for Selecting Proportions for Normal, Heavyweight and Mass Concrete, ACI 211.1-91, American Concrete Institute, Farmington Hills, Michigan, Reapproved 2002, table 6.3.3. Used with permission.

TABLE 4.13

Reduction of Water Content for Non-Angular Coarse Aggregates

Aggregate Shapes	Reduction in Water lb/yd³ (kg/m³)
Sub-angular	20 (12)
Gravel with crushed particles	35 (21)
Round gravel	45 (27)

Data Tabulated from the recommendations by the PCA (2013) (pg. 238).

environmental conditions, slump, and the water to cementing materials ratio, as well as the aggregate size, shape, and texture. An increase in the air content, aggregate size, rounded aggregates, water-reducing admixtures, or fly ash causes a reduction in the water demand. On the other hand, increased temperatures, cement contents, slump, water–cement ratio, aggregate angularity, and a decrease in the proportion of coarse aggregate to fine aggregate, causes an increase in the water demand.

Step 7. Determination of the Cement Content Requirements

Once the water content is determined from Step 6, the cement content can be calculated using the selected W/C ratio from Step 2 as shown in Eq. 4.11.

$$\text{Cement weight} = \frac{\text{Water Weight}}{W/C} \tag{4.11}$$

Then, the calculated cement content is checked with the minimum cement content listed in Table 4.14 for exterior use.

The minimum cement content requirements serve to ensure the satisfactory durability and finishability, to improve the wear resistance of slabs, and to guarantee a

TABLE 4.14

Minimum Requirements of Cementing Materials for Different Aggregate Sizes

NMAS in. (mm)	Cement Content lb/yd³ (kg/m³)
1.5 (37.5)	470 (280)
1.0 (25)	520 (310)
3/4 (19)	540 (320)
1/2 (12.5)	590 (350)
3/8 (9.5)	610 (360)
Severe exposure, i.e., deicer exposure	564 (335)

Courtesy of American Concrete Institute, ACI 302, Farmington Hills, Michigan, table 6.2. Used with permission.

suitable appearance of vertical surfaces. This is important even though the strength requirements may be satisfied at a lower cementing materials content. However, excessively large amounts of cementing materials should be avoided to maintain economy in the mixture, and to not adversely affect the workability and other properties.

Step 8. Evaluation of the Admixture Needs

Water-reducing admixtures are added to concrete to reduce the water–cementing materials ratio, reduce cementing materials, reduce water content, reduce paste content, or to improve the workability of a concrete without changing the water–cementing materials ratio. Water reducers usually decrease the water content by 5–10% and some also increase the air content by ½ to 1 percentage point. Retarders may also increase the air content. High-range water reducers (plasticizers) reduce water contents between 12% and 30%, and some can simultaneously increase the air content up to 1 percentage point; others can reduce or not affect the air content.

Step 9. Estimation of the Fine Aggregate Mass

In previous steps, the coarse aggregate, water and cement weights (actually masses for metric units) have been determined per unit volume of the mix. These weights per unit volume of the mix are converted into volume of the respective component per unit volume of the mix as follows:

$$V = \frac{\text{Weight}}{\text{Unit Weight}} = \frac{W}{\gamma} = \frac{W}{G\gamma_w} \qquad (4.12)$$

Therefore,

$$V = \frac{W}{G\gamma_w} \qquad (4.13)$$

where,

V = Volume of component per unit volume of the mix (ft^3/yd^3 or m^3/m^3)

G = Bulk specific gravity of the component (coarse aggregate, cement, etc.)

W = Weight of component per unit volume of the mix (lb/yd^3 or kg/m^3)

γ = Unit weight of component (pcf or kg/m^3)

γ_w = Unit weight of water (62.4 pcf or 1,000 kg/m^3)

Note that the units should be consistent. If you use weight (W) in kg/m^3, the unit weight of water (γ_w) should be in 1,000 kg/m^3.

The calculated volumes of components and the air volume per unit volume of the mix are added together to calculate the subtotal volume of the mix ($V_{sub\text{-}total}$) except the fine aggregate as shown in Eq. 4.14.

$$V_{\text{sub-total}} = V_{\text{coarse}} + V_{\text{water}} + V_{\text{cement}} + V_{\text{air}} \qquad (4.14)$$

where,

V_{coarse} = Volume of coarse aggregate per unit volume of the mix (ft^3/yd^3 or m^3/m^3)

V_{water} = Volume of water per unit volume of the mix (ft^3/yd^3 or m^3/m^3)

V_{cement} = Volume of cement per unit volume of the mix (ft^3/yd^3 or m^3/m^3)

V_{air} = Volume of air per unit volume of the mix (ft^3/yd^3 or m^3/m^3)

Then, the fine aggregate volume per unit volume (V_{fine}) can be calculated from the total volume using Eq. 4.15.

$$V_{fine} = V_{total} - V_{sub\text{-}total} \tag{4.15}$$

where,

V_{total} = total volume of the mix (1.0 yd³ or 1.0 m³ is assumed)

The volume of fine aggregate per unit volume of the mix is then converted to weight per volume as shown in Eq. 4.16.

$$W = VG\gamma_w \tag{4.16}$$

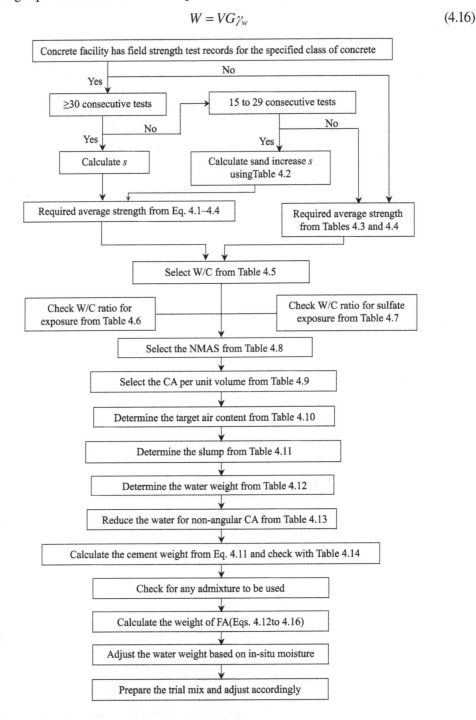

FIGURE 4.6   Summary of the mix design procedure.

Step 10. Determination of the Moisture Corrections

The contents of the coarse and fine aggregate determined in previous steps are dry weights. If the aggregate to be used has moisture in it, then the required water content is to be adjusted. This is because mix design assumes that water to be used to hydrate the cement is the free water available in the aggregate surfaces. The water absorbed by aggregates does not take part in the hydration of cement. Upon adding water, some amount of the water is absorbed by the aggregate if it is dry. Water lost to saturate the aggregates (absorbed) must be compensated (added). Similarly, if aggregate has excess water beyond the absorption capacity, this amount must be deducted from the water content.

Step 11. Trial Mix Procedures

After the mix design, a trial batch is prepared, and fresh concrete is tested for slump and air content. If the slump and air content requirements are satisfied, then three 6-in. diameter and 12-in. height samples are prepared, cured for 28 days, and then tested for compressive strength. If the samples fulfill the requirement of strength, the mix design is complete. Otherwise, the mixture is adjusted accordingly. Additional batches are to be made and tested in order to find out the most desired and economical mixture.

The total mix design procedure is summarized using the flow chart in Figure 4.6. The mix design method presented is also called the *absolute volume method*. Now, let us practice some worked-out examples for a better understanding of the concrete mix design.

Example 4.1 Average Compressive Strength of Concrete

For a project, the structural design engineer specifies the concrete strength of 4,000 psi. Determine the required average compressive strength of concrete to be prepared for the following conditions:

a) Standard deviation of the concrete strength is unknown
b) Standard deviation the concrete strength is 350 psi for 22 test results available from the past
c) Standard deviation of the concrete strength is 290 psi from the intensive history of the plant

Solution

Note: This is not a complete mix design procedure. This example describes how the desired strength of concrete is determined from the laboratory test data, which is the first step of the mix design process.

a) If fewer than 15 samples are tested, or no standard deviation data is available, then for a specified concrete strength of 4,000 psi, the required strength (from Eq. 4.6):

$$f'_c = f'_c + 1,200 = 4,000 + 1,200 = 5,200 \text{ psi}$$

b) If the standard deviation (s) is calculated using more than 30 samples, then the required average strength is the larger of the following (from Eqs. 4.1 and 4.2):

$$f'_{cr} = f'_c + 1.34s$$

and

$$f'_{cr} = f'_c + 2.33s - 500$$

As there are 22 test results available from the past, the s is to be multiplied by the modification factor (F). By interpolation from Table 4.2:
$F = 1.06$ for 22 test results

$$f'_{cr} = f'_c + 1.34(1.06s) = 4,000 + 1.34(1.06)(350) = 4,497 \text{ psi}$$

$$f'_{cr} = f'_c + 2.33(1.06s) - 500 = 4,000 + 2.33(1.06)(350) - 500 = 4,364 \text{ psi}$$

The larger of the above two results is $4,497 \approx 4,500$ psi

c) If the standard deviation (s) is calculated using more than 30 samples (intensive history), then, the required average strength is the larger of the following (from Eqs. 4.1 and 4.2):

$$f'_{cr} = f'_c + 1.34s = 4,000 + 1.34(290) = 4,389 \text{ psi}$$

$$f'_{cr} = f'_c + 2.33s - 500 = 4,000 + 2.33(290) - 500 = 4,176 \text{ psi}$$

The larger of the above two results is $4,389$ psi $\approx 4,390$ psi
Answers:

a)   *5,200 psi*
b)   *4,500 psi*
c)   *4,390 psi*

Example 4.2 Nominal Maximum Aggregate Size

A concrete mix is to be designed for casting a reinforced-concrete beam, as shown in Figure 4.7. The beam has 3 reinforcement bars (#8, or 1.0-in. diameter) placed 3 in. apart on the center. The bars are located 3 in. from the edge and bottom of the beam. Determine the Nominal Maximum Aggregate Size (NMAS) possible.

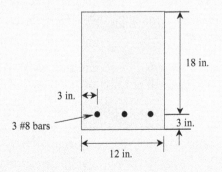

FIGURE 4.7   A reinforced concrete section for Example 4.2.

TABLE 4.15

Calculation of NMAS for Example 4.2

Situation	NMAS
Form dimension = 12 in.	1/5 (12 in.) = 2.4 in.
Clear space between reinforcements = 3 in. − ½ (1.0 in.) − ½ (1.0 in.) = 2.0 in.	¾ (2.0 in.) = 1.5 in.
Clear space between reinforcement and form (or clear cover) = 3 in. − ½ (1.0 in.) = 2.5 in.	¾ (2.5 in.) = 1.875 in.

Solution

Note: This example shows how the NMAS is determined. The NMAS is required to select the coarse-aggregate proportion in the third step of the mix design process. Let us use Table 4.8 to calculate the NMAS as shown in Table 4.15.

The NMAS is the smallest of the values obtained in Table 4.15, i.e., 1.5 in.

Answer: 1.5 in.

Example 4.3 Mix Design Calculations

A concrete mix design yields a water–cement ratio of 0.5, with 2,100 lb/yd³ of dry coarse aggregate, 4% air content, and 235 lb/yd³ of water. The bulk specific gravity of the cement used is 3.15. The coarse aggregate has a bulk specific gravity of 2.65, in-situ moisture content of 2%, and an absorption of 4%. The fine aggregate (sand) has a bulk specific gravity of 2.45, in-situ moisture content of 2.5%, and an absorption of 2.0%. Determine the weight of cement, wet coarse aggregate, wet fine aggregate, and amount of water for a 1 yd³ trial mix.

Solution

Note: This example shows only the Steps 9 and 10 of the mix design process, and how to perform computations to determine the amount of fine-aggregate, once the proportions of dry-coarse aggregate, cement, and water are known.

Weights:
Dry coarse aggregate = 2,100 lb/yd³
Water = 235 lb/yd³
Water–cement ratio, W/C = 0.5

From Eq. 4.11:

$$\text{Cement weight n} = \frac{\text{Water Weight}}{\text{W/C}} = \frac{235 \text{ lb/yd}^3}{0.5} = 470 \frac{\text{lb}}{\text{yd}^3}$$

Fine Aggregate Volume:

Let us assume that the total volume of the mix = 1 yd^3 = (1 yd^3) × (27 ft^3/yd^3) = 27 ft^3

From Eq. 4.13:

$$V = \frac{W}{G\gamma_w}$$

$$\text{Water Volume} = \frac{W}{G\gamma_w} = \frac{235 \dfrac{\text{lb}}{\text{yd}^3}}{1 \times 62.4 \text{ pcf}} = 3.766 \text{ ft}^3/\text{yd}^3$$

$$\text{Cement Volume} = \frac{W}{G\gamma_w} = \frac{470 \dfrac{\text{lb}}{\text{yd}^3}}{3.15 \times 62.4 \text{ pcf}} = 2.391 \text{ ft}^3/\text{yd}^3$$

Air Volume = 4% = 0.04 yd^3/yd^3 = (0.04 yd^3/yd^3) × (27 ft^3/yd^3) = 1.080 ft^3/yd^3

$$\text{Coarse Aggregate Volume} = \frac{W}{G\gamma_w} = \frac{2{,}100 \dfrac{\text{lb}}{\text{yd}^3}}{2.65 \times 62.4 \text{ pcf}} = 12.699 \text{ ft}^3/\text{yd}^3$$

From Eq. 4.14: Subtotal Volume = 3.766 + 2.391 + 1.080 + 12.699 = 19.936 ft^3/yd^3

From Eq. 4.15: Fine Aggregate = Total Volume – Subtotal volume

= 27 ft^3/yd^3 – 19.936 ft^3/yd^3 = 7.064 ft^3/yd^3

From Eq. 4.16: Fine Aggregate Dry Weight = $VG\gamma_w$ = 7.064 ft^3/yd^3(2.45 × 62.4 pcf) = 1,080 lb/yd^3

Moisture Correction:

Dry weight of the coarse aggregate = 2,100 lb/yd^3

Moist-weight of the coarse aggregate

= Dry weight (1 + Moisture Content)

= 2,100 (1 + 0.02)

= 2,142 lb/yd^3

Dry weight of the fine aggregate = 1,080 lb/yd^3

Moist-weight of the fine aggregate

= Dry weight (1 + Moisture Content)

= 1,080 (1 + 0.025)

= 1,107 lb/yd^3

The moisture content of the coarse aggregate is less than the absorption, and the moisture content of the fine aggregate is greater than the absorption. Therefore, the water quantity may need to be adjusted.

Shortage of moisture in the coarse aggregate to reach the absorption

= Dry weight (Absorption – Moisture Content)

= 2,100 (0.04 – 0.02)

= 42 lb/yd^3

Excess of moisture in the fine aggregate above the absorption =
   = Dry weight (Moisture Content – Absorption)
   = 1,080 (0.025 – 0.02)
   = 5.4 lb/yd³
Therefore, the additional moisture required = Shortage of moisture – Excess of moisture
   = 42 – 5.4 = 36.60 lb/yd³
Total water required = 235 + 36.6 = 271.6 lb/yd³
Answers:

Materials	Amount, lb/yd³
Water	272
Moist Coarse Aggregate	2,142
Moist Fine Aggregate	1,107
Cement	470

Note: It is recommended not to round off in every step of the mix design. The final results should be rounded off to a whole number for simplicity.

Example 4.4 Mix Design

Given Information:
   Coarse Aggregates
      Gravel with crushed particles
      NMAS = ¾ in.
      Bulk oven-dry specific gravity = 2.890
      Oven-dry rodded density = 105 pcf (2,835 lb/yd³)
      Moisture content = 0.4%
      Absorption = 2.0%
   Fine Aggregates
      Natural sand
      Bulk oven-dry specific gravity = 2.598
      Moisture content = 3.0%
      Absorption = 1.0%
      Fineness modulus = 2.80
   Cement
      Specific Gravity = 3.15
Strength Information:
   The specified strength of concrete = 4.5 ksi (Old data says the standard deviation of compressive strength is 300 psi when more than 30 samples are tested).
Project Environment:
   Concrete is to be used in an interior beam (width of 12 in.) with both a clear space between bars and a clear cover of 3 in.

Air Entrainer:
    None – concrete will not be exposed to freeze–thaw conditions and deicing salts.
Design the mix of the concrete using the ACI 318 absolute volume method.

Solution

Note: This is a complete mix design problem of a concrete sample. Follow all the steps of the mix design described in this chapter. Also, practice how to select values from different tables based on the requirements. In real life, the job conditions change from project to project, site to site, and so on.

Step 1. Strength Requirements
Given
The specified strength of concrete, $f'_c = 4,500$ psi
The standard deviation of compressive strength, $s = 300$ psi for more than 30 samples.
    From Table 4.1, the required average strength is the larger of the following (from Eqs. 4.1 and 4.2):

$$f'_{cr} = f'_c + 1.34s = 4,500 + 1.34(300) = 4,902\,\text{psi}$$

$$f'_{cr} = f'_c + 2.33s - 500 = 4,500 + 2.33(300) - 500 = 4,699\,\text{psi}$$

Therefore, the required average strength is 4,902 psi $\approx$ 4.9 ksi
Step 2. Water–Cement Ratio Requirements
From Table 4.5, $\dfrac{W}{C} = 0.49$ (by interpolation) for the required strength of 4.9 ksi for non-air-entrained concrete.
    Tables 4.6 and 4.7 are not applicable as there is no exposure or sulfates problem mentioned.
Step 3. Coarse-Aggregate Requirements
The permissible NMAS is the smallest of the following:
    1/5 of the minimum form dimension = 1/5 (12.0 in.) = 2.40 in.
    ¾ of the minimum clear cover = ¾ (3.0 in.) = 2.25 in.
    ¾ of the minimum clear space = ¾ (3.0 in.) = 2.25 in.
    The minimum value is 2.25 in.
The NMAS of ¾ in. is good as it is less than the minimum value of 2.25 in.
    From Table 4.9, for the NMAS of ¾ in. and the FM of 2.8, the coarse aggregate factor is 0.62.
    Therefore, the dry weight of coarse aggregate = 0.62 (2,835 lb/yd³) = 1,758 lb/yd³
Step 4. Air Entrainment Requirements
From Table 4.10, the target air content is 2% for non-air-entrained concrete.
Step 5. Workability Requirements
From Table 4.11, the slump range for a beam = 1–4 in. (let us use 3 in.)
Step 6. Water Content Requirements
From Table 4.12, for a 3-in. slump and ¾ in. NMAS, the water content is 340 lb/yd³.
For gravel with crushed particles, the reduction of water is 35 lb/yd³ from Table 4.13.

Required water = 340 − 35 = 305 lb/yd³

Step 7. Cement Content Requirements

From Step 2, $\dfrac{W}{C} = 0.49$

From Step 6, required water = 305 lb/yd³

From Eq. 4.11:

$$\text{Cement weight n} = \frac{\text{Water Weight}}{\text{W/C}} = 305 \text{ lb/yd}^3 / 0.49 = 622.45 \text{ lb/yd}^3$$

There is no need to check the minimum cement content, as the mix is to be used inside a building.

Step 8. Admixture Requirements

None.

Step 9. Fine Aggregate Requirements

Let us assume that the total volume of the mix = 1 yd³ = (1 yd³) × (27 ft³/yd³) = 27 ft³

From Eq. 4.13:

$$V = \frac{W}{G\gamma_w}$$

$$\text{Water Volume} = \frac{W}{G\gamma_w} = \frac{305\,\dfrac{\text{lb}}{\text{yd}^3}}{1 \times 62.4 \text{ pcf}} = 4.888 \text{ ft}^3/\text{yd}^3$$

$$\text{Cement Volume} = \frac{W}{G\gamma_w} = \frac{622.45\,\dfrac{\text{lb}}{\text{yd}^3}}{3.15 \times 62.4 \text{ pcf}} = 3.167 \text{ ft}^3/\text{yd}^3$$

Air Volume = 2% = 0.02 yd³/yd³ = (0.02 yd³/yd³) × (27 ft³/yd³) = = 0.54 ft³/yd³

$$\text{Coarse Aggregate Volume} = \frac{W}{G\gamma_w} = \frac{1{,}758\,\dfrac{\text{lb}}{\text{yd}^3}}{2.890 \times 62.4 \text{ pcf}} = 9.748 \text{ ft}^3/\text{yd}^3$$

From Eq. 4.14: Sub-total Volume = 4.888 + 3.167 + 0.54 + 9.748 = 18.343 ft³/yd³

From Eq. 4.15: Fine Aggregate = Total Volume − Subtotal volume
= 27 ft³/yd³ − 18.343 ft³/yd³ = 8.657 ft³/yd³

From Eq. 4.16: Fine Aggregate Dry Weight = $VG\gamma_w$ = 8.657 ft³/yd³(2.598 × 62.4 pcf) = 1,403.4 lb/yd³

Step 10. Moisture Correction

Dry weight of the coarse aggregate = 1,758 lb/yd³

Moist-weight of the coarse aggregate
= Dry weight (1 + Moisture Content)
= 1,758 (1 + 0.004) = 1,765 lb/yd³

Dry weight of the fine aggregate = 1,403.9 lb/yd³

Moist-weight of the fine aggregate
= Dry weight (1 + Moisture Content)
= 1,403.4 (1 + 0.03) = 1,446 lb/yd³

The moisture content of coarse aggregate is less than the absorption, and the moisture content of fine aggregate is greater than the absorption. Therefore, the water quantity may need to be adjusted.

Shortage of moisture in the coarse aggregate to reach the absorption

= Dry weight (Absorption − Moisture Content)

= 1,758 (0.02 − 0.004) = 28.13 lb/yd³

Excess of moisture in the fine aggregate above the absorption

= Dry weight (Moisture Content − Absorption)

= 1,403.4 (0.03 − 0.01) = 28.07 lb/yd³

Therefore, the additional moisture required

= Shortage of moisture − Excess of moisture

= 28.13 − 28.07 = 0.06 lb/yd³

Total water required = 305 + 0.06 = 305.06 lb/yd³

Answers:

Materials	Amount, lb/yd³
Water	305
Moist Coarse Aggregate	1,765
Moist Fine Aggregate	1,446
Cement	623

Now, you can adjust your material quantity based on the amount of mix to be prepared.

The next example shows the mix design process in SI units. The job conditions are also different from the previous problem, for further practice.

Example 4.5 Mix Design

Given Information:
    Materials
    Coarse Aggregates

    Rounded river gravel
    NMAS = 19 mm
    Bulk oven-dry specific gravity = 2.55
    Oven-dry rodded density = 1,761 kg/m³
    Moisture Content = 2.0%
    Absorption = 4.0%

Fine Aggregates
    Natural sand
    Bulk oven-dry specific gravity = 2.66
    Moisture content = 2%

Absorption = 4%
Fineness modulus = 2.47

Cement
Specific Gravity = 3.15
Strength Information
The specified strength of concrete = 27.6 MPa (Old data says the standard deviation of compressive strength is 2.1 MPa when more than 30 samples are tested).
Project Environment
A building frame with a minimum dimension of 150 mm, a minimum clear space between bars of 50 mm, and a minimum clear cover of 40 mm with mild exposure conditions.
Air Entrainer
6.3 ml/1% air/100 kg cement
Design the mix of the concrete using the ACI 318 absolute volume method.

Design:

Step 1. Strength Requirements
Given,
The specified strength of concrete, f'_c = 27.6 MPa
The standard deviation of compressive strength, s = 2.1 MPa for more than 30 samples.
The required average strength for f'_c = 27.6 MPa is the larger of the following (from Eqs. 4.1 and 4.4):

$$f'_{cr} = f'_c + 1.34s = 27.6 + 1.34(2.1) = 30.4 \text{ MPa}$$

$$f'_{cr} = f'_c + 2.33s - 3.45 = 27.6 + 2.33(2.1) - 3.45 = 29.0 \text{ MPa}$$

Therefore, the required average strength is 30.4 MPa

Step 2. Water–Cement Ratio Requirements
From Table 4.5, for air-entrained concrete, $\dfrac{W}{C}$ = 0.45 (by interpolation) for the required strength of 30.4 MPa.
Table 4.6 shows the maximum W/C value of 0.45 for exposure to freezing and thawing (mild exposure). Table 4.7 is not applicable as there is no sulfates problem. Therefore, $\dfrac{W}{C}$ = 0.45 is okay.

Step 3. Coarse-Aggregate Requirements
The permissible NMAS is the smallest of the following:
  1/5 of the minimum form dimension = 1/5 (150 mm) = 30 mm
  ¾ of the minimum clear cover = ¾ (50 mm) = 37.5 mm
  ¾ of the minimum clear space = ¾ (40 mm) = 30 mm
  The smallest value is 30 mm.
Therefore, the given NMAS of 19 mm is good as it is less than 30 mm.
From Table 4.9, for the NMAS of 19 mm and FM of 2.47, the coarse aggregate factor is about 0.65 by visual inspection.
Dry weight of the coarse aggregate = 0.65(1,761 kg/m³) = 1,145 kg/m³

Step 4. Air Entrainment Requirements

From Table 4.10, the target air content is 3.5% for air-entrained concrete with mild exposure.

Step 5. Workability Requirements

From Table 4.11, the slump range for a frame (say, beam) = 25–100 mm. Let us use 50-mm slump.

Step 6. Water Content Requirements

From Table 4.12, for 50-mm slump, 19 mm NMAS, and air-entrained concrete, the water content is 168 kg/m^3.

Reduction in the water content because of round gravel is 27 kg/m^3.

Required water content = 168 – 27 = 141 kg/m^3

Step 7. Cement Content Requirements

From Step 2, $\dfrac{W}{C} = 0.45$

From Step 6, required water = 141 kg/m^3

From Eq. 4.11:

$$\text{Cement weight} = \frac{\text{Water Weight}}{\text{W/C}} = 141/0.45 = 313 \text{ kg/m}^3$$

By Table 4.14, the minimum cement content is 320 kg/m^3.

Therefore, the cement content = 320 kg/m^3

Step 8. Admixture Requirements

3.5% air

Cement = 320 kg/m^3

Admixture = 6.3 ml/1% air/100 kg cement

$\qquad$ = (6.3 ml) × (3.5 air) × (320/100) = 71 ml/m^3

Step 9. Fine Aggregate Requirements

Let us assume that the total volume of the mix = 1 m^3

From Eq. 4.13: $V = \dfrac{W}{G\gamma_w}$

$$\text{Water Volume} = \frac{W}{G\gamma_w} = \frac{141\dfrac{\text{kg}}{\text{m}^3}}{1 \times 1{,}000\dfrac{\text{kg}}{\text{m}^3}} = 0.141\,\text{m}^3/\text{m}^3$$

$$\text{Cement Volume} = \frac{W}{G\gamma_w} = \frac{320\dfrac{\text{kg}}{\text{m}^3}}{3.15 \times 1{,}000\dfrac{\text{kg}}{\text{m}^3}} = 0.102\,\text{m}^3/\text{m}^3$$

Air Volume = 3.5% = 0.035 m^3/m^3

$$\text{Coarse Aggregate Volume} = \frac{W}{G\gamma_w} = \frac{1{,}145\dfrac{\text{kg}}{\text{m}^3}}{2.55 \times 1{,}000\dfrac{\text{kg}}{\text{m}^3}} = 0.449\,\text{m}^3/\text{m}^3$$

From Eq. 4.14: Subtotal Volume = 0.141 + 0.102 + 0.035 + 0.449 = 0.727 m^3/m^3

From Eq. 4.15: Fine Aggregate = 1 m³/m³ – 0.727 m³/m³ = 0.273 m³/m³

From Eq. 4.16: Fine Aggregate Dry Weight = $VG\gamma_w$ = 0.273 m³/m³ (2.66×1,000 kg/m³) = 726 kg/m³

Step 10. Moisture Correction

Dry weight of the coarse aggregate = 1,145 kg/m³

Moist-weight of the coarse aggregate

    = Dry weight (1 + Moisture Content)

    = 1,145 (1 + 0.02) = 1,168 kg/m³

Dry weight of the fine aggregate = 726 kg/m³

Moist-weight of the fine aggregate

    = Dry weight (1 + Moisture Content)

    = 726 (1 + 0.02) = 740.5 kg/m³

The moisture contents of the coarse and fine aggregates are less than the absorption. Therefore, the water quantity may need to be adjusted.

Shortage of moisture in the coarse aggregate to reach the absorption

    = Dry weight (Absorption – Moisture Content)

    = 1,145 (0.04 – 0.02) = 22.9 kg/m³

Shortage of moisture in the fine aggregate above the absorption

    = Dry weight (Absorption – Moisture Content)

    = 726 (0.04 – 0.02) = 14.52 kg/m³

Therefore, the additional moisture required = 22.9 + 14.52 = 37.42 kg/m³

Total water required = 141 + 37.42 = 178.42 kg/m³

Answers:

Materials	Amount
Water	178 kg/m³
Moist Coarse Aggregate	1,168 kg/m³
Moist Fine Aggregate	741 kg/m³
Cement	320 kg/m³
Admixture	71 ml

4.4 MIX DESIGNS FOR SMALL JOBS

The mix design procedure described above is good for large jobs. For smaller jobs, which do not require as large a quantity of concrete, the full effort of this mix design procedure is not very economical. For smaller jobs, Tables 4.16 and 4.17 can be used to select the proportions of ingredients of concrete.

While selecting ingredients by volume, the required volume should be increased by about 50% (or multiplied by 3/2). This is because the combined volume is approximately two-thirds of the sum of the original bulk volumes of the components, as the water and fines fill the voids of the coarse materials.

TABLE 4.16

Relative Components of Concrete for Small Jobs by Weight

NMAS in. (mm)	Air-Entrained Concrete				Non-Air-Entrained Concrete			
	Cement	Wet Fine Aggregate	Wet Coarse Aggregate[a]	Water	Cement	Wet Fine Aggregate	Wet Coarse Aggregate[a]	Water
3/8 (9.5)	0.210	0.384	0.333	0.073	0.200	0.407	0.317	0.076
1/2 (12.5)	0.195	0.333	0.399	0.073	0.185	0.363	0.377	0.075
3/4 (19)	0.176	0.296	0.458	0.070	0.170	0.320	0.442	0.068
1.0 (25)	0.169	0.275	0.493	0.063	0.161	0.302	0.470	0.067
1.5 (37.5)	0.159	0.262	0.517	0.062	0.153	0.287	0.500	0.060

[a] If crushed stone is used, decrease coarse aggregate by 50 kg and increase fine aggregate by 50 kg for each m³ of concrete (or decrease coarse aggregate by 3 lb and increase fine aggregate by 3 lb for each ft³ of concrete).

Adapted from Kosmatka, Steven H. and Wilson, Michelle L., 2016. Design and Control of Concrete Mixtures, EB001, 16th edition, Portland Cement Association, Skokie, IL, table 15.16. Used with permission.

TABLE 4.17

Relative Components of Concrete for Small Jobs by Bulk Volume[a]

NMAS in. (mm)	Air-Entrained Concrete				Non-Air-Entrained Concrete			
	Cement	Wet Fine Aggregate	Wet Coarse Aggregate	Water	Cement	Wet Fine Aggregate	Wet Coarse Aggregate	Water
3/8 (9.5)	0.190	0.429	0.286	0.095	0.182	0.455	0.272	0.091
1/2 (12.5)	0.174	0.391	0.348	0.087	0.167	0.417	0.333	0.083
3/4 (19)	0.160	0.360	0.400	0.080	0.153	0.385	0.385	0.077
1.0 (25)	0.154	0.346	0.423	0.077	0.148	0.370	0.408	0.074
1.5 (37.5)	0.148	0.333	0.445	0.074	0.143	0.357	0.429	0.071

[a] The combined volume is approximately 2/3 of the sum of the original bulk volumes.

Adapted from Kosmatka, Steven H. and Wilson, Michelle L., 2016. Design and Control of Concrete Mixtures, EB001, 16th edition, Portland Cement Association, Skokie, IL, table 15.17. Used with permission.

Example 4.6 Mix Design for Small Jobs by Volume

Determine the required volumes of components required to make a batch of 0.25 m³ of non-air-entrained concrete mix with the NMAS of 12.5 mm.

Solution

From the note of Table 4.17, the combined volume is approximately 2/3 of the sum of the original bulk volumes. Therefore, the sum of the original bulk volume of the components = 0.25 m³ × 3/2 = 0.375 m³ (water and fines occupy the voids of coarse aggregates)

From Table 4.17, the relative components of cement, fine aggregate, coarse aggregate and water are 0.167, 0.417, 0.333 and 0.083 respectively.

Volume of cement = 0.167×0.375 m³ = 0.063 m³
Volume of wet fine aggregate = 0.417×0.375 m³ = 0.156 m³
Volume of wet coarse aggregate = 0.333×0.375 m³ = 0.125 m³
Volume of water = 0.083×0.375 m³ = 0.031 m³

Answers:
Volume of cement = 0.063 m³
Volume of wet fine aggregate = 0.156 m³
Volume of wet coarse aggregate = 0.125 m³
Volume of water = 0.031 m³

Example 4.7 Mix Design for Small Jobs by Weight

Determine the required weights of components to make a batch of 2,500 lb of air-entrained concrete mix with the NMAS of 19 mm.

Solution

From Table 4.16, the relative components of cement, fine aggregate, coarse aggregate and water are 0.176, 0.296, 0.458 and 0.070 respectively.
Weight of cement = $0.176 \times 2,500$ lb = 440 lb
Weight of wet fine aggregate = $0.296 \times 2,500$ lb = 740 lb
Weight of wet coarse aggregate = $0.458 \times 2,500$ lb = $1,145$ lb
Weight of water = $0.070 \times 2,500$ lb = 175 lb

Answers:
Weight of cement = 440 lb
Weight of wet fine aggregate = 740 lb
Weight of wet coarse aggregate = 1,145 lb
Weight of water = 175 lb

4.5   CONCRETE MIX ADMIXTURES

Admixtures are the natural or manufactured chemicals used in PCC prior to or during the mixing process which do not include aggregate, Portland cement, or water. Admixtures are added to attain special properties to fresh or solid concrete, such as workability, setting time, strength, or durability.

Accelerators reduce the initial setting time of concrete and give the concrete its strength earlier than expected. They speed up the setting process and the rate of strength gain and thus make concrete stronger to resist damage from the freezing cold. Some other examples of usages of accelerators are the fast-track construction (requiring quick removal of the form), exposure to traffic, structures underwater, etc. On the other hand, *retarders* are chemicals that delay the initial setting of concrete and allow more time for placing and finishing. These are particularly helpful in hot weather, where the rapid setting causes heat generation.

Air-entraining admixtures are used to produce microscopic air bubbles throughout the whole concrete specimen uniformly. These bubbles improve the concrete's capacity against the freezing and thawing, and also deicing salt exposure. This addition also reduces the

TABLE 4.18
Concrete Admixtures

Type	Desired Effect on Concrete Mix	Materials
Accelerators	Accelerate setting and early strength gain	Calcium chloride, triethanolamine, sodium thiocyanate, calcium formate, calcium nitrite, and calcium nitrate
Retarders	Delay setting time	Lignin, borax, sugars, tartaric acid, and salts
Air-Entrainment	Improve durability against freeze–thaw, deicers, sulfate, and alkali; reduce bleeding, segregation, shrinkage, and laitance; and improve workability	Salts of wood resins, lignin, petroleum acids, proteinaceous material or sulfonated hydrocarbons, fatty and resinous acids and their salts, and alkylbenzene sulfonates
Air Detrainers	Decrease air content	Tributyl phosphate, dibutyl phthalate, octyl alcohol, water-insoluble esters of carbonic and boric acid, silicones
Water Reducers	Reduce water demand by 5%	Lignosulfonates, hydroxylated carboxylic acids, and carbohydrates
Superplasticizers	Reduce water–cement ratio by 12%	Sulfonated melamine formaldehyde condensates, sulfonated naphthalene formaldehyde condensates, and lignosulfonates
Workability Agents	Improve workability	Air-entraining admixtures, cementitious materials, natural pozzolans, and inert minerals (except silica fume)
Alkali-Reactivity Reducers	Reduce alkali-reactivity expansion	Pozzolans, blast-furnace slag, salts of lithium and barium, and air-entraining agents
Bonding	Increase bond strength	Rubber, polyvinyl chloride, polyvinyl acetate, acrylics, and butadiene styrene copolymers
Corrosion Inhibitors	Reduce steel corrosion activity in a chloride environment	Calcium nitrite, sodium nitrite, sodium benzoate, certain phosphates or fluorosilicates, and fluoroaluminates
Damp Proofing	Retard moisture penetration.	Soaps of calcium or ammonium stearate or oleate, butyl stearate, and petroleum products
Inert minerals	Improve workability filler	Marble, dolomite, quartz, and granite
Natural pozzolans	Improve workability and sulfate resistance; reduce alkali reactivity, permeability, and heat of hydration	Diatomaceous earth, opaline cherts, clays, shales, volcanic tuffs, pumicites, fly ash, and silica fume
Permeability reducers	Reduce permeability	Silica fume, fly ash, ground slag, natural pozzolans, and latex
Pumping aids	Improve pumpability	Organic and synthetic polymers, organic flocculants, organic emulsions of paraffin, coal tar, asphalt, acrylics, bentonite and pyrogenic silicas, natural pozzolans, fly ash, and hydrated lime

water requirement by improving the workability, and reduces the bleeding and segregation of aggregates in the mix. The fine air bubbles can reasonably be considered to form part of the paste, and thus, increase plasticity and concrete volume. The coarse air bubbles may perform like fine aggregate particles, which move easily. Air bubbles, fine or coarse, improve workability. No air entrainment is necessary for interior concrete members, as these are not subjected to freezing and thawing, or deicing salt. *Water reducers* are used to lower the water content requirements, which causes an increase in the strength and obtains a higher slump without adding water. Water reducers disperse the cement particles in concrete, making it more workable. They are particularly helpful in hot weather to offset the increased water demand. Some admixtures, their desired effects, and examples are listed in Table 4.18.

4.6 HANDLING FRESH CONCRETE

Similar to the handling of concrete, certain precautions should be taken for better quality concrete. Segregation and compaction are the two most important parameters to consider.

4.6.1 AVOIDING SEGREGATION

Segregation is the separation of coarse and fine portions of the mixture. It produces non-homogeneous concrete, meaning weaker areas may be developed. The most important rules for avoiding segregation during the placing of concrete are:

- Concrete should be placed vertically and as near as possible to its final position.
- Concrete should not flow to its final position; concrete should be shoveled into positions as needed.

Other techniques for avoiding segregation during placement depend on the type of element being constructed and on the type of distribution equipment being used. For walls and columns (i.e. deep, narrow forms), problems occur when the concrete is dropped from a height and ricochets off the reinforcement and form-faces, resulting in segregation.

4.6.2 AIDING COMPACTION

To attain adequate compaction of the concrete, concrete should be placed little by little in layers of small thicknesses, which can then be compacted properly with the available compaction equipment. It is virtually impossible to adequately compact deep layers, and thus, entrapped air may be left behind, causing voids and blow holes in the surface of the concrete. This flaw subsequently produces less durable and weaker concrete.

4.7 CURING OF CONCRETE

4.7.1 CONCEPT OF CURING

As discussed earlier, hydration takes place once water is added to cement. Hydration is the reaction of chemical compounds in cement with the water, causing concrete to begin hardening. A certain amount of heat is produced during hydration. Some amount of water available in the concrete (during its fresh condition) starts decreasing due to the heat of hydration jump-starting evaporation. Thus, the freshly placed concrete is kept moist by adding more water to let the hydration continue and gain strength. *Curing* is the method to keep concrete moist, facilitate the hydration, and gain the desired strength. In other words, curing is the process in which the concrete is protected from the loss of moisture and kept within a reasonable temperature range. The summary of the curing objectives can be written as follows:

a) To supplement the water lost due to the evaporation, so that continuous hydration can take place in the cement particles in concrete
b) To keep the temperature at the surface low, so that shrinkage, and hence, thermal cracks, do not occur

4.7.2 IMPORTANCE OF CURING

The result of the curing process is increased strength and decreased permeability. Curing is also a key component of mitigating cracks in the concrete, which severely impact the durability. If proper curing is not performed, due to the decrease in moisture, the concrete shrinks and cracks in block shapes. These types of cracks are called *shrinkage cracks*, shown in Figure 4.8.

Figure 4.9 shows the strength of a batch of different concretes with different levels of curing. Concrete samples which completed the entire curing period have the highest strength,

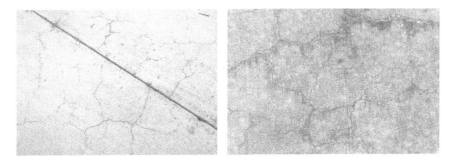

FIGURE 4.8 Shrinkage cracks due to improper curing. Photos taken at the Colorado State University–Pueblo.

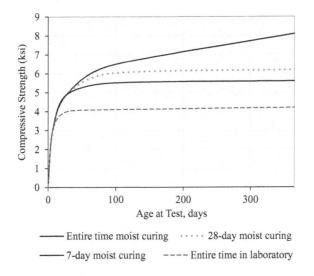

FIGURE 4.9 Compressive strength with age of curing.

while the samples with entire curing time spent in laboratory air have the least strength, for the same mix design. This means that the amount of strength desired is related to the duration of curing as well. After concrete is placed, it increases in strength very quickly for a period of 3–7 days. Concrete which is moist and cured for 7 days is about 50% stronger than the uncured concrete. The strength gain is very fast, up to 28 days, from the initial placing of concrete. Thus, it is very important to cure concrete for about 28 days. Ideally, the curing period depends upon the type of cement used, mixture proportions, required strength, size and shape of member, ambient weather, future exposure conditions, and method of curing.

4.7.3 CURING METHODS

There are three basic ways of curing concrete which are commonly used. These are:

a) Water curing – keeping the surface of the concrete moist by the use of ponding, sprinkling, damp sand, etc.
b) Sheet curing – covering the concrete with polythene sheeting or leaving the formwork in place
c) Spray curing – spraying or applying curing compounds

4.7.3.1 Water Curing

Water curing is the most efficient method of curing, and sometimes is the most appropriate method to be used for horizontal types of work such as floors, pavements, footpaths, etc. Ways to implement water curing include ponding, sprinkling, and wet coverings.

Ponding. On horizontal surfaces, such as pavements, footpaths, and floors, concrete can be cured by ponding. An enclosed boundary is formed by using the earth or sand dykes around the perimeter of the concrete surface, and some water is 'ponded' within this boundary.

Sprinkling. Continuous or periodic sprinkling of the concrete with water is also an excellent method of curing. This method is particularly appropriate for vertical walls, or any areas where ponding is difficult. If periodic sprinkling is conducted, intervals must be selected to prevent the concrete from drying between two consecutive sprinklings. This method is costly, as it requires an adequate supply of water, a sufficient amount of pressure to reach the extreme point, and supervision.

Wet Coverings. Wet coverings which use moisture-retaining fabrics are also common for curing. The coverings must be kept wet continuously to ensure a continuous film of water on the concrete surface. While placing coverings, care should be taken to avoid any concrete surface damage.

4.7.3.2 Sheet Curing

As an alternative to applying water, polythene sheet coverings can be applied to concrete. In addition, ensuring formwork remains in place can limit the need for water in the curing process. However, these methods may not be as effective but are usually adequate for most jobs.

Polythene Sheet. Polythene sheet is a very thin material with negligible weight that is easy to handle. If a layer of polythene is applied on freshly placed concrete, it may protect the concrete from moisture loss at the surface. The polythene should be placed into position as soon as possible – on vertical surfaces within half an hour of the removal of the formwork.

Formwork Protection. If the formwork is left in position, it protects concrete from the loss of moisture, and thus, allows curing to proceed. If any part of the concrete is exposed, it must be cured following any applicable method. If the formwork is made with wood, care should be taken, as possible defects (knots, shake, cracks), as well as joints in the wood may contribute to moisture loss. This method is effective in humid areas and interior environments if the formwork is used for at least 4 days.

4.7.3.3 Spray Curing

In this method, appropriate types of curing compounds are sprayed on the surface, forming a thin continuous film or membrane which seals in moisture and protects the concrete from most of the wind and sun actions for several weeks. A super-grade compound with white or aluminized pigment is recommended for external paving. A lower, non-pigmented grade is satisfactory for structural concrete. In tropical climates, the higher efficiency grade should be used in all jobs. Surfaces exposed to heavy daylight are not recommended to be treated with a curing compound. Curing compounds are generally non-toxic, but approval should be obtained before using them on structures.

4.8 STRENGTH BEHAVIOR OF CONCRETE

Fresh concrete gains strength upon the hydration of chemical components of cement. Hydration is the chemical reaction(s) of cement chemicals with water to form a slowly developing, cementitious structure that adheres to and binds together the fine and coarse aggregate, forming hardened concrete. As the age increases, the degree of hydration also increases and, consequently, the strength increases. The strength of concrete is commonly referred to as the *unconfined compression strength of concrete.*

Portland cement consists of four major compounds, namely tricalcium silicate, dicalcium silicate, tricalcium aluminate, and tetracalcium aluminoferrite. When water is added, each of these compounds undergoes the hydration process and contributes to the final, solid concrete. Only the calcium silicates contribute to the strength. Tricalcium silicate is responsible for most of the early strength (first 7 days), while dicalcium silicate reacts slower than tricalcium silicate and contributes only to the strength at later times.

It is assumed that the water–cement ratio is the primary factor affecting the strength of concrete. This is because the effects of other vital factors, such as porosity and aggregate structures, whose effect cannot be accurately determined. The strength of concrete increases when a small proportion of water is used to prepare the concrete. For a particular age and temperature of a fully compacted concrete, the strength is inversely proportional to the water–cement ratio. The hydration reaction itself consumes some amount of water. Concrete is actually mixed with more water than is needed for the hydration reactions. This extra water is added to give concrete sufficient workability. Flowing of concrete is desired to achieve proper filling and composition of the forms. The water not consumed in the hydration reaction remains in the microstructure pore space. These pores make the concrete weaker due to the lack of strength-forming calcium silicate hydrate bonds.

Closely behind the water–cement ratio, the degree of compaction, temperature, and age are also very important for affecting the strength of concrete. Some secondary factors affecting the strength of concrete are the aggregate–cement ratio and quality of aggregate, which includes grading, surface texture, shape, size of aggregate, and so on.

Incomplete compaction produces a very low, early strength, as shown in Figure 4.10a. Loose concrete has a very high porosity, and consequently a very low strength. For the density increase of 75% to full-compaction level, the strength increases by 90% (Glanville et al. 1950).

Temperature also affects the strength gain of concrete (Figure 4.10b). Fresh concrete with a higher temperature than normal results in more rapid hydration of cement, which leads to the rapid setting and the rapid strength gain. However, due to the less uniform development of the gel, the long-term strength gain is less. If the high temperature is accompanied by low humidity, then a rapid loss of water causes a rapid loss of workability, and a higher amount of plastic shrinkage. High temperature is also detrimental for placing a large amount of concrete because of the greater temperature differentials. Rapid heat production due to the hydration and subsequent cooling cause tensile stress inside the mix, and consequently may cause thermal cracking. The cold temperature problem arises from the freezing action on the fresh concrete. If the fresh concrete is allowed to freeze, the water mixed in the concrete freezes, and the overall concrete volume increases. When this occurs, no water is left for the hydration process, and thus, the setting and hardening of the concrete is delayed. During the thawing period, concrete hydrates and hardens in its expanded state, which leaves large pores, and consequently, yields a lower strength. If freezing occurs after the concrete has set, but before it has gained some strength, then the expansion associated with the formation of ice causes a disruption and irreparable loss of strength. If the temperature is cold, but not freezing, the strength gain is delayed, but will yield a higher strength in the long run.

As concrete ages, strength increases due to hydration. The exact strength gaining pattern cannot be specified since the hydration depends on curing, temperature, etc. As a rough estimation, concrete attains 60–65% of its ultimate strength in 7 days for full-moist curing. About 75% of its ultimate strength is achieved in 14 days, 90% in 28 days, and the remaining strength takes many years to achieve.

Regarding aggregate, the larger the aggregate-cement ratio, the larger the strength attained for a particular water–cement ratio. This is because if the paste represents a small proportion of the volume of the concrete, then the total porosity of the concrete is lower, and hence, its strength is higher. Smooth aggregate shows cracking at a lower stress than the rough and angular aggregate. Thus, concrete with angular aggregates has a higher flexural strength than that with rounded aggregates. However, the rounded aggregates require less water to attain certain workability than do the angular aggregates. Therefore, both types of aggregates are considered to have similar flexural characteristics.

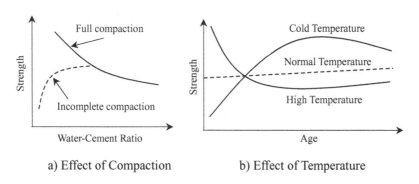

a) Effect of Compaction   b) Effect of Temperature

FIGURE 4.10 Effects of compaction and temperature on the strength of concrete.

4.9 READY MIXED CONCRETE

Ready mixed (or ready-mix) concrete refers to concrete that is produced according to the specification of the contractor in a central plant (Figure 4.11) and delivered to the job site in a timely manner, using a large truck (Figure 4.12). Ready-mix concrete is better in quality compared to the conventional concrete, as the plant uses skilled laborers, advanced equipment, and consistent methods. There is a more well-defined process in a plant compared to a job site, and thus, the ready-mix concrete is more appreciated by the contractors. The ready-mix plant also uses more mechanical devices instead of manual labor. Thus, it has a smaller amount of human error in its operations. Better mixing practices reduce the consumption of cement and produce less dust, as ready-mix concrete uses a bulk concrete instead of bags of cement.

The United States uses about 240 million cubic yards of ready mixed concrete each year, and about 68.8% of total cement consumption comes from ready mixed concrete (PCA 2013, Kosmatka et al. 2016). The United States Census Bureau (72013) shows that the United States had about 5,500 ready mixed concrete plants in 2011, many of which produce concrete for paving. The value of ready-mix concrete produced in the United States was estimated to be $34.7 billion in 2007, with a value of about $1.7 billion based on 5% of the cement used (United States Census Bureau 2007).

FIGURE 4.11   A ready-mix plant. Courtesy of Fujian South Highway Machinery, Co Ltd., Fengze District, Quanzhou City, Fujian, China. Used with permission.

FIGURE 4.12   Ready-mix concrete trucks and delivery. Photos taken in Colorado Spring.

Even with all of its advantages, ready-mix concrete has some disadvantages as well, especially if the job site is located far away or takes a long time to reach. The prolonged time to reach the site may cause a loss of the concrete's workability and can result in setting of concrete. These potential risks demand additional water or admixtures to maintain the workability and delay the setting period. In addition, a large amount of ready-mix concrete is delivered at a time. This means that a large construction area must be prepared in a timely manner, for the concrete pouring.

There are broadly three types of ready-mix concrete, depending upon the mixing of the various ingredients. These are mentioned below:

a) Transit-mixed concrete
b) Shrink-mixed concrete
c) Central-mixed concrete

4.9.1 Transit-Mixed Concrete

Transit-mixed concrete, also known as truck-mixed materials, is proportioned in a central plant, completely or partially mixed in the truck while in transit, and finished at the job site (if not completed in transit). While mixing in transit, the water is kept separate from the cement and aggregates, which allows the water to be mixed before the placement at the job site.

This method eliminates the unexpected loss of the workability and setting of concrete during the transportation of the mix. In addition, this way of mixing allows concrete to be transported to a large distance from the plant. However, the capacity of a truck used for transit-mix is smaller than that of the same truck containing the central-mixed concrete. Three types of transit-mixed concrete are possible. These are listed below:

Concrete mixed at job site. While being transported to the destination, the drum is revolved at a slow or agitating speed of 2 rpm. Upon reaching the site, just before discharging the material, the drum is revolved at a maximum speed of 12–15 rpm for nearly 70–100 revolutions to ensure homogeneous mixing.

Concrete mixed in transit. The drum is rotated at medium speed during the transit time, i.e. approximately 8 rpm for about 70 revolutions. After 70 revolutions, it is slowed down to the agitating speed of 2 rpm until the concrete is discharged.

Concrete mixed in the yard. The drum is turned at a high speed of 12–15 rpm for about 50 revolutions in the yard itself. The concrete is then agitated slowly during transit.

4.9.2 Shrink-Mixed Concrete

Shrink-mixed concrete is partially mixed in the plant mixer and partially mixed in the truck's mounted drum mixer during transit time. The amount of mixing conducted in transit depends on the extent of mixing done in the central mixing plant. Shrink-mixed concrete is used to increase the truck's load capacity and retain the advantages of transit-mixed concrete.

4.9.3 Central-Mixed Concrete

Central-mixed concrete is completely mixed at the plant and loaded into the truck mixer. The truck mixer agitates the mix only while transporting the concrete to the job site. In the case that the workability requirement is low, non-agitating units or dump trucks can also be used. This type of mix is commonly used for large project and short-distance job sites.

4.10 EARLY-AGE CRACK CONTROL

At the early age of concrete, cracks may occur for a variety of reasons. Cracking in concrete is inevitable if proper care is not taken. The main reasons are shrinkage due to moisture loss, thermal-contraction due to a decrease in temperature, and brittleness. Some other reasons for this occurrence are the settlement of the concrete foundation (such as soil) and the movement of the formwork before the concrete has fully hardened. Cracks which form before concrete has fully hardened are known as the *early-age cracks*, or prehardening cracks. There are three main types of early-age cracks:

a) Plastic shrinkage cracks
b) Plastic settlement cracks
c) Cracks caused by formwork movement

All of these early-age cracks occur due to improper construction conditions and/or poor practices. Sometimes, the faulty formwork design may lead to a complete collapse. Early-age cracks are often avoidable, following some good construction practices.

4.10.1 PLASTIC SHRINKAGE CRACKS

Plastic shrinkage cracks are formed in the concrete surface when the mix is still plastic, meaning before it has set and begun to harden. These cracks may not be visible until they age. These cracks occur due to the rapid shrinkage of the concrete's surface upon too rapid a loss of moisture from the surface of the concrete. This rapid loss of moisture commonly happens during hot, dry, and windy conditions. Figure 4.13 shows some plastic shrinkage cracks. Usually, these cracks do not show any regular pattern, and may range from 1.0 in. (25 mm) to 80 in. (2 m) in length. They are fairly straight, and vary from a hairline to perhaps one-tenth of 1.0 in. (2.5 mm) in width.

FIGURE 4.13 Plastic shrinkage cracks in a concrete slab. Photo taken at the Colorado State University–Pueblo.

The most effective way to reduce the risk of plastic shrinkage cracking is to prevent rapid loss of moisture from the surface of the concrete. This can be achieved in different ways, such as:

- Using aliphatic alcohols sprayed over the surface, prior to and after finishing, but before curing
- Drying any excess water in the place where concrete is to be poured
- Using cool aggregates to lower the temperature of the fresh concrete
- Adding polypropylene fibers to the concrete mix

4.10.2 PLASTIC SETTLEMENT CRACKS

After placing at a job site, most concrete bleeds. Bleeding of concrete occurs due to water within the concrete rising to the surface as the solid particles settle. Once the bleed water evaporates, there is a loss of volume from the concrete. This loss of volume is called the *plastic settlement* of concrete. There is a very slight lowering of the surface due to the decrease in volume, provided there is no restraint such as reinforcement. If there are any restraints near the surface, the lowering of the concrete surface may be resisted, and potential cracks may develop in the nearby concrete which is not restrained. Settlement cracks follow the pattern of the reinforcement. Some factors contributing to the plastic settlement include:

- High slump
- Rate of bleeding
- Depth of reinforcement (relative to total thickness)
- Total time of settlement
- Constituents of the mix

Plastic settlement cracks may be closed by revibrating the concrete after the settlement is complete (after half an hour to one hour). Vibrating before the concrete has begun to stiffen may allow the cracks to reopen, while vibrating too late, i.e., after the concrete has hardened, may damage the bond with reinforcement or reduce its strength. Other procedures which may help reduce the plastic settlement cracking include:

- Using lower slump mixes
- Using more cohesive mixes
- Using an air entrainer to improve cohesiveness and reduce bleeding
- Increasing cover to top bars
- Avoiding retarders in cold weather

4.10.3 FORMWORK MOVEMENT CRACKS

Concrete cracks if it is not fully hardened whenever there is a movement of the formwork, intentional or unintentional. Freshly placed concrete requires some time to gain enough strength to support its own weight. This type of crack follows the movement of the formwork. The formwork should be constructed carefully, with a strong base and tight connections. The formwork should not be touched until the concrete has gained sufficient strength to support itself. If supplementary cementitious materials, such as fly ash, are used, it takes a longer time to gain strength. Forms should be properly aligned, clean, tight, adequately braced, and constructed of materials that impart the desired off-the-form finish to the hardened concrete. Formwork for concrete can be constructed of lumber, metal, hardboard, or plastic. The forms should be straight and free from warping, and have sufficient strength to resist concrete pressure without deforming.

4.11 WEATHER CONSIDERATION

Weather conditions at a job site may be vastly different from the optimum concrete placement conditions assumed at the time a concrete mixture is designed. Longer duration projects require changes to the concrete mixture as the seasonal weather changes. Hot weather conditions can adversely influence concrete quality, primarily by accelerating the rate of evaporation/moisture loss and the rate of cement hydration. Many difficulties with hot weather concreting, including increased water demand and accelerated slump loss, lead to the addition of water on site. Adding water to concrete at the job site can adversely affect the strength and serviceability properties of the hardened concrete. Before concrete is placed, certain precautions should be taken during the hot weather times to maintain or reduce concrete's temperature:

- Mixers, chutes, conveyor belts, hoppers, pump lines, and other equipment for handling concrete should be shaded, painted white, or covered with wet burlap to reduce the effect of solar heating
- Forms, reinforcing steel, and subgrade should be wetted with cool water just before the concrete is placed
- Fogging cools the contact surfaces and increases humidity to minimize rate of evaporation
- Moistening of subgrade for slabs on ground
- Ensuring there is no standing water or puddles when concrete is placed
- During extremely hot periods, concrete placements may be improved by restricting pouring times to early morning, late evening, or nighttime hours

A set retarding admixture may be beneficial in delaying the setting time in hot weather concreting, despite the potential for an increased rate of slump loss resulting from its use. A hydration control admixture or set stabilizer can be used to stop the cement hydration and setting. Hydration is resumed with the addition of a special accelerator.

Cold weather is also detrimental to concrete. If concrete freezes before it gains sufficient strength (about 500 psi), ice forms within the concrete, resulting in the disruption of the cement paste, as well as a loss of strength as high as 50% (Kosmatka et al. 2016). Fresh concrete should not be allowed to freeze within the first 24 hours of placing it. In addition, the rapid cooling or heating of concrete may cause cracks due to the differential thermal gradient (exterior is cold and interior is hot).

4.12 CHAPTER SUMMARY

The concrete mix design procedure has been discussed with mathematical examples. Engineers are directly responsible for design of a mix which is durable, satisfies the requirement, and is workable. However, simply designing a proper mix cannot guarantee a good mix. Proper placing, compaction, and curing are essential to ensure the quality of the mix. Some parts of mix design, such as admixtures, slump tests, and air tests, are discussed in detail in the next chapter. The majority of jobs are constructed with traditional concrete. With the improvements in modern technology, special or alternative concretes are becoming more common every day. The next chapter discusses these methods, in addition to the common laboratory and field testing procedures performed in reality.

Concrete mix design is the process of determining the proportions of cement, aggregates, and water, considering the required strength, service environment, and construction conditions, such as construction underwater, etc. The following steps are followed while designing a mix:

Step 1. Strength requirements – The strength requirement is slightly more than the specified strength (the strength that is used in designing the member). An appropriate increase in strength must be made.

Step 2. Determination of the W/C – The water–cement ratio (ratio of the weight of water to the weight of cement) is a very important factor while mix designing. A lower water–cement ratio yields higher strength and durability. At the same time, low water may make the mix too dry and difficult to work with. The water–cement ratio varies from 0.32 to 0.82 based on different strength requirement and whether air entrainment is used or not. Concrete exposed to sulfates in soil or water uses the water–cement ratio of 0.40 to 0.50 based on the severity.

Step 3. Estimation of the coarse aggregate mass – The maximum size of the aggregate is selected based on the concrete member's dimensions and reinforcement spacing. The volume of coarse aggregate per unit volume of concrete is determined using the maximum aggregate size and the fineness modulus of cement.

Step 4. Air entrainment requirements – Air entrainment is provided if the proposed concrete is expected to be exposed to freeze–thaw conditions and deicing salts, as well as for workability.

Step 5. Workability needs – The slump test is conducted on freshly mixed concrete to check the adequacy of workability. A high slump value means there is too much water, but concrete is easy to work, and vice-versa. An optimum slump is adopted based on the type of construction, such as reinforced foundations, walls, or columns, etc. The minimum of 1.0 in. and the maximum of 4.0 in. slump are used.

Step 6. Estimation of the water content – The amount of water needed to produce the desired slump is determined using the NMAS of coarse aggregate and the air-entrainment requirements.

Step 7. Determination of the cement content requirements – Once the amount of water is determined, the cement content can be determined using the preselected water–cement ratio.

Step 8. Evaluation of admixture needs – Different admixtures may be mixed as desired. For example, water-reducing admixtures are added to concrete to reduce the water–cementing materials ratio. Retarders may be used to delay the setting of concrete.

Step 9. Estimation of the fine aggregate mass – The volume of fine aggregate is determined by subtracting the volumes of the coarse aggregate, water, and cement from the total volume.

Step 10. Determination of the moisture corrections – The moisture volume per batch must be adjusted depending on the in-situ moisture content of the aggregate. The contents of coarse aggregate and fine aggregate are determined considering dry weights. If the in-situ aggregate is dry, additional water is added to saturate the aggregate; otherwise some part of the added water is absorbed by aggregate, resulting in less water available for the reaction with cement.

Step 11. Trial mix procedures – After the mix design has been established, a trial batch is prepared. The fresh concrete is then tested for quality control and quality assurance.

For smaller jobs, the above lengthy and expensive procedures are not followed. The ACI recommended weight or volume methods are commonly used.

The strength and performance of concrete are dependent, not only on the mix design but also the handling and curing as well. Segregation must be avoided while placing fresh concrete. Concrete must be compacted properly to attain the desired strength. Curing is also imperative to attain the desired strength and avoid hydration-related cracks. Concrete commonly has 50% less strength if curing is not performed. Curing can be done in three different ways: water curing, sheet curing, and spraying chemicals. Early-age cracking, such as plastic shrinkage cracks, plastic settlement cracks, or cracks due to formwork movement, should also be mitigated while placing concrete.

Ready-mix concrete is another improvement of the concrete mix design process. Ready-mix concrete is produced in a central plant and delivered to the job site. Ready-mix concrete is better in quality, as the plant uses skilled laborers, up to date equipment, and consistent methods. However, ready-mix concrete is more appropriate for large projects.

ORGANIZATIONS DEALING WITH CONCRETE MIX

American Concrete Institute (ACI). 'The American Concrete Institute (ACI) is a non-profit, technical society and standards-developing organization. ACI is a leading authority and resource worldwide for the development, distribution, and adoption of consensus-based standards, technical resources, educational programs, and proven expertise for individuals and organizations involved in concrete design, construction, and materials, who share a commitment to pursuing the best use of concrete.'

Location: Farmington Hills, Michigan, USA.
Website: www.concrete.org

National Ready Mixed Concrete Association (NRMCA). 'Founded in 1930, the National Ready Mixed Concrete Association is the leading industry advocate. Its mission is to provide exceptional value for our members by responsibly representing and serving the entire ready mixed concrete industry through leadership, promotion, education, and partnering, to ensure ready mixed concrete is the building material of choice.'

Location: Silver Spring, MD
Website: https://www.nrmca.org
PCA is also a resource for concrete mix design.

REFERENCES

ACI 211.1-91. 2002. Standard Practice for Selecting Proportions for Normal, Heavyweight and Mass Concrete, ACI 211.1-91, American Concrete Institute, Farmington Hills, Michigan, Reapproved 2002.

ACI Committee 318. 2002. Building Code Requirements for Structural Concrete, ACI 318-02, and Commentary, ACI 318R-02, American Concrete Institute, Farmington Hills, Michigan.

ACI. 2011. Committee 318, Building Code Requirements for Structural Concrete, ACI 318-11, and Commentary, American Concrete Institute, Farmington Hills, Michigan.

PCA. 2013. *Design and Control of Concrete Mixtures, EB001*, 15th Edition. Portland Cement Association, Skokie, IL.

Kosmatka, Steven H. and Wilson, Michelle L. 2016. *Design and Control of Concrete Mixtures, EB001*, 16th Edition. Portland Cement Association, Skokie, IL, p. 632.

Glanville, W., Collins, A. and Matthews, D. 1950. The grading of aggregate and workability of concrete, Road Research Tech, Paper No. 5, London.

United States Geological Survey (USGS). 2013. *2011 Minerals Yearbook, Cement. Advance Release*. US Department of the Interior, United States Geological Survey, Reston, VA.

United States Census Bureau. 2007. *Manufacturing: Industry Series: Detailed Statistics by Industry for the United States: 2007 Economic Census. Ready Mixed Concrete Manufacturing*. United States Census Bureau, Washington, DC.

FUNDAMENTALS OF ENGINEERING (FE) EXAM STYLE QUESTIONS

FE Problem 4.1

The maximum aggregate size possible for a 6-in. unreinforced slab is closest to:

A. 1.2 in.
B. 4.5 in.
C. 2.0 in.
D. 2.4 in.

Solution: C

The maximum aggregate size possible for the unreinforced slab = 1/3 (6 in.) = 2.0 in.

FE Problem 4.2

Which of the following types of aggregates requires a greater amount of water while preparing the concrete mix?

A. Sub-angular
B. Gravel with crushed particles
C. Round gravel
D. Angular

Solution: D

The angular aggregate requires more water when preparing the concrete mix. If the aggregate is sub-angular, gravel with crushed particles, or round gravel, the water content is decreased.

FE Problem 4.3

The minimum amount of cement required for severe exposure, i.e., deicer exposure, conditions is closest to:

A. 335 kg/m^3
B. 564 kg/m^3
C. 280 kg/m^3
D. No minimum criteria

Solution: A

The minimum amount of cement required for severe exposure, i.e., deicer exposure, conditions is 335 kg/m^3 (564 lb/yd^3).

PRACTICE PROBLEMS

Problem 4.1

Given Information

Coarse Aggregates
  Gravel with crushed particles
  NMAS = ¾ in.
  Bulk oven-dry specific gravity = 2.890
  Oven-dry rodded density = 105 pcf (2,835 lb/yd^3)
  Moisture content = 0.4%
  Absorption = 4.0%

Fine Aggregates
  Natural sand
  Bulk oven-dry specific gravity = 2.598
  Moisture content = 2.0%
  Absorption = 1.0%
  Fineness modulus = 2.80

Cement
  Specific gravity = 3.15

Strength Information

Specified strength of concrete=4.5 ksi (old data shows the standard deviation of the compressive strength is 300 psi when more than 30 samples are tested).

Project Environment

Concrete is to be used in an interior beam (width of 12 in.) with both clear space between bars and clear cover of 3 in.

Air Entrainer

None, as the concrete is not expected to be exposed to freeze–thaw conditions and deicing salts.
  Design the concrete mix using the absolute volume method.

PROBLEM 4.2

Given Information

Coarse Aggregates

  Crushed stone (angular)
  NMAS=2 in.
  Bulk oven-dry specific gravity=2.573
  Oven-dry rodded density=120 pcf
  Moisture content=1.0%
  Absorption=2.80%

Fine Aggregates

  Natural sand
  Bulk oven-dry specific gravity=2.540
  Moisture content=4.5%
  Absorption=3.4%
  Fineness modulus=2.68

Cement

  Specific gravity=3.15

Strength Information

Specified strength of concrete=3.0 ksi (old data says the standard deviation of the compressive strength is 250 psi when more than 30 samples are tested).

Project Environment

Concrete is to be used in a pavement slab, 12 in. thick, in the City of Bozeman, Montana (cold climate).

Air Entrainer

0.15 fl. oz/1% air/100 lb cement
  Design the concrete mix using the absolute volume method.

PROBLEM 4.3

For a project, the structural design engineer specifies a concrete strength of 5,200 psi. Determine the required average compressive strength of concrete to be prepared for the following conditions:

a) The standard deviation of the concrete strength is unknown.
b) The standard deviation of the concrete strength is 250 psi for 17 test results available from the past.
c) The standard deviation of the concrete strength is 210 psi for the intensive history of the plant.

PROBLEM 4.4

For a project, the structural design engineer specifies a concrete strength of 2,800 psi. Determine the required average compressive strength of concrete to be prepared for the following conditions:

a) The standard deviation of the concrete strength is unknown.
b) The standard deviation of the concrete strength is 410 psi for 27 test results available from the past.
c) The standard deviation of the concrete strength is 270 psi for 82 test results available from the past.

PROBLEM 4.5

Determine the nominal maximum aggregate size possible if:

- The minimum dimension of the concrete member = 8 in.
- 0.75-in. diameter steel is placed in the middle at 4-in. center to center, with clear cover of 1.0 in.

PROBLEM 4.6

Determine the nominal maximum aggregate size possible for a 4-in. thick, reinforced concrete pavement slab. The temperature reinforcement of 0.5-in. diameter steel is placed in the middle of the beam at 12-in. center to center.

PROBLEM 4.7

A concrete mix design yields a water–cement ratio of 0.42, 1,900 lb/yd³ of dry coarse aggregate, 2% air content, and 225 lb/yd³ of water. The bulk specific gravity of cement used is 3.15. The coarse aggregate has a bulk specific gravity of 2.65, in-situ moisture content of 2.5%, and an absorption of 4.5%. The fine aggregate (sand) has a bulk specific gravity of 2.42, in-situ moisture content of 0.5%, and an absorption of 2.5%.

Determine the weight of cement, wet coarse aggregate, wet fine aggregate, and water for a 1 yd³ trial mix.

PROBLEM 4.8

A concrete mix design yields a water–cement ratio of 0.46, 2,300 lb/yd³ of dry coarse aggregate, 4% air content, and 255 lb/yd³ of water. The bulk specific gravity of cement used is 3.15. The coarse aggregate has a bulk specific gravity of 2.67, in-situ moisture content of 0.5%, and an absorption of 4%. The fine aggregate (sand) has a bulk specific gravity of 2.55, in-situ moisture content of 3.5%, and an absorption of 1.0%.

Determine the weight of cement, wet coarse aggregate, wet fine aggregate, and water for a 1 yd³ trial mix.

PROBLEM 4.9

Determine the required volumes of components required to make a batch of 0.20 m³ of an air-entrained concrete mix with the NMAS of 12.5 mm.

PROBLEM 4.10

Determine the required weights of components to make a 2,500-lb batch of a non-air-entrained concrete mix with the NMAS of 19 mm.

PROBLEM 4.11

Determine the required volumes of components required to make a batch of 0.20 yd³ of an air-entrained concrete mix with the NMAS of 1.0 in.

PROBLEM 4.12

Determine the required weights of components to make a 3,200-kg batch of an air-entrained concrete mix with the NMAS of 0.375 in.

PROBLEM 4.13

List the effect(s) of fly ash on a concrete mix.

PROBLEM 4.14

List some admixtures which can decrease the water demand in a concrete mix.

PROBLEM 4.15

Differentiate among workability, slump, and consistency.

5 Portland Cement Concretes

This chapter discusses the behavior, properties, and characterization procedures of hardened (solid) concrete. Once the mix design is complete, trial specimens are prepared and tested to verify the mix design. With the improvement of technology, some non-conventional concrete products have become more available recently. This chapter also discusses several non-conventional concretes.

5.1 BACKGROUND

Portland cement concrete (PCC), shown in Figure 5.1, is produced by mixing Portland cement, aggregate, water, and admixtures (optional) as described in the previous chapter. The compacted concrete is then cured to the best quality possible to have the desired strength and performance. Care should also be taken in their early stages to avoid the plastic shrinkage cracks, plastic settlement cracks, and damages during the formwork removals. Laboratory-prepared specimens are treated according to the appropriate standards and tested for quality control and quality assurance.

PCC is one of the most versatile and widely used civil engineering materials in the world. Sometimes, admixtures are added to concrete to attain some special properties. Some inherent advantages of concrete are:

* Low material cost
* Low maintenance
* Great resistance to fire and water
* Ability to be cast into any shape
* Very importantly, the high compression strength

Two major disadvantages of PCC are low tensile strength and heavy weight. The disadvantage of low tensile strength can be overcome by adding reinforcing bars (rebar). The problem of heavy weight can be turned into an advantage by using PCC in footings, floors, basements, piers, rigid pavements, and similar jobs. *Rigid pavement* is the pavement where the surface layer is predominantly Portland cement concrete. If both asphalt concrete and Portland cement concrete are used, the contribution of Portland cement concrete must be predominant so that, upon loading, a major part of the slab deflects, not only under the loading. About four tons of concrete are produced per person per year worldwide and about 1.7 tons of concrete are produced per person per year in the United States (PCA 2016).

5.2 HARDENED PCC BEHAVIOR

5.2.1 EARLY VOLUME CHANGE

Concrete undergoes volume changes due to physical and chemical reactions with its internal and external environment. All concrete goes through a phase of expansion during the

a) Laboratory Compacted   b) Field Cored

FIGURE 5.1   Hardened cement concretes. *Photos taken at the Colorado State University–Pueblo.*

plastic state, followed by a phase of drying shrinkage during the hardened state. Unless the shrinkage is compensated by cement, the shrinkage always results in a final volume that is less than the original volume. Even with the use of shrinkage-compensating cement, the final volume is less than the volume of the expansion phase. Understanding the nature and cause of these volume changes can help in understanding the proper design and installation of coatings and polymer flooring systems on concrete.

5.2.2   Creep Properties

During the hardening process of concrete, creep may occur due to the application of a sustained load on it, such as any ponding water for curing, self-weight, or any other imposed load. Creep is the gradual increase in the strain with time due to a sustained loading. It can be of the same order of magnitude as the drying shrinkage. If the sustained load is removed, the strain which decreases immediately is the elastic strain, at the given age. Then, the gradual decrease in the strain, called the creep recovery starts. When the hydrated cement is completely dried, little to no creep occurs. Creep continues for a very long time, and the detectable amount of creep can be observed after as long as 30 years. However, the creep rate is very fast at the early age of concrete and decrease with time. A rough measurement is that about 75% of the 20-year creep occurs during the first year.

5.2.3   Permeability

Permeability is defined as the rate of flow (flow volume per unit time) of a fluid through a porous solid. Permeability can be related to the ability to resist the weathering action, chemical attack, abrasion, or any process of deterioration. It is difficult for water and corrosive materials to penetrate the less permeable concrete. Thus, concrete permeability influences durability because it controls the rate that moisture can penetrate into it. Low permeable concretes are desired for applications exposed outdoors without aggressive chemicals. Figure 5.2 shows a permeable concrete specimen.

Some factors affecting the permeability of concrete are mentioned below:

- Inadequate compaction leaves enormous voids inside the mix, resulting in highly porous concrete.
- The W/C ratio is directly related to the permeability, as the presence of large amounts of water in the mix leaves large, unfilled spaces after the water is consumed by hydration or evaporation.

FIGURE 5.2 Permeable concretes. *Photos taken at the Colorado State University–Pueblo.*

- Wet curing reduces the permeability, as evaporation of mix water is prevented.
- Some admixtures, such as silica fume, latex emulsions, and high-range water reducers, allow placement of the highly impermeable concrete.
- Hydration over a long period of time (years) requires moisture consumption and porosity. Coarse cement particles require longer hydration times.

To avoid high permeability, a portion of the Portland cement can be replaced with pozzolans or slag, in addition to considering the above factors affecting permeability.

5.2.4 Stress–Strain Behavior

Typical stress–strain curves for the normal-strength concretes under the compressive load are presented in Figure 5.3. The salient features of the concrete stress–strain curves under compressive load can be sequentially presented as:

- The stress–strain curves become non-linear after about 30–40% of the ultimate stress. The non-linearity occurs due the coalescence of microcracks at the paste-aggregate interface. The coalesced microcracks start connecting with each other, and a large crack network is formed when the ultimate stress is reached.
- The strain corresponding to the ultimate stress is usually around 0.002 for most normal-strength concretes, as shown in Figure 5.3. The stress–strain behavior in

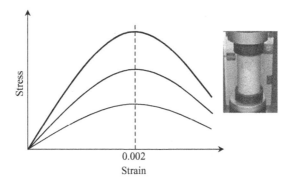

FIGURE 5.3 Stress–strain behavior of normal-strength concrete.

tension is similar to that in compression. The ultimate stress is presented as f'_c and considered as the *unconfined compression strength*, or simply compression strength, or even strength of concrete.

- The descending portion of the stress–strain curve (the post-peak response) of the concrete is commonly not recorded as concrete exhibits catastrophic failure.
- Using a displacement-controlled machine, the post-peak behavior can be recorded slowly. The strain at failure is typically around 0.003–0.005 for normal-strength concrete.
- The post peak behavior is actually a function of the stiffness of the testing machine, in relation to the stiffness of the test specimen and the rate of strain. With the increasing strength of concrete, its brittleness also increases, and this is shown by a reduction in the strain at failure.

Cement paste and aggregates individually have linear stress–strain relationships. However, the behavior of concrete is non-linear. This is due to the microcracking created at the bonding between aggregate and cement.

It can also be seen that the concrete having a higher ultimate compression strength (f'_c) has a higher *modulus of elasticity* (the initial slope of the stress–strain curve). The modulus of elasticity of concrete can be empirically calculated from the following relationship:

$$E_c = 33w_c^{1.5} \sqrt{f'_c} \tag{5.1}$$

where,
  E_c = Modulus of elasticity of concrete, psi
  w_c = Unit weight of concrete, 90–155 pcf
  f'_c = Compression strength of concrete, psi

For the normal weight concrete weighing 145 pcf, the following simplified equation can be used:

$$E_c = 57,000\sqrt{f'_c} \tag{5.2}$$

Eqs. 5.1 and 5.2 are empirically obtained and thus the only possible units are as mentioned.
    Note: 1 MPa = 145 psi. If the unit of f'_c is given in MPa, it must be converted into psi to use Eq. 5.2.

Example 5.1 Modulus of Elasticity

Determine the modulus of elasticity of a concrete with f'_c of 4 ksi and a unit weight of 150 pcf.

Solution

Given,
    Unit weight of concrete, w_c = 150 pcf
    Compression strength of concrete, f'_c = 4,000 psi

From Eq. 5.1: Modulus of elasticity

$$E_c = 33w_c^{1.5}\sqrt{f_c'} = 33(150 \text{ pcf})^{1.5}\sqrt{4,000 \text{ psi}}$$

$$= 3,834,253 \text{ psi}$$

$$= 3,834 \text{ ksi}$$

Answer: 3,834 ksi

5.3 PREPARATION OF LABORATORY/FIELD SPECIMENS

Preparation of specimens is crucial before characterizing concrete specimens. Two types of specimens are commonly prepared:

a) Cylindrical specimens for compressive or splitting tensile strength testing
b) Beam specimens for flexural strength testing

ASTM C 192 describes the specimen preparation in laboratory and ASTM C 31 covers for the field specimen preparation. Compressive or splitting tensile strength specimens are cylindrical (Figure 5.4) and allowed to set in an upright position. The length is twice the diameter. The cylinder diameter should be at least three times greater than the NMAS of the coarse aggregate. When the NMAS of the coarse aggregate exceeds 2 in. (50 mm), the concrete specimen should be treated by wet sieving through a 2-in. (50 mm) sieve. For acceptance testing for specified compressive strength, cylinders should be 6 by 12 in. (150 by 300 mm) or, when specified, 4×8 in. (100×200 mm). The mix is placed in the molds in layers one by one. Each layer is rodded before adding a new layer. The number of layers and number of rodding per layer depend on the diameter of the mold and are shown in Table 5.1. After each layer is rodded, tap the outsides of the mold lightly 10–15 times with the mallet, so as to close any holes left by rodding and to release any large air bubbles that may have been trapped. Once the final layer is rodded, trim any extra material above the mold using a bar.

Flexural strength specimens are beam shaped (Figure 5.5) and hardened in the horizontal position. The length should be at least 2 in. (50 mm) greater than three times the depth.

Weighing   Mixing   Compacting   Finishing

FIGURE 5.4 Preparing concrete cylinder. *Photos taken at the Farmingdale State College of the State University of New York.*

The ratio of the width to the depth as molded should not exceed 1.5. The standard beam cross-section should be 6 in. by 6 in. (150 by 150 mm) and should be used for concrete with the NMAS of coarse aggregate up to 2 in. (50 mm). When the NMAS of the coarse aggregate exceeds 2 in. (50 mm), the smaller cross-sectional dimension of the beam should be at least three times the nominal maximum size of the coarse aggregate. Beams made in the field should not have a width or depth of less than 6 in. The mix is placed in the molds in layers and rodded before adding a new layer. The number of layers and number of rodding per layer are shown in Tables 5.2 and 5.3. After each layer is rodded, the outsides of the mold are lightly tapped 10–15 times, similarly to cylindrical specimens. Once the final layer is rodded, any extra material is trimmed above the mold.

TABLE 5.1
Rodding requirements for concrete cylinders

Diameter, in. (mm)	Number of Layers of Equal Depth	Number of Rodding per Layer
4 (100)	2	25
6 (150)	3	25
9 (225)	4	50

| Batching | Mixing | Compacting | Finishing |

FIGURE 5.5 Preparing concrete beam specimen. *Photos taken at the Farmingdale State College of the State University of New York.*

TABLE 5.2
Layer and Rodding Requirements for Concrete Beam

Depth, in. (mm)	Number of Layers of Equal Depth
Up to 8 (200)	2
>8 (200)	3 or more

TABLE 5.3
Number of Rodding per Layer

Top Surface Area, in.^{2} (mm^2)	Diameter of Rod, in. (mm)	No. of Rodding/layer
25 (15,625) or less	3/8 (9.375)	25
26–49 (16,250–30,625)	3/8 (9.375)	One per 1.0 in.^{2} (625 mm^2)
50 (31,250) or more	5/8 (9.375)	One per 2.0 in.^{2} (625 mm^2)

Once the specimens are prepared, they are stored for a period up to 24 hours at a temperature between 60 and 80°F (16 and 27°C), and in an environment preventing moisture loss from the specimens. Upon completion of the initial curing and within 30 minutes after removing the molds, cure specimens with free water maintained on their surfaces at all times, at a temperature of 73 ± 3°F (23 ± 2°C) using water storage tanks (saturated-lime-water) or moist rooms. Water by itself may leach out calcium hydroxide from concrete specimens and reduce their strength, so the water must be saturated with added calcium hydroxide (hydrated lime). Beams are to be cured the same way as cylinders, except that they should be stored in water saturated with calcium hydroxide at 73 ± 3°F (23 ± 2°C) at least 20 hours prior to testing.

5.4 MECHANICAL CHARACTERIZATIONS OF HARDENED PCC

5.4.1 Elastic Modulus and Poisson's Ratio

The elastic modulus of PCC can be determined using the ASTM C 469 test standard. A uniaxial compressive load is applied within 40% of the ultimate concrete strength and the resulting axial and lateral deformations are measured, as shown in Figure 5.6. The stress is calculated by dividing the load by the cross-sectional area and the strain is calculated by dividing the deformation by the original dimension. Using the equations provided in chapter 1, the modulus of elasticity and Poisson's ratio of PCC are calculated.

The Poisson's ratio of new PCC typically ranges between 0.11 and 0.21, and values between 0.15 and 0.18 are typically considered for the PCC design. The elastic modulus of concrete ranges from 2,000 to 6,000 ksi (14 to 40 GPa).

FIGURE 5.6 Modulus of elasticity and Poisson's ratio of concrete testing. *Photo taken at the Colorado State University–Pueblo.*

5.4.2 Unconfined Compression Strength

The ASTM C 39 test method covers determination of compressive strength of cylindrical concrete specimens, such as molded cylinders and drilled cores. This test consists of applying a compressive axial load to molded cylinders or cores, at a rate which is within a prescribed range, until failure occurs (Figure 5.7). The compressive strength of the specimen is calculated by dividing the maximum load attained during the test by the cross-sectional area of the specimen. Test procedures can be summarized as follows:

1) Carefully place and align the axis of the specimen with the center of the loading frame.
2) Apply the load continuously and without shock, at a rate of approximately 20–50 psi/s.
3) Apply the load until the specimen fails and record the maximum load.
4) Calculate the compressive strength by dividing the maximum load by the average cross-sectional area.
5) If the specimen length to diameter ratio (L/D) is 1.75 or less, correct the result obtained by the appropriate correction factor listed in Table 5.4.

The commonly used normal-weight concrete has a compressive strength between 3,000 and 6,000 psi (20 and 40 MPa).

John Hendrickson at the SUNY –Farmingdale

| Prepared Specimens | Testing | Failed Specimen |

FIGURE 5.7 Testing concrete cylinder under compression. *Photos taken at the Farmingdale State College of the State University of New York.*

TABLE 5.4
Specimen Length to Diameter Correction Factor

L/D	1.75	1.50	1.25	1.0
Correction Factor	0.98	0.96	0.93	0.87

5.4.3 Flexural Strength or Modulus of Rupture

Flexural strength is typically used in PCC mix design for pavements because it best simulates slab flexural stresses as they are subjected to loading. For the normal-weight concrete, the flexural strength or modulus of rupture (R) can be approximated as follows (Mamlouk and Zaniewski 2014):

For the US customary unit: $R = (7.5 \text{ to } 10)\sqrt{f_c'}$ (psi)   (5.3)

For the Metric unit : $R = (0.62 \text{ to } 0.83)\sqrt{f_c'}$ (MPa)   (5.4)

Note: Eqs. 5.3 and 5.4 gives a range of output. For example, R ranges from 7.5 to 10 times of $\sqrt{f_c'}$ for the US customary unit. Consider the average value of 7.5 and 10 for normal use, the lowest value of 7.5 for risky structures and the highest value of 10 for non-important structure.

There are two basic flexural tests: the third-point loading and the center-point loading. The beam is supported on each end and is loaded at its third points (for the third-point loading test) or at the middle (for the center-point loading test) until the failure occurs.

5.4.3.1   Using Center-Point Loading

The ASTM C 293 test method describes the testing procedure of the flexural strength of concrete specimens by the use of a simple beam with the center-point loading (Figure 5.8). The specimen is positioned symmetrically, and a load is applied on the specimen continuously and without shock, at a rate between 125 and 175 psi/min (0.9 and 1.2 MPa/min). The load is applied until the failure, and the peak load is recorded. The flexural strength or modulus of rupture (R) is calculated using the following equation:

$$R = \frac{Mc}{I} = \frac{\left(\dfrac{PL}{4}\right)\left(\dfrac{d}{2}\right)}{\dfrac{bd^3}{12}} = \frac{3}{2}\frac{PL}{bd^2}$$ (5.5)

where,
R = Modulus of rupture (psi or MPa)
M = Developed maximum moment (lb.in. or N.mm)
P = Maximum applied load indicated by the testing machine (lb, or N)
c = The distance from the neutral axis to the extreme tensile fiber, half of the depth for rectangular section (in.)
I = Moment of inertia of the beam section, = $bd^3/12$ for rectangular section (in.⁴ or mm⁴)
b = Width of the beam at the fracture (in. or mm)
d = Height of the beam at the fracture (in. or mm) (h is also used instead of d)

5.4.3.2   Using Third-Point Loading

The third-point loading test (Figure 5.8) is preferred because, ideally, in the middle third of the span the specimen is subjected to pure moment force with zero shear. The center-point loading test gives results about 15% higher than that of the center-point loading. According to the ASTM C 78 test method, the specimen is positioned symmetrically, and a load is applied on the specimen continuously and without shock. The load should be applied at a constant rate to the breaking point, at a rate that constantly increases the extreme fiber stress between 125 and 175 psi/min (0.86 and 1.21 MPa/min) until rupture occurs. The peak load is recorded and used in the flexural strength calculation.

a) Center-Point Loading

b)Third-Point Loading

c) A Concrete Beam Failed Out of the Middle-Third

FIGURE 5.8 Concrete modulus of rupture testing. *Photos taken at the Farmingdale State College of the State University of New York.*

The modulus of rupture is calculated as follows:

a) If the fracture initiates in the tension surface *within* the middle third of the span length, calculate the flexural strength or modulus of rupture (R) as follows:

$$R = \frac{Mc}{I} = \frac{\left(\dfrac{PL}{6}\right)\dfrac{d}{2}}{\dfrac{bd^3}{12}} = \frac{PL}{bd^2} \tag{5.6}$$

where,
R = Modulus of rupture (psi or MPa)
P = Maximum applied load indicated by the testing machine (lb, or N)
L = Span length (in. or mm)
b = Average width of specimen at the fracture (in. or mm)
d = Average depth of specimen at the fracture (in. or mm)

b) If the fracture occurs in the tension surface *outside* of the middle third of the span length by *no more* than 5% of the span length, calculate the flexural strength or modulus of rupture (R) as follows:

$$R = \frac{3Pa}{bd^2} \tag{5.7}$$

where,

  a = Average distance between line of fracture and the nearest support measured on the tension surface of the beam (in. or mm).

c) If the fracture occurs in the tension surface *outside* of the middle third of the span length by *more* than 5% of the span length, discard the test.

Example 5.2 Flexural Strength

Determine the flexural strength of a concrete with f'_c of 30 MPa.

Solution

Given,

  Compression strength of concrete, $f'_c = 30$ MPa

  From Eq. 5.4: $R = (0.62 \text{ to } 0.83)\sqrt{f'_c}$

  Average of 0.62 and 0.83 = (0.62 + 0.83)/2 = 0.725

  Therefore,

$$R = 0.725\sqrt{f'_c} = 0.725\sqrt{30 \text{ MPa}} = 3.97 \text{ MPa}$$

  Answer: 3.97 MPa

Example 5.3 Flexural Strength

A 6×6×20 in. concrete beam is being tested applying the center-point loading for flexure, as shown in Figure 5.9. The beam fails when the load reaches 3.0 kip. Calculate the flexural strength of the beam.

Solution

Given,

  Peak load, $P = 3,000$ lb
  Span of beam, $L = 20$ in.
  Width of beam, $b = 6.0$ in.
  Depth of beam, $d = 6.0$ in.

From Eq. 5.5: Flexural strength, $R = \dfrac{3}{2}\dfrac{PL}{bd^2} = \dfrac{3}{2}\dfrac{(3,000 \text{ lb})(20 \text{ in.})}{(6 \text{ in.})(6 \text{ in.})^2} = 417$ psi

  Answer: 417 psi

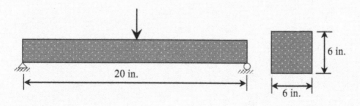

FIGURE 5.9 Center-point loading for Example 5.3.

5.4.4   SPLITTING TENSION TEST

Tensile strength of PCC is important for structures experiencing tensile stress upon load-ing. For example, the concrete slab in rigid pavement undergoes tensile stress for differ-ent loading and temperature conditions. Tensile stresses are typically measured indirectly by one of two means: a *splitting tension test* or a *flexural strength test*. According to the AASHTO T 198 and ASTM C 496, a splitting tension test uses a standard 6-in. (150-mm) diameter, 12 in. (300 mm) long test cylinder, laid on its side as shown in Figure 5.10. A diametric compressive load is then applied along the length of the cylinder until it fails. The splitting tensile strength or fracture strength, *T* is calculated as:

$$T = \frac{2P}{\pi l D} \tag{5.8}$$

where,

   T = Tensile strength (psi or MPa)
   P = Peak force needed to crack the specimen diagonally (lb or N)
   D = Diameter of the specimen (in. or mm)
   l = Length of the specimen (in. or mm)

Because PCC is much weaker in tension than compression, the cylinder typically fails due to horizontal tension and not vertical compression. The tensile strength of concrete varies from 8% to 15% of its compressive strength, and commonly ranges from 360 to 450 psi (2.5 to 3.1 MPa).

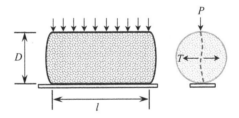

FIGURE 5.10   Concrete splitting tension testing.

Example 5.4 Tensile Strength Test

In a splitting tensile strength test, the peak load required to fail a specimen of 6.0-in. diameter and 12-in.thickness is 45,000 lb. Calculate the indirect tensile strength of the specimen.

Solution

Given:

   Diameter, $D = 6.0$ in.
   Length/thickness, $l = 12.0$ in.
   Peak load, $P = 45,000$ lb

From Eq. 5.8: Indirect tensile strength,

$$T = \frac{2P}{\pi l D} = \frac{2(45,000\ \text{lb})}{\pi(12.0\ \text{in.})(6.0\ \text{in.})} = 398\ \text{psi}$$

Answer: 398 psi

5.4.5 COEFFICIENT OF THERMAL EXPANSION

The change in length per unit length (thermal strain) due to the temperature change is called the thermal expansion. This property is required for designing pavement with concrete slabs. The developed thermal stress in concrete is dependent on the coefficient of thermal expansion/contraction and the decrease/increase in temperature, as discussed in Chapter 1. The coefficient of thermal expansion/contraction can be determined using the AASHTO TP 60 test protocol. The coefficients of thermal expansion/contraction of PCC vary typically between 3.8×10^{-6} and 6.7×10^{-6} per °F (7×10^{-6} and 12×10^{-6} per °C).

5.4.6 REBOUND HAMMER TEST

The *rebound hammer test* (ASTM C 805) is based on the principle that the rebound of an elastic mass depends on the hardness of the concrete surface against which the mass strikes. A spring-driven steel hammer impacts with the concrete surface, and the distance that the hammer rebounds is measured as shown in Figure 5.11. The instrument is held firmly perpendicular to the concrete surface. Gradually the instrument is pushed toward the test surface until the hammer impacts. After impact, the rebound number is read on the scale to the nearest whole number and is recorded. The rebound reading is then correlated with the compressive strength of concrete.

FIGURE 5.11 Rebound hammer test. *Courtesy of James Instruments, Inc., 3727 N. Kedzie, Chicago, IL. Used with permission.*

5.4.7 PENETRATION RESISTANCE TEST

Penetration resistance determines the resistance of a hardened concrete to penetration of a steel probe or pin following the ASTM C 803 test standard. A driver delivers a known

FIGURE 5.12   Penetration resistance testing. *Courtesy of James Instruments, Inc., 3727 N. Kedzie, Chicago, IL. Used with permission.*

amount of energy to either a steel probe or pin, as shown in Figure 5.12. The penetration resistance of the concrete to this probe or pin is determined by measuring either the exposed lengths of probes that have been driven into the concrete, or by measuring the depth of the holes created by the penetration of the pins into the concrete. This resistance is empirically correlated with the in-place strength.

5.4.8   ULTRASONIC PULSE VELOCITY TEST

The strength and quality of concrete can be assessed by measuring the velocity of an ultrasonic pulse passing through a concrete structure or natural rock formation, as shown in Figure 5.13. According to the ASTM C 597 protocol, the test is conducted by passing a pulse of ultrasonic wave through the concrete to be tested, and measuring the time taken by the pulse to get through the structure. Higher velocities indicate good quality and continuity

FIGURE 5.13   Ultrasonic pulse velocity testing. *Courtesy of James Instruments, Inc., 3727 N. Kedzie, Chicago, IL. Used with permission.*

of the material, while lower velocities may indicate concrete with many cracks or voids. The ultrasonic testing equipment includes a pulse generation circuit and a pulse reception circuit that receives the signal. The quality of concrete in terms of strength, homogeneity, trapped air, internal flaws, cracks, segregation, honeycombing, compaction, workmanship, and durability can be concluded from this test.

5.4.9 CONCRETE MATURITY TEST

The maturity test determines how mature the placed concrete is. To perform the test, a strength-maturity relationship is developed by laboratory testing on the concrete mixture to be used. The temperature history of the field concrete, for which strength is to be estimated, is recorded from the time of concrete placement to the time when the strength estimation is desired. The recorded temperature history is used to predict the maturity status of the field concrete. The maturity level is expressed either in terms of the temperature-time factor, or in terms of the equivalent age at a specified temperature. Using the predicted maturity index and the strength-maturity correlation, the strength of the field concrete can also be determined. The ASTM C 1074 test method covers the estimation of concrete strength by means of the maturity method. This practice can be used to determine the in-place strength of concrete to perform construction activities, such as:

- Removal of formwork or any support
- Post-tensioning of tendons
- Termination of cold weather protection
- Opening of the roadways to traffic

5.5 NON-CONVENTIONAL CONCRETES

The above discussed concrete is the traditional concrete commonly used in civil and construction engineering. With the improvement in technology, some non-conventional concretes are being used and are gaining popularity for special jobs. Some of them are commercially available nationwide. Several common non-conventional concretes are discussed in this section.

5.5.1 SELF-COMPACTING CONCRETE

Self-consolidating concrete, or *self-compacting concrete*, (SCC) mixes have a low yield stress, high deformability, good segregation resistance, and moderate viscosity. This type of mix has uniform suspension of solid particles during transportation, placement, and thereafter, until the concrete sets. Without external compaction, the solid particles stay in a suspended condition. SCC flows very easily within and around formwork, obstructions, and corners, as shown in Figure 5.14. It does not require any vibration or tamping after pouring and takes the shape of the mold after hardening. SCC gains its fluid properties from the high proportion of fine aggregate (typically 50%), combined with the superplasticizers and viscosity-enhancing admixtures. SCC is commonly used for heavily reinforced sections and complex shapes of formwork where vibration is difficult.

FIGURE 5.14 Self-compacting concrete. *Courtesy of Roadstone.*

5.5.2 SHOTCRETE

In common practice, *shotcrete* refers to concrete or mortar conveyed through a hose and pneumatically thrown at a high velocity onto a surface as an innovative construction technique, as shown in Figure 5.15. Specifically, shotcrete is the wet mix, previously prepared or a ready-mixed concrete. Gunite is the mixture of dry ingredients, with which water is mixed just before leaving the hose. In summary:

- Shotcrete is the wet mix previously prepared or ready-mixed concrete
- Gunite is the mixture of dry ingredients water is mixed with just before applying

FIGURE 5.15 Applying shotcrete. *Courtesy of Shotcrete Services Ltd., Cranbrook, Kent, United Kingdom. Used with permission.*

Both can be used with conventional construction techniques, such as being reinforced by steel rods, steel mesh, or fibers. However, gunite is more popular for swimming-pool construction. Gunite swimming pools are more durable compared to conventional materials, and allow for customization in shape, depth, and style. Shotcrete is placed and compacted at the same time, due to the force with the nozzle. It can be sprayed onto any type or shape of surface, including vertical and overhead areas. Gunite involves taking the dry ingredients into a hopper and then conveying them pneumatically through a hose to the nozzle. The nozzle man adds water at the nozzle. The water and the dry mixture are partially mixed at this point but become fully mixed when the mixture hits the receiving surface. Thus, this method requires a skilled nozzle man. Advantages of the dry mix process are that the water content can be adjusted by the nozzle man, allowing more effective placement in overhead and vertical applications without using accelerators. The dry mix process is useful in repair applications when it is necessary to stop frequently.

5.5.3 FLOWABLE FILL

Flowable fill, or simply *flow fill*, is a mixture of cement, water, fine aggregate, fly ash, and/ or slag. Due to the nature of the product, many materials that do not meet the quality standards for use in concrete have been used successfully in the flow fill mixtures. Chemical admixtures specifically developed for use in flow fill mixtures have also become more commonly used as a result of their unique benefits on the performance properties of the mixture. A common application of flow fill is around the pipelines under roadways. The mix flows all around the pipelines and provides rigid protection from the traffic impact to the pipelines, as shown in Figure 5.16. On top of the flow fill, conventional pavement materials (soil, aggregate base, and asphalt layer) are placed and compacted.

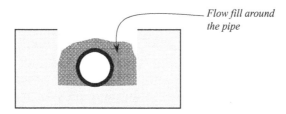

Flow fill around
the pipe

FIGURE 5.16 Flowable fill.

5.5.4 LIGHTWEIGHT CONCRETE

Lightweight concretes (Figure 5.17) can either be lightweight aggregate concrete, foamed concrete, or Autoclaved Aerated Concrete (AAC). Lightweight aggregate concrete can be produced using a variety of lightweight aggregates. Lightweight aggregates originate from different sources, and include:

FIGURE 5.17 Light-weight concrete specimens. *Photo taken at the Colorado State University–Pueblo.*

- Natural materials, like volcanic pumice
- The thermal treatment of natural, raw materials, like clay, slate or shale
- Manufacture from industrial by-products, such as fly ash
- Processing of industrial by-products, such as pelletized expanded slabs

Lightweight concretes are often used for high thermal insulation, but little in structural requirement areas. Foamed concrete is a highly workable and low-density material which can incorporate up to 50% entrained air. It has self-leveling, self-compacting, and 'pumpability' properties. Foamed concrete is ideal for difficult-to-access sewer systems, pipelines, and culverts. It is also good for the reinstatement of temporary road trenches. As mentioned, foamed concrete is used for thermal insulation, and thus suitable for sub-screeds and filling under-floor voids. The benefits of using lightweight aggregate concrete include:

- Reduction in dead loads, providing savings in foundations, reinforcement, and formwork
- Improved thermal insulations
- Improved fire resistance
- Easy transporting and handling

5.5.5 HEAVYWEIGHT CONCRETE

Heavyweight concrete (Figure 5.18) uses heavy, natural aggregates such as barites, magnetite or steel slug, etc. This is mainly used for radiation shielding (medical or nuclear), ballasting for pipelines, and similar types of structures. Cement contents and water–cement ratios are similar to those for normal concretes. However, the aggregate–cement ratios are significantly higher because of the higher densities of the aggregates. Heavyweight concrete can be prepared, transported, and compacted using the conventional equipment, however, higher heat generation may occur during the hydration. Heavyweight concrete has a density of more than 175 pcf (2,800 kg/m^3), whereas lightweight concrete has a density less than 125 pcf (2,000 kg/m^3). The normal-weight concrete has a density in between 125–175 pcf (2,000–2,800 kg/m^3).

FIGURE 5.18   Heavyweight concrete. *Courtesy of Sika.*

5.5.6 HIGH-STRENGTH CONCRETE

There is no sharp separation point between the high-strength concrete and normal-strength concrete. ACI defines the high-strength concrete as the concrete with the compressive strength greater than 6,000 psi (41 MPa). The United States uses concrete with compressive strength of 19,000 psi (131 MPa). By carrying higher loads than the normal-strength concrete, high-strength concrete reduces the amount of material and the sizes of members. The requirements for production of the high-strength concrete are high-quality Portland cement; high-quality, optimum size aggregates; and optimum proportions of cement, water, aggregates, and admixtures. Pozzolans, such as fly ash and silica fume, are commonly used admixtures for high-strength concrete. These materials impart additional strength to the concrete by reacting with Portland cement hydration products to create additional strength enhancing C-S-H gel. Chemical admixtures such as superplasticizer are also important to produce the high-strength concrete mixtures. The superplasticizer causes an increase in the workability at a low water–cement ratio and thus, yields greater strength. Water-reducing retarders may be used to slow down the hydration of the cement and offer more time to use the concrete.

5.5.7 SHRINKAGE-COMPENSATING CONCRETE

Shrinkage-compensating concrete is made with either an expansive cement, or a normal cement with an expansive component that is added separately. It is proportioned so that the concrete increases in volume after the setting and during the early-age hardening. The expansion must be restrained by internal reinforcement. This restriction may produce a compressive stress in the concrete. When the concrete dries later in service, the resulting shrinkage reduces the compressive stress instead of producing a tensile stress. The benefit of shrinkage-compensating concrete is that it offers greater joint spacing.

5.5.8 FIBER-REINFORCED CONCRETE

Fiber-reinforced concrete (FRC) is concrete which contains short, discrete fibers that are uniformly distributed and randomly oriented, as shown in Figure 5.19. There are different

FIGURE 5.19 Fiber-reinforced concrete specimen. *Photo taken at the Colorado State University–Pueblo.*

types of fibers, such as steel fibers, glass fibers, synthetic fibers, and natural fibers, which include sisal and jute fibers. Fibers in concrete help control cracking due to the plastic shrinkage and drying shrinkage, reduce the permeability, crack-width, bleeding of water, and can even produce greater abrasion resistance in concrete. Synthetic fibers are used in a small amount (about 0.1% by the volume of concrete), whereas metal fibers are used in a large amount (0.3% or more by the volume of concrete). The drawbacks of using fibers in concrete are the decreased workability, longer mixing time, and careful preparation required.

5.5.9 POLYMER CONCRETE

Polymer concretes (Figure 5.20) use polymers to replace the lime-type cements as a binder. In some cases, the polymer is used in addition to Portland cement to form the *polymer cement concrete* or the *polymer modified concrete*. Both thermoplastic and thermosetting resins are used as the polymer component due to their high thermal stability and resistance to many chemicals. Polymer concrete is also composed of superior quality and dust-free aggregates that include silica, quartz, granite, limestone, etc. Polymer concrete has an adhesive property and allows for repair of both polymer and cement-based concretes. The low permeability and corrosive resistance of polymer concrete allows it to be used in swimming pools, sewer structure applications, manholes, drainage channels, and other structures that contain liquids or corrosive chemicals.

FIGURE 5.20   Polymer concrete. *Courtesy of Gomaco.*

5.5.10 ROLLER-COMPACTED CONCRETE

Roller-compacted concrete (RCC), or the rolled concrete (rollcrete), has the same ingredients as conventional concrete, but in different proportions, with partial substitution of fly ash for the Portland cement and a smaller amount of water. The produced mix is drier and has negligible or no slump. RCC is placed in a manner similar to paving, i.e., the material is delivered by dump trucks, spread by small bulldozers, and then compacted by vibratory rollers, as shown in Figure 5.21. Having less water, RCC is stronger than the conventional concrete (Harrington et al. 2010). However, it requires a tremendous compaction effort that can be supplied only by a machine roller.

FIGURE 5.21   Roller-compacted concrete. *Courtesy of Myers Concrete.*

5.5.11   HIGH-PERFORMANCE CONCRETE

Ultra-high-performance concrete (UHPC), also known as the *reactive powder concrete* (RPC), is a high-strength, ductile concrete formulated by combining Portland cement, silica fume, quartz flour, fine silica sand, high-range water reducer, steel, or organic fibers. The material provides compressive strength up to 29,000 psi, flexural strength up to 7,000 psi, and a modulus of elasticity up to 7,300 ksi (Zia et al. 1991). The materials are usually supplied in a three-component premix: pre-blended powders (Portland cement, silica fume, quartz flour, and fine silica sand), superplasticizers, and organic fibers. This concrete is very ductile, and the use of it is simplified by the elimination of reinforcing steel and the ability of the material to be virtually self-placing or dry cast.

5.5.12   HIGH-PERMEABLE CONCRETE

Pervious concrete has a high porosity, contains inter-connected pores, and allows water to move from one side of the concrete to other. It is made with large aggregates, with little to no fine aggregates present. The concrete paste then coats the aggregates, producing large voids which allow water to pass through the concrete slab and groundwater recharge. Pervious concrete is traditionally used in parking areas, areas with light traffic, pedestrian walkways, and similar applications. However, it should not be used in freeze–thaw-prone areas; ice lenses formed inside the pores apply pressure to surrounding particles, causing concrete to break easily.

5.6   CHAPTER SUMMARY

Portland cement concrete is the compacted concrete which has started hardening (gaining strength) or is already hardened (gained strength). PCC has numerous advantages, including its low cost, low-maintenance, great resistance to fire and water, ability to be cast into any shape, and very importantly, its high compression strength. Two major disadvantages of PCC, its low tensile strength and heavy weight, can be avoided by adding reinforcing bars (rebar) and using it in footings, respectively.

During the early life of PCC, it shrinks a lot due to physical and chemical reactions with its internal and external environment. PCC also suffers from the creep deformation due to the sustained loads it experiences, such as self-weight, construction loads, etc.

Permeability is a very important physical property of PCC which influences the PCC's durability. It is the trapped air pockets from incomplete compaction, which create porous spaces in the hardened concrete. It also occurs due to the loss of mixing water by evaporation. There are some factors that adversely affect the permeability of PCC, such as inadequate compaction, high W/C ratio, and warm-curing.

PCC has a high compressive strength, but a very low tensile strength. Compressive strength is determined by applying a uniaxial compression load until the failure occurs. The peak load is used to calculate the compressive strength. The stress–strain curves of concretes of different strengths show that the curve becomes non-linear after about 30–40% of the ultimate stress. The ultimate stress occurs around the stain of 0.002 for most of the normal-strength concretes. The strain at failure is typically around 0.003–0.005 for the normal-strength concrete. Concrete having a higher ultimate compression strength has a higher modulus of elasticity.

Cylindrical or beam specimens are prepared to characterize the PCC. The basic procedure of preparing specimens is to mix the ingredients thoroughly first. The liquid mix is poured into the mold (cylindrical or rectangular), layer by layer; each layer is rodded before adding a new layer. The number of layers and the number of rodding for each layer depend on the dimensions of the mold. The produced specimens are then cured properly.

The modulus of elasticity and Poisson's ratio of concrete are determined by using uniaxial compression loading. The loading is conducted slowly enough to record the axial and lateral deformation with the loading. To determine the unconfined compression strength, only the peak load is required. Thus, a uniaxial compression load is applied rapidly until the failure occurs. The peak load is recorded and used in the calculations. Flexural strength is measured by applying the third-point loading or center-point loading test on a beam. The beam is loaded until failure occurs. The peak load and the beam's dimensions are used to calculate the maximum bending stress (flexural strength). Tensile strength is determined by a splitting test in which the load is applied diagonally on a cylindrical specimen, and the peak load is used to calculate the tensile strength of the specimen.

Self-compacting concrete flows by itself without any external vibration. It has high deformability and good segregation resistance, but low yield stress. Shotcrete is a wet mix, previously prepared or a ready-mixed concrete, and can be pneumatically thrown at high velocity onto a surface. Gunite is the mixture of dry ingredients and water and is mixed just before leaving the hose. Flowable fill is a mixture of cement, water, fine aggregate, fly ash, and/or slag. This mix can flow with the action of gravity and surface tension. High-strength concrete can be made by proper selection of aggregates, the optimum size of the aggregate, the bond between the cement paste and the aggregate, and the surface characteristics of the aggregate. Pozzolans, such as fly ash and silica fume, are the most commonly used mineral admixtures in high-strength concrete. Shrinkage-compensating concrete is made with either expansive cement, or with normal cement and a separately added expansive component. Fiber-reinforced concrete is concrete containing short, discrete fibers that are uniformly distributed and randomly oriented. Fibers include steel fibers, glass fibers, synthetic fibers, and natural fibers. Polymer concretes use thermoplastic polymers to replace lime or Portland cements as a binder. Polymer concretes also use superior quality aggregates including silica, quartz, granite, and limestone. Roller-compacted concrete is a mix of cement/fly ash, water, sand, aggregate, and common additives but contains much less water than traditionally used. The produced mix is drier with no slump. This concrete is stronger than the conventional concrete since it has a smaller amount of water. Ultra-high-performance concrete is a high-strength, ductile material formulated by combining Portland cement, silica fume, quartz flour, fine silica sand, high-range water reducer, water, and steel or organic fibers. Pervious concrete is made using large aggregates, with little to no fines, and thus has high permeability. It allows water to pass directly from a site, allowing groundwater recharge.

ORGANIZATIONS DEALING WITH CONCRETE

American Shotcrete Association (ASA). The American Shotcrete Association (ASA) is a non-profit organization of contractors, suppliers, manufacturers, designers, engineers, owners, and others with a common interest in advancing the use of shotcrete. The mission of the ASA is to provide knowledge, resources, qualification, certification, education, and leadership to increase the acceptance, quality, and safe practices of the shotcrete process.

Location: Farmington Hills, Michigan
Website: https://www.shotcrete.org/
ACI, ASTM International, NRMCA, and PCA are also resources for PCC materials.

REFERENCES

Harrington, D., F. Abdo, W. Adaska, and C. Hazaree. 2010. *Guide for Roller-Compacted Concrete Pavements*. SN298. Portland Cement Association, Skokie, IL.

Mamlouk, M. and Zaniewski, J. 2014. *Materials for Civil and Construction Engineers*, 4th edition. Pearson, Upper Saddle River, NJ.

PCA. 2016. *Design and Control of Concrete Mixtures, EB001*, 16th edition. Portland Cement Association, Skokie, IL.

Zia, P., Leming, M., and Ahmad, S. 1991. *High Performance Concretes: State of the Art Report, SHRP-C/FR-91-103*, Washington, DC.

FUNDAMENTALS OF ENGINEERING (FE) EXAM STYLE QUESTIONS

FE PROBLEM 5.1

The strength of concrete does not increase with

A. Age
B. Water–cement ratio
C. Curing
D. None of the above

Solution: B

The strength of concrete does not increase with the water–cement ratio. It increases with age and curing but decreases with the water–cement ratio.

FE PROBLEM 5.2

In the flexural testing of concrete using the third-point loading, the test result is discarded if

A. The failure occurs inside the middle third
B. The failure occurs outside the middle third by no more than 5% of the span
C. The failure occurs outside the middle third by more than 5% of the span
D. Never, unless mesh occurs

Solution: C

In the flexural testing of concrete using the third-point loading, the test result is discarded, if the fracture occurs in the tension surface outside of the middle third of the span length by more than 5% of the span.

FE Problem 5.3

The stress reaches the ultimate point in the unconfined compression test of concrete specimen at the strain level of about...

A. 0.001
B. 0.002
C. 0.003
D. 0.004

Solution: B

Regardless of strength, the ultimate stress of concrete occurs at the strain of 0.002 in the compression testing.

FE Problem 5.4

While preparing a concrete cylinder or beam, after each layer is rodded, the outsides of the mold are tapped lightly 10–15 times with the mallet. The outside of the mold is tapped... [select all that apply]

A. to close any holes left by rodding
B. to release any large air bubbles that may have been trapped
C. to make the surface level
D. To raise the inside moisture up

Solution: A and B

While preparing a concrete cylinder or beam, after each layer is rodded, tap the outsides of the mold lightly 10–15 times with the mallet to close any holes left by rodding and to release any large air bubbles that may have been trapped.

FE PROBLEM 5.5

The curing of a laboratory-prepared cylinder or beam is performed in saturated lime water...

    A. To maintain temperature
    B. To prevent leaching out of calcium hydroxide from the concrete
    C. To insert calcium hydroxide inside the concrete specimens
    D. To anti-freeze the water

Solution: B

Water by itself may leach out calcium hydroxide from the concrete specimens and reduce their strength, so the water must be saturated with added calcium hydroxide (hydrated lime).

PRACTICE PROBLEMS

PROBLEM 5.1

In a splitting tensile strength test, the peak load required to fail a specimen of 5.5-in. diameter and 10-in. thickness is 80,000 lb. Calculate the tensile strength of the specimen.

PROBLEM 5.2

In an unconfined compression strength test, the peak load required to fail a specimen of 6.0-in. diameter and 9-in. length is 80,000 lb. Calculate the unconfined compression strength of the specimen.

PROBLEM 5.3

In an unconfined compression strength test, the peak load required to fail a specimen of 6.0-in. diameter and 10.50-in. length is 120,000 lb. Calculate the unconfined compression strength of the specimen.

PROBLEM 5.4

In an unconfined compression strength test, the peak load required to fail a specimen of 6.0-in. diameter and 12-in. length is 130,000 lb. Calculate the unconfined compression strength of the specimen.

Problem 5.5

A center-point loading flexural test was performed on a concrete beam of 8-in. span. The beam's cross-section is 2 in. square. The beam failed at the load of 400 lb. Calculate the flexural strength of the concrete beam.

Problem 5.6

A center-point loading flexural test was performed on a concrete beam of 18-in. span. The beam's cross-section is 6 in. square. The beam failed at the load of 8,000 lb. Calculate the flexural strength of the concrete beam.

Problem 5.7

A third-point loading flexural test was performed on a concrete beam of 18-in. span. The beam's cross-section is 6 in. square. A crack appears at the middle of the beam when the load is 8,000 lb. Calculate the flexural strength of the concrete beam.

Problem 5.8

A third-point loading flexural test was performed on a concrete beam of 18-in. span. The beam's cross-section is 6 in. square. A crack appears at 5.5-in. from the end of the beam when the load is 9,000 lb. Calculate the flexural strength of the concrete beam.

Problem 5.9

A concrete of 155 pcf unit weight has an average compressive strength of 4 ksi. Determine its modulus of elasticity.

Problem 5.10

A normal-weight concrete has an average compressive strength of 4 ksi. Determine its flexural strength.

PROBLEM 5.11

A compressive force of 15 kip is applied on a 6-in. diameter and 12 in. tall long cylinder. After applying the load, the diameter of the rod increases to 6.15 in., and the length decreases to 11.9998 in. Assuming no permanent deformation occurs in the material, calculate the modulus of elasticity. [Hint: You may need to use equations from chapter 1]

6 Asphalt Binders

As cement is used in concrete, asphalt is used in asphalt concrete to bind aggregates together. The performance of asphalt concrete is primarily dependent on the amount and type of asphalt binder used. This chapter discusses the source, properties, grading, and laboratory characterization procedures of asphalt binder. Some non-conventional forms of asphalt binders, such as cutbacks, emulsion, and foamed asphalt, are also discussed.

6.1 BACKGROUND

Asphalt concrete (AC) is the general name of hardened asphaltic material mixed with aggregates and/or fines. It is the most common material for asphalt pavement design. It broadly consists of asphalt binder, coarse aggregate, and fine aggregate, as shown in Figure 6.1.

Asphalt is one of the bituminous materials, and its generic flow can be described as shown in Figure 6.2. Both asphalt and tar are black in color, and both have very good waterproofing and adhesion properties. Asphalt binder is obtained from crude petroleum residue or natural lakes. There are three natural asphalt lakes in world: La Brea Tar Pits in Los Angeles, Lake Guanoco in Venezuela, and the Pitch Lake in Trinidad and Tobago. The Pitch Lake is the largest natural deposit of asphalt in the world. The lake covers about 100 acres and 250 feet deep. This lake contains about 50–57% of all asphalt in the world (Wallace and Martin 1967).

On the other hand, tar is obtained from crude petroleum vapor and from the bituminous coal destructive distillation. Asphalt material is primarily used in asphalt pavement road construction, whereas tar is primarily used for waterproofing membranes in roofs and pavement treatment, especially in parking lots where fuel spills may dissolve the asphalt. Tar is no longer used for road construction because of its health hazards (such as eye and skin irritation) and high-temperature susceptibility. Carbon and hydrogens are the main elements present in asphalt binder (Peterson 1984). Some other minor elements present are sulfur, nitrogen, oxygen, and occasionally vanadium and nickel are found (Halstead 1985).

In the United States, there are about 4,000 asphalt processing plants which generate about 525 million tons of asphalt, with the total value of over $3 billion. There are about 300,000 employees working in this industry. There is a total of 8.7 million miles of pavement (2.5% interstate and 97.5% non-interstate), of which, interstate comprises of 65% concrete and 35% asphalt pavements. Regarding non-interstate pavements, 94% are made with asphalt, while 6% use concrete (Rivera et al. 2017). Thus, the importance of asphalt binder is enormous. The United States produced about 452 million tons in 2007 (NAPA 2012) and about 396 million tons in 2010 (Hansen and Newcomb 2011). The country used approximately 23 million tons ($7.7 billion) of asphalt binder in 2011 (EIA 2011; U.S. Census Bureau 2007). Roads and highways constitute the largest sole use of asphalt, at 85% of the total (Asphalt Institute, 2001). Approximately 83% of asphalt binder used in the United

Asphalt Binder and Aggregate   Asphalt Pavement

FIGURE 6.1 Asphalt binder, aggregate and asphalt pavement. *Photos taken in Pueblo, Colorado.*

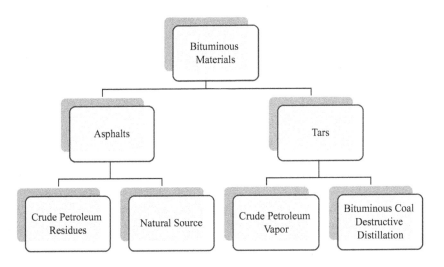

FIGURE 6.2 Classification of asphalt binder.

States in 2011 was used for paving purposes (Grass 2012). In the United States, more than 92% of all paved roads and highways are surfaced with asphalt products.

6.2   ASPHALT BINDER

Asphalt is a dark brown to black, highly viscous, hydrocarbon produced from petroleum distillation residue. This distillation can occur naturally, resulting in asphalt lakes, or occur in a petroleum refinery using crude oil. Asphalt functions as a waterproof, thermoplastic, viscoelastic adhesive. By weight, asphalt generally accounts for between 4 and 8% of AC mixes and makes up about 30% of the cost of AC pavement structures, depending upon type and quantity. In addition to the common asphalt, three other types of asphalts are produced. These are mentioned below:

a) Asphalt emulsion
b) Asphalt cutback
c) Foamed asphalt

6.2.1 ASPHALT EMULSION

Asphalt emulsion is the suspension of small asphalt cement in water, assisted by an emulsifying agent (such as soap). The emulsifying agent imparts an electrical charge to the surface of the asphalt cement globules so that they do not coalesce. Emulsions are used because they effectively reduce the asphalt viscosity for low temperature uses.

6.2.2 CUTBACK ASPHALT

Cutback asphalt is simply the combination of asphalt cement and petroleum solvent. Like emulsions, cutbacks reduce the asphalt viscosity for low temperature uses (tack coats, fog seals, slurry seals, and stabilization material). Like emulsified asphalts, after a cutback asphalt is applied, the petroleum solvent evaporates, leaving behind asphalt cement residue on the surface to which it was applied. Cutback asphalt cures as the petroleum solvent evaporates away. The use of cutback asphalts is decreasing because

* Cutback asphalts contain volatile chemicals that pollute the environment
* The petroleum solvents used to produce cutbacks require high amounts of energy to manufacture and are expensive.

6.2.3 FOAMED ASPHALT

Foamed asphalt is formed by combining hot asphalt binder with a small amount of cold water. When the cold water comes in contact with the hot asphalt binder, it turns into steam, which becomes trapped in tiny asphalt binder bubbles. The result is a thin-film, high-volume asphalt foam, with about 10 times more coating potential than the asphalt binder in its normal liquid state. This high-volume foam state lasts for a few minutes, after which the asphalt binder resumes its original properties. Foamed asphalt can be used as a binder in soil or base course stabilization and is often used as the stabilizing agent in full-depth asphalt reclamation.

6.2.4 RECYCLED ASPHALT

Recycled (or reclaimed) asphalt binder is extracted from the recycled (or reclaimed) asphalt pavement (RAP) milling process. Solvent extraction following the AASHTO T 164 and ASTM D 2172 protocols uses a chemical solvent (trichloroethylene, 1,1,1-trichloroethane or methylene chloride) to separate the asphalt binder from the aggregate. The milled asphalt is submerged in trichloroethylene for some time. The asphalt binder/solvent and aggregate are then separated using a centrifuge (Figure 6.3). The mixture of binder/solvent is further processed to separate the binder and the solvent. The solvent extraction method is only sparingly used due to the hazardous nature of the solvents.

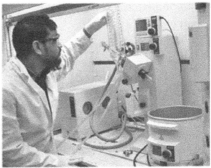

FIGURE 6.3 Extraction of asphalt binder from RAP by the author. *Photos taken at the University of New Mexico.*

6.3 PHYSICAL PROPERTIES OF ASPHALT BINDER

Asphalt can be classified by its chemical composition and physical properties. The pavement industry typically relies on physical properties for performance characterization, although asphalt's physical properties are a direct result of its chemical composition. Typically, the most important physical properties are:

Age. Asphalt binder continuously reacts with oxygen and ages if exposed to air. Lighter constituents of asphalt binder also volatize. As asphalt binder ages, its viscosity increases, and it becomes stiffer and brittle.

Viscosity. Asphalt binder is a viscous material. Its deformation behavior is dependent on the duration or frequency of loading. The longer the load is sustained on asphalt or the smaller the frequency, the larger the deformation and damage. In other words, a faster vehicle produces less damage in asphalt material compared to a slow moving, similar vehicle.

Safety. Asphalt cement volatilizes (gives off vapor) when heated. At extremely high temperatures (well above those experienced in the manufacture and construction of AC), asphalt cement can release enough vapor to increase the volatile concentration immediately above the asphalt cement, to a point where it ignites (causes a flash) when exposed to a spark or open flame. For safety reasons, the flash point of asphalt cement is tested and controlled.

Purity. Asphalt cement, as used in AC paving, should consist of almost pure bitumen. Impurities are not active cementing constituents and may be detrimental to the asphalt performance.

Temperature Dependency. Asphalt cement is largely dependent on temperature, as shown in Figure 6.4. Its viscosity decreases with the increase in temperature and vice-versa. Similarly, asphalt's mechanical properties and strength are affected by temperature.

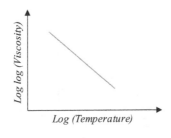

FIGURE 6.4 Variation of viscosity with temperature.

6.4 GRADING OF ASPHALT BINDER

Asphalt binder can be graded broadly in three systems:

a) Penetration grading
b) Viscosity grading
c) Performance grading

6.4.1 PENETRATION GRADING

Penetration grading (simply termed pen grade) is primarily developed based on the penetration test on asphalt binder following the ASTM D 946 method. In this test, a standard needle penetrates an asphalt binder specimen when placed under a 100-g load for 5 seconds, as shown in Figure 6.5. The test is simple and easy to perform, but it does not measure any fundamental parameter and can only characterize asphalt binder at one temperature (77°F). Penetration grades are listed as a range of penetration units (one penetration unit = 0.1 mm of penetration by the standard needle), such as 40–50 if the penetration ranges 4–5 mm, by Table 6.1. The higher the penetration, the softer the asphalt binder is. Typical asphalt binders used in the United States are 65–70 and 85–100 pen grade.

Conditioning specimen in water   Penetrating an asphalt binder specimen

FIGURE 6.5 Asphalt penetration testing. *Photos taken at the Farmingdale State College of the State University of New York.*

TABLE 6.1
Penetration Grading for Asphalt Binder

	Penetration Unit at 77°F (25°C)	
Penetration Grade	1 Unit = 0.10 mm	
	Minimum Penetration	**Maximum Penetration**
40–50	40	50
60–70	60	70
85–100	85	100
120–150	120	150
200–300	200	300

Some advantages of penetration grading are (Brown et al. 2009):

- Penetration grading is based on penetration at 77°F (25°C), which is very close to the normal pavement temperature.
- This method is relatively cheap considering time, equipment, and skillset.

Some disadvantages of penetration grading are (Brown et al. 2009):

- Penetration is an empirical and physical test; it does not dictate the performance of asphalt binder. In addition, this test is conducted at 77°F (25°C), which does not simulate asphalt behavior in summer and winter months.
- Viscosity information of asphalt binder is required to determine the mixing and compaction temperature. This method does not provide that.

6.4.2 VISCOSITY GRADING

Viscosity grading is a superior grading system compared to that of penetration grading, but it does not test low-temperature asphalt binder rheology. This grading system is based on the absolute viscosity of virgin binder and Rolling Thin-Film Oven (RTFO) test aged binder. The RTFO test simulates the effects of short-term aging that occurs during the mixing and compaction of an asphalt mixture by heating a film of asphalt binder in an oven for 5 hours at 163°C (325°F). Absolute viscosity, or simply viscosity, is defined as the resistance to flow of a fluid. The absolute viscosity test is discussed later in this chapter.

The basic absolute viscosity test (ASTM D 2171 and AASHTO T 202) measures the time required for a fixed volume of asphalt binder to be drawn up through a capillary tube, as shown in Figure 6.6. The drawing of asphalt is conducted by means of a closely controlled

FIGURE 6.6 Capillary viscometer. *Courtesy of Raysky Scientific Instruments, Labfreez Group, Guangdong, China. Used with permission.*

vacuum at 60°C. This temperature is selected for being approximately the maximum asphalt surface temperature during the summer in most areas. The *viscometer* is a U-tube with a reservoir where the asphalt is introduced, and a section with a calibrated diameter and timing marks. A vacuum is applied at one end, and the time during which the asphalt flows between two timing marks on the viscometer is measured. The flow time (in seconds) is multiplied by the viscometer calibration factor to obtain the absolute viscosity in poises.

Grading of virgin binder is denoted by two letters (AC) and a number. AC means asphalt cement here, and the numerical value indicates viscosity at 60°C (140°F) in hundreds of poises. Both ASTM D 3381 and AASHTO M 226 use this procedure. Table 6.2 lists the AC grading for virgin binder, where the AC grades are listed in hundreds of poises (cm-g-s = dyne-second/cm^2). The lower the number of poises, the lower the viscosity, meaning the more easily a substance flows.

Grading on RTFO aged binder is dented by the two-letter AR grading and a number. AR means *asphalt residue*, and the numerical value indicates viscosity at 60°C (140°F) in hundreds of poises by the AASHTO M 226 protocol (ASTM D 3381 uses the viscosity in poise), as listed in Table 6.2. For example, according to AASHTO M 226, AR-40 means the RTFO aged binder has an absolute viscosity of 4,000 poise. ASTM D 3381 expresses this binder as AR-4000. Common asphalt binders used in the United States are AC-10, AC-20, AC-30, AR-4000, and AR-8000.

Some advantages of viscosity grading include (Brown et al. 2009):

- Viscosity is a fundamental test and provides one of the major properties of asphalt binder. Thus, this grading system is better than the penetration grading.
- The viscosity is tested at 60°C (140°F), which is the maximum pavement temperature in most areas during the summer. Thus, this grading system works well for high-temperature performance of asphalt binder, such as permanent deformation along the wheel paths which occurs predominantly at high temperatures.

TABLE 6.2
Standard Viscosity Graded Binders

AASHTO M 226	ASTM D 3381	Absolute Viscosity at 60°C (140°F) (poises)
Grading Based on Original Asphalt (AC)		
AC-2.5	AC-2.5	250 ± 50
AC-5	AC-5	500 ± 100
AC-10	AC-10	$1,000 \pm 200$
AC-20	AC-20	$2,000 \pm 400$
AC-30	AC-30	$3,000 \pm 600$
AC-40	AC-40	$4,000 \pm 800$
Grading Based on Aged Residue (AR)		
AR-10	AR-1000	$1,000 \pm 200$
AR-20	AR-2000	$2,000 \pm 400$
AR-40	AR-4000	$4,000 \pm 800$
AR-80	AR-8000	$8,000 \pm 1,600$
AR-160	AR-16000	$16,000 \pm 3,200$

However, as the viscosity is tested at 60°C (140°F), the performance at low or average temperature is not measured in this grading system. Low temperature is critical for thermal cracking, and average temperature is critical for fatigue (back of an alligator-shaped) cracking.

6.4.3 PERFORMANCE GRADING (PG)

The *performance grading* (PG) system was developed as part of the Superpave research effort to more accurately and fully characterize asphalt binders for use in asphalt pavements. This method tests asphalt binder at an upper-performance temperature and at a lower-performance temperature. This method also tests virgin asphalt, RTFO aged asphalt residue, and Pressure-Aging Vessel (PAV) aged asphalt residue. The PAV provides simulated, long-term aged asphalt binder for in-service aging over a 7–10-year period. The basic PAV procedure calls for obtaining the RTFO aged asphalt binder specimens, placing them in stainless steel pans, and then aging them for 20 hours in a heated vessel, pressurized to 305 psi (2.10 MPa) at 90°C or 100°C.

Superpave performance grading is reported using two numbers: the first being the average 7-day maximum pavement temperature in °C, and the second being the single-day minimum pavement design temperature likely to be experienced in °C. Thus, a PG 64-28 is intended for use where the average 7-day maximum pavement temperature is 64°C, and the expected single-day minimum pavement temperature is −28°C. Notice that these numbers are pavement temperatures and not air temperatures. The PG grading system requirement is listed in Table 6.3.

Example 6.1 PG Binder

In a region, the average 7-day maximum pavement design temperature is 50°C, and the minimum pavement design temperature is −12°C. Determine the recommended asphalt binder for this region.

Solution

From Table 6.3, for the average 7-day maximum pavement design temperature less than 52°C, PG 52 is recommended. Then, for the minimum pavement design temperature is −12°C, the second number is −16 (as −12 > −16). Therefore, the PG 52-16 binder is recommended.

Example 6.2 Direct Tension Test of PG Binder

For the asphalt binder, PG 64-28, determine the recommended design temperature for the direct tension test.

Solution

From Table 6.3, the recommended design temperature for PG 64-28 binder for the direct tension test is −18°C.

TABLE 6.3
Performance-Graded (PG) Binder Grading System

Performance Grade (PG)	PG 52							PG 58					PG 64				
	-10	-16	-22	-28	-34	-40	-46	-16	-22	-28	-34	-40	-16	-22	-28	-34	-40
Average 7-Day Maximum Pavement Design Temperature, °C	<52							<58					<64				
Minimum Pavement Design Temperature, °C	>-10	>-16	>-22	>-28	>-34	>-40	>-46	>-16	>-22	>-28	>-34	>-40	>-16	>-22	>-28	>-34	>-40
Original Binder																	
Flash Point Temp, T48: Minimum °C	230																
Viscosity, ASTM D 4402; Maximum, 3 Pa-s (3,000 cP), Test Temp. °C	135																
Dynamic Shear, TP5: $G^*/sin\delta$, Min. 1.00 kPa Test Temperature @ 10 rad/sec., °C	52							58					64				
Rolling Thin Film Oven (T 240) or Thin Film Oven (T 179) Residue																	
Mass Loss, Max. %	1.00																
Dynamic Shear, TP5: $G^*/sin\delta$, Min. 2.20 kPa Test Temp @ 10 rad/sec., °C	52							58					64				
Pressure Aging Vessel (PAV) Residue																	
PAV Aging Temperature, °C	90							100					100				
Dynamic Shear, TP5: $G^*/sin\delta$, Min. 5,000 kPa Test Temp @ 10 rad/sec., °C	25	22	19	16	13	10	7	25	22	19	16	13	28	25	22	19	16
Physical Hardening	Report																
Creep Stiffness, TP1: S, Max. 300 MPa m-value, Min. 0.300 Test Temp, @ 60 sec., °C	0	-6	-12	-18	-24	-30	-36	-6	-12	-18	-24	-30	-6	-12	-18	-24	-30
Direct Tension, TP3: Failure Strain, Min. 1.0% Test Temp @ 1.0 mm/min, °C	0	-6	-12	-18	-24	-30	-36	-6	-12	-18	-24	-30	-6	-12	-18	-24	-30

Adapted from Superpave Fundamentals Reference Manual. NHI Course #131053, Federal Highway Administration (FHWA), 2017, Washington DC.

The following tests are required for the PG grading system:

Determining Temperature. The pavement surface temperatures for the pavement site can be determined using the LTPPBind 3.0/3.1 software. It is available at the Federal Highway Administration (FHWA) website (https://infopave.fhwa.dot.gov/).

Rolling Thin-Film Oven Test. The RTFO procedure following AASHTO T 240 and ASTM D 28723 provides the simulated, short-term aging during the mixing and placement of asphalt concrete. It also provides a quantitative measure of the volatiles lost during the aging process. This test requires virgin (unaged) asphalt binder specimens in cylindrical glass bottles to be placed in a rotating carriage within an oven (Figure 6.7). The carriage rotates within the oven while the 325°F (163°C) temperature ages the specimens for 85 minutes.

Pressure Aging Vessel. The Pressure Aging Vessel (PAV) provides simulated, long-term aging that occurs during the service life of over a 7–10-year period. The basic PAV procedure, following the AASHTO R 28, requires that RTFO aged asphalt binder specimens are placed in stainless steel pans and then aged for 20 hours in a heated vessel, pressurized to 305 psi (2.10 MPa or 20.7 atmospheres) at 90°C or 100°C, as shown in Figure 6.8. Specimens are then stored for use in physical property tests.

Exterior View   Interior View

FIGURE 6.7   Rolling thin-film oven test apparatus. *Photos taken at the University of New Mexico.*

FIGURE 6.8   Pressure aging vessel test apparatus. *Photos taken at the University of New Mexico.*

FIGURE 6.9 Dynamic shear rheometer testing by the author. *Photos taken at the University of New Mexico.*

Dynamic Shear Test. The Dynamic Shear Rheometer (DSR) is used to characterize the viscous and elastic behavior of asphalt binders at medium to high temperature. The testing temperature is used as the actual temperatures anticipated in the area where the asphalt binder is to be used. The DSR measures the shear modulus (G^*) and phase angle (δ). The complex shear modulus (G^*) can be considered the specimen's total resistance to deformation when repeatedly sheared, while the phase angle (δ), is the lag between the applied shear stress and the resulting shear strain. The larger the phase angle (δ), the more viscous the material is. A zero-degree phase angle means an elastic material, whereas, a 90-degree phase means a pure viscous material. Asphalt material is in between these two values.

According to AASHTO T 315, a small specimen of asphalt binder is sandwiched between two plates, shown in Figure 6.9. The test specimen is kept at near constant temperature as desired. The top plate oscillates at 10 rad/sec (1.59 Hz) in a sinusoidal waveform, while the equipment measures the maximum applied stress, the resulting maximum strain, and the time lag between them. The software then automatically calculates the complex modulus (G^*) and phase angle (δ).

Bending Beam Rheometer (BBR) Creep Stiffness. Creep stiffness is the stiffness or modulus measured using creep (sustained) loading. The Bending Beam Rheometer (BBR) test provides a measure of the low-temperature creep stiffness and the relaxation properties of asphalt binders. These parameters give an indication of an asphalt binder's ability to resist the low-temperature cracking.

According to ASTM D 6648 or AASHTO T 313, an asphalt beam 4.0 in. (102 mm) long, 0.5 in. (12.5 mm) wide, and 0.25 in. (6.25 mm) tall is prepared by pouring the heated binder in a mold. After cooling, the beam is then kept in the test bath for an hour. Then, a load of 100 g (980 mN) is applied at the center of the beam for a total of 240 seconds, as shown in Figure 6.10. The deflection of the beam is recorded during this loading period. Using the classical strength of materials equation for a center-point loaded beam, the stiffness after 60 seconds of loading is calculated as shown by Eq. 6.1.

$$S(t) = \frac{PL^3}{4bh^3 \delta(t)}$$

(6.1)

where,

$S(t)$ = Creep stiffness at time, $t = 60$ sec

P = Applied constant load, 100 g (980 mN)

Specimen Preparation   BBR Testing by the Author   Inside view of the Equipment

FIGURE 6.10 Bending beam rheometer (BBR) testing by the author. *Photos taken at the University of New Mexico.*

L   = Beam span, 4.0 in. (102 mm)
b   = Beam width, 0.5 in. (12.5 mm)
h   = Beam height, 0.25 in. (6.25 mm)
$\delta(t)$ = Deflection at time, $t = 60$ sec

The *m*-value is the slope of the "Log *S* versus Log-time" curve at 60 seconds of loading. The *m*-value indicates the rate of change of stiffness with loading time. The PG binder specification requires that the *m*-value be equal to or greater than 0.30 at 60 seconds of loading (McGennis et al. 1994).

Example 6.3 BBR Test of PG Binder

In a BBR testing, an asphalt beam of 12.5 mm wide, 6.25 mm high and 102 mm long is used. After applying a load of 100 g for 60 sec, a deflection of 0.904 mm is recorded. Calculate the creep stiffness of the specimen.

Solution
Given,
P   = Applied constant load, 100 g = 0.1 kg (9.81 m/sec²) = 0.981 N
L   = Beam span = 102 mm = 0.102 m
b   = Beam width = 12.5 mm = 0.0125 m
h   = Beam height = 6.25 mm = 0.00625 m
$\delta(t)$ = Deflection at 60 sec = 0.904 mm = 0.000904 m

From Eq. 6.1: Creep stiffness,

$$S(t) = \frac{PL^3}{4bh^3\delta(t)} = \frac{0.981\,\text{N}(0.102\,\text{m})^3}{4(0.0125\,\text{m})(0.00625\,\text{m})^3(0.000904\,\text{m})}$$

$$= 94,338,950\,\text{Pa}$$

$$= 94.3\,\text{MPa}$$

Answer: 94.3 MPa

Direct Tension Test. The Direct Tension Test (DTT) test, which measures the tensile strength of asphalt binder at a critical cracking temperature, is conducted following the

AASHTO T 314 test protocol at different temperatures (commonly 10°C higher than the low temperature grade of the binder). The effective length of the specimen is slightly more than 1.5 in. (40 mm) and the effective area of the cross section of the specimen is about 1.5 in. (36 mm) square. The specimen is submerged under an alcoholic bath and is pulled at a constant strain rate of 3%/minute. The variations of stress and strain are recorded. The DTT test setup and three failed specimens are shown in Figure 6.11.

  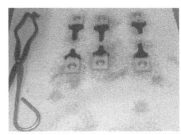

Specimen
Preparation   Testing   Three Failed Specimens

FIGURE 6.11   Direct tension test set and failed specimens. *Photos taken at the University of New Mexico.*

6.5   OTHER TESTS ON ASPHALT BINDER

6.5.1   KINEMATIC VISCOSITY

The *kinematic viscosity* of a liquid is the absolute (or dynamic) viscosity divided by the density of the liquid at the temperature of measurement. The 135°C (275°F) measurement temperature is selected to simulate the mixing and compaction temperatures commonly used. The basic kinematic viscosity test (ASTM D 2170 and AASHTO T 201) measures the time it takes for a fixed volume of asphalt binder to flow through a Zeitfuchs Cross-Arm viscometer under the closely controlled head and temperature. The kinematic viscosity in centistokes is obtained by multiplying the time taken by the calibration factor of the viscometer, provided by the manufacturer. Absolute viscosity can be obtained from this kinematic viscosity by multiplying it by the density of asphalt binder as follows:

$$\text{Absolute Viscosity (poises)} = \text{Kinematic Viscosity (stokes)} \times \text{Specific Gravity} \quad (6.2)$$

Example 6.4 Viscosity of Asphalt Binder

An asphalt binder has a kinematic viscosity of 1,200 centistokes. If its specific gravity is 0.98, determine its absolute viscosity in poise.

Solution

1 stoke = 100 centistokes
Therefore, 1,200 centistokes = 12 stokes

$$\text{Absolute Viscosity (poises)} = \text{Kinematic Viscosity (stokes)} \times \text{specific Gravity}$$

$$= 12 \times 0.98$$

$$= 11.76 \text{ poise}$$

Answer: 11.76 poise

6.5.2 BROOKFIELD VISCOSITY

The Brookfield viscometer, or Rotational viscometer (RV), is used to determine the viscosity of asphalt binders in the high-temperature range. The RV test can be conducted at various temperatures. The test for the PG binder specification is always conducted at 275°F (135°C) to simulate the mixing and paving temperature. About 11 g of asphalt binder is poured into the RV chamber. The test measures the torque required to maintain a rotational speed (20 revolutions per minute, rpm) of a cylindrical spindle while submerged in the asphalt binder, shown in Figure 6.12. This torque is then converted to a viscosity and displayed automatically by the RV in the unit of centipoise (1,000 centipoise = 1 Pa.s). The standard test methods are ASTM D 4402 and AASHTO T 316.

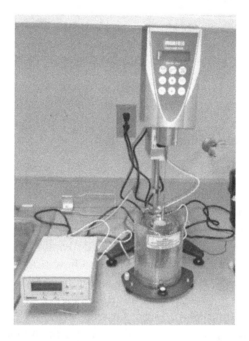

FIGURE 6.12   Rotational (Brookfield) viscometer. *Courtesy of Dr. Andrew Braham, University of Arkansas, Fayetteville, Arkansas, www.andrewbraham.com. Used with permission.*

6.5.3 SPECIFIC GRAVITY

Specific gravity tests are useful in making volume corrections based on temperature, as the specific gravity of asphalt binder changes with temperature. The specific gravity at 15.6°C (60°F) is commonly used when buying/selling asphalt cements. The typical specific gravity for asphalt is around 1.03. The standard test method is AASHTO T 228.

6.5.4 FLASH POINT TEMPERATURE

A typical flash point test involves heating a small specimen of asphalt binder in a test cup. The temperature of the specimen is increased and, at specified intervals, a test flame is passed across the cup. The flash point is the lowest liquid temperature at which application of the test flame causes the vapors of the specimen to ignite.

6.5.5 DUCTILITY

The ductility at 25°C (77°F) test measures asphalt binder ductility by stretching a standard-sized briquette of asphalt binder to its breaking point. The stretched distance in centimeters at breaking is then reported as the ductility. Like the penetration test, this test has limited use since it is empirical and conducted only at 25°C (77°F).

6.5.6 SOLUBILITY IN TRICHLOROETHYLENE

Asphalt cement, as used for asphalt paving, should consist of almost pure bitumen. Impurities are not active cementing constituents and may be detrimental to asphalt cement performance. Mineral impurities can be quantified by dissolving a specimen of asphalt cement in trichloroethylene or 1,1,1 trichloroethane through a filter mat. Anything remaining on the mat is considered an impurity.

6.6 GRADING OF CUTBACKS

Asphalt cutbacks are manufactured by blending liquefied asphalt and petroleum solvents. There are three types of cutback asphalts based on the relative rate of evaporation of the solvent:

a) Rapid-curing (RC), curing time of 4–8 hours
b) Medium-curing (MC), curing time of 12–24 hours
c) Slow-curing (SC), curing time of 48–60 hours

Rapid-curing (RC) cutback asphalt is designed to react quickly primarily for spray applications, such as bond/tack coats, aggregate chips seals, sand seals, and similar surface treatments.

Medium-curing (MC) cutback asphalt is designed for mixing with aggregates. These grades do not break immediately upon contact with aggregate; mixes can remain workable for an extended period and lend themselves to cold-mix stockpiles.

The rate of evaporation determines the type of asphalt cutback in the mixture. Gasoline, or naphtha, is a highly volatile material and mixing this with asphalt produces the rapid-cure cutback (RC) with a curing time of 4–8 hours. Similarly, kerosene (medium volatility) produces a medium-curing cutback (MC) with a curing time of 12–24 hours. A fuel oil (low volatility) produces a slow-curing cutback (SC) with a curing time of 48–60 hours. Table 6.4 shows the percentage of components by grade for different types of asphalt cutbacks.

TABLE 6.4
Asphalt Cutbacks Composition (% of Total Volume)

		Grades				
Type	Solvent	30	70	250	800	3,000
Rapid Curing (RC)	Asphalt Cement	–	65	75	83	87
	Gasoline or Naphtha	–	35	25	17	13
Medium Curing	Asphalt Cement	54	64	74	82	86
(MC)	Kerosene	46	36	26	18	14
Slow Curing (SC)	Asphalt Cement	–	50	60	70	80
	Fuel Oil	–	50	40	30	20

6.7 GRADING OF EMULSIONS

Emulsions are classified primarily by their ionic charge. Some other parameters, such as setting time, viscosity level, and type of asphalt base are also considered. Cationic emulsions begin with a "C." If there is no C, the emulsion is usually an anionic type (Table 6.5). The charge is important when designing an emulsion for compatibility with certain aggregates. After the charge designation, the next set of letters describes how quickly an emulsion sets or coalesces to a continuous asphalt mass. The standard terms are:

- RS (Rapid set, 5–10 minutes)
- MS (Medium set, several hours)
- SS (Slow set, few months)

RS emulsions break rapidly and have little or no ability to mix with aggregates. MS emulsions are designed to mix with aggregates, and are often called the mixing grade emulsions. MS emulsions are used in cold recycling, cold and warm dense-graded aggregate mixes, patch mixes, and other mixes. SS emulsions are designed to work with the fine aggregates to allow for maximum mixing time and extended workability. They are the most stable emulsions, and can be used in dense-graded aggregate bases, slurry seals, soil stabilization, asphalt surface courses, and some recycling. SS emulsions can be diluted with water to reduce their viscosity, so they can be used for tack coats, fog seals, and dust palliatives. SS emulsions are also used as driveway sealers.

An "HF" that precedes the setting-time designation indicates a *High Float emulsion*. HF emulsions are designed so the emulsifier forms a gel structure in the asphalt residue. The thicker asphalt film allows these emulsions to perform in a wider temperature range. High Floats are used in chip seals, cold mixes, and road mixes.

After the set designation, there is a series of numbers and letters that further describe the characteristics of the emulsions. The number 1 or 2 designates the viscosity of the emulsion, with the number 1 meaning lower viscosity and 2 meaning higher viscosity. If there is an "h" or "s" at the end of the name, the "h" indicates a harder base, and the "s" a softer asphalt base. For example, SS-1h is a slow setting emulsion with a lower viscosity made from a relatively hard base asphalt. A "P" may be added to the set designation to show the presence of polymer in the emulsion. An "L" indicates the presence of latex polymer. For example, CRS-2P is a cationic, rapid setting emulsion having a higher viscosity and containing some polymer. Polymers and latex are used to add strength, elasticity, adhesion, and durability to the pavement. Polymer asphalt emulsions can be less brittle at low temperatures to resist cracking, and stiffer at high temperatures to resist rutting and bleeding. Polymers permit the application of micro surfacing in wheel-path ruts and other locations where multiple stone depths are required.

TABLE 6.5
Grading of Emulsions

First Term	Second Term	Dash	Third Term	Fourth Term
C = Cationic	RS = Rapid Setting	–	1 = Lower Viscosity	h = Harder asphalt base
No Letter = Anionic	MS = Medium Setting			s = Softer asphalt base
	SS = Slow Setting		2 = Higher Viscosity	P = Presence of polymer
				L = Presence of latex polymer

6.8   CHAPTER SUMMARY

Asphalt binder is the asphalt cement that is used to bind fine and coarse aggregates. It is largely used in road construction. Asphalt can be obtained from two sources: crude petroleum residue and natural lakes, such as the Pitch Lake in Trinidad and Tobago. During destructive distillation of crude petroleum, the residue is collected and refined to produce the asphalt. Asphalt binder should not be confused with tar, which is the crude petroleum vapor.

In addition to the conventional asphalt, asphalt emulsions, asphalt cutbacks, and foamed asphalts are also used. Asphalt emulsion is the suspension of small asphalt cement in water with the help of an emulsifying agent (such as soap). A cutback asphalt is the combination of asphalt cement and petroleum solvent. Foamed asphalt is produced by combining hot asphalt binder with small amounts of cold water. Nowadays, RAP materials are used with virgin aggregate and asphalt binder.

The performance grading (PG) system is the most recent and the most robust asphalt binder grading system. Penetration and viscosity grading systems are older and have grown less popular. Penetration grading is primarily developed based on the penetration test, in which a 100-g needle is allowed to penetrate for 5 seconds. Depending on the amount of penetration, the binder is graded. Viscosity grading is based on the absolute viscosity of asphalt binder and asphalt residue (aged), tested at a high temperature (60°C). The performance grading system characterizes asphalt for high, low, and normal temperatures. Low temperature properties are creep stiffness (S and m values) and direct tensile strain. The dynamic shear test is conducted both at normal temperature and high temperature. Viscosity and flash point temperature are also measured for high temperature.

RTFO and PAV are two methods to simulate the asphalt binder up to the initial construction and long-term age of service life, respectively. The aging occurred during the mixing and compaction is simulated using the RTFO test by oven-aging some asphalt binder at 325°F (163°C) temperature for 85 minutes. The long-term service life aging is simulated using the PAV test, in which asphalt binder is exposed to heat for 20 hours in a heated vessel, pressurized to 305 psi (2.10 MPa or 20.7 atmospheres) at 90°C or 100°C.

DSR testing is conducted to determine the shear modulus (G^*) and phase angle (δ) of asphalt binder. A small specimen of asphalt binder is sandwiched between two plates, shear stress is applied, and the resulting strain and the time lag between the stress–strain are measured. The complex modulus (G^*) is the ratio of the maximum stress to the maximum strain.

The BBR test is conducted to determine the low-temperature creep stiffness (S-value) and the m-parameter. The m-value is the slope of the "Log S versus Log-time" curve at 60 seconds of loading. The m-value indicates the rate of change of stiffness with loading time.

The DTT test measures the tensile strength of asphalt binder at a critical cracking temperature by pulling the specimen at a constant strain rate of 3%/minute under an alcoholic bath.

Using the capillary viscometer, the absolute viscosity is determined by measuring the time required by an asphalt specimen to flow from one point of the capillary tube to another, by means of applied vacuum suction. The flow time (in seconds) is multiplied by the viscometer calibration factor to obtain the absolute viscosity in poises. The Rotational Viscometer (RV) measures the torque required to maintain a constant rotational speed of 20 rpm of a cylindrical spindle while submerged in an asphalt binder at a constant temperature. This torque is then converted to a viscosity.

The flash point temperature test involves heating a small specimen of asphalt binder in a test cup and measuring the lowest liquid temperature causing vapors to ignite due to the test flame. The ductility test measures the asphalt binder ductility by stretching a standard-sized asphalt binder to its breaking point at 25°C (77°F). The presence of mineral impurities can be separated and quantified by dissolving asphalt binder in trichloroethylene or 1,1,1 trichloroethane through a filter mat.

Asphalt cutback is a mixture of asphalt binder and a petroleum solvent. Depending on the volatility of the solvent used, asphalt cutback can be of three types: rapid-curing (RC), medium-curing (MC) and slow-curing (SC). Mixing of gasoline, or naphtha, with asphalt produces a rapid-cure cutback with a curing time of 4–8 hours. Similarly, kerosene produces a medium-curing cutback with a curing time of 12–24 hours; and fuel oil produces a slow-curing cutback with a curing time of 48–60 hours.

Emulsions are classified by their ionic charge, created by the emulsifying agent that is added while producing asphalt emulsions. Some other parameters, such as setting time, viscosity level, and type of asphalt base, are also added with its grading.

ORGANIZATIONS DEALING WITH ASPHALT

Asphalt Institute (AI)

The Asphalt Institute promotes the safe use, benefits, and quality performance of petroleum asphalts in a unified voice for its membership. Founded in 1919, the Institute's members represent 90% of the liquid asphalt produced in North America and an increasing percentage in international markets.

Location: Lexington, KY
Website: http://www.asphaltinstitute.org

The Transportation Research Board (TRB). The Transportation Research Board (TRB) provides innovative, research-based solutions to improve transportation. TRB manages transportation research by producing publications and online resources.

Location: Washington, DC
Website: http://www.trb.org
AASHTO and ASTM International are also resources for asphalt materials.

REFERENCES

Asphalt Institute. 2001. *SUPERPAVE Mix Design, Series No. 2 (SP-2)*. Asphalt Institute Research Center, Lexington, KY.

Brown, E. R., Kandhal, P. S., Roberts, F. L., Kim, Y. R., Lee, D., and Kennedy, T. W. 2009. *Hot-Mix Asphalt Materials, Mixture Design, and Construction*, 3rd edition. Second Printing 2016, NAPA Research and Education Foundation, Maryland.

Energy Information Agency (EIA). 2011. *Table F1: Asphalt and Road Oil Consumption, Price, and Expenditure Estimates*. Energy Information Agency, Washington, DC.

Grass, P. T. 2012. "Asphalt Consumption Trends." *Technical Presentation, South East Asphalt User/ Producer Group Annual Meeting*. South East Asphalt User/Producer Group, Ridgeland, MS.

Halstead, W. 1985. Relation of asphalt chemistry to physical properties and specifications. *Proceedings of Association of Asphalt Paving Technologists (AAPT)*, St Paul, MN, No. 54, 1985.

Hansen, K. and D. Newcomb. 2011. *Asphalt Pavement Mix Production Survey on Reclaimed Asphalt Pavement, Reclaimed Asphalt Shingles, and Warm-Mix Asphalt Usage. 2009–2010*. Information Series 138. National Asphalt Pavement Association, Lanham, MD.

McGennis, R., Shuler, S. and Bahia, H. 1994. Background of Superpave Asphalt Binder Test Methods, FHWA Report No. FHWA-SA-94-069, 1994.

National Asphalt Pavement Association (NAPA), European Asphalt Pavement Association (EAPA). 2012. *The Asphalt Paving Industry, a Global Perspective, Production, Use, Properties, and*

Occupational Exposure Reduction Technologies and Trends. Global Series 101, 3rd edition. National Asphalt Pavement Association, Lanham, MD.

Peterson, J. 1984. Chemical composition of asphalt as related to asphalt durability – State of the art. *Journal of Transportation Research Record*, Vol. 999, 1984.

Rivera, J. Islam, M. R. and Mincic, M. 2017. Statistical Analysis of Pavement System in the United States. *Proceedings of the CSU–Pueblo Student Symposium*, Pueblo, CO, April 7, 2017.

U.S. Census Bureau. 2007. *EC0731SX1: Manufacturing: Subject Series: Industry-Product Analysis: Industry Shipments by Products: 2007 Economic Census.* U.S. Census Bureau, Washington, DC.

Wallace, H. and Martin, J. 1967. *Asphalt Pavement Engineering.* McGraw Hill, New York.

FUNDAMENTALS OF ENGINEERING (FE) EXAM STYLE QUESTIONS

FE Problem 6.1

Asphalt binder can be obtained from the

A. Crude petroleum vapor
B. Crude petroleum residue
C. Bituminous coal destructive distillation
D. All of the above

Solution: B

Asphalt binder is the crude petroleum residue; the other two options for true for tars.

FE Problem 6.2

Which of the following statements is true?

A. Asphalt emulsion is a suspension of small asphalt cement in water
B. Cutback asphalt is a combination of asphalt cement and petroleum solvent
C. Foamed asphalt is formed by combining hot asphalt with cold water
D. All of the above

Solution: D

FE Problem 6.3

In a region, the average 7-day maximum pavement design temperature is 65°C, and the minimum pavement design temperature is −23°C. The recommended PG asphalt binder for this region is:

A. PG 64-22
B. PG 70-22
C. PG 70-28
D. PG 76-34

Solution: C

PG 70-28 binder is recommended; PG 76-34 is over-designed.

PRACTICE PROBLEMS

PROBLEM 6.1

What are the differences between tar and asphalt binder?

PROBLEM 6.2

Which grading system of asphalt binder best characterizes the asphalt binder?

PROBLEM 6.3

The average 7-day maximum design temperature of a pavement is 51°C, and the minimum pavement design temperature is −18°C. What is most likely the recommend asphalt binder?

PROBLEM 6.4

For the asphalt binder, PG 64-22, what is the recommended design temperature for the dynamic shear test of binder?

PROBLEM 6.5

In a BBR testing, an asphalt beam 12.5 mm wide, 6.25 mm high and 102 mm long is used. After applying a load of 100 g, the deflection data presented in Table 6.6 is recorded. Calculate the creep stiffness of the specimen.

TABLE 6.6
BBR Test Data for Problem 6.5

Time (sec)	Deflection (mm)
0	0
10	0.355
30	0.605
40	0.705
50	0.782
60	0.825

7 Asphalt Mix Design

Asphalt mix design is performed with due attention to the climate of the region, temperature, traffic volume, and serviceability requirement. Detailed volumetric and performance analyses are required to achieve the desired asphalt mix. This chapter discusses the two popular asphalt mix design procedures, *Marshall* and *Superpave*, using step-by-step solved examples.

7.1 BACKGROUND

Asphalt mix design determines the following:

- Which aggregate to use
- Which grade of asphalt binder to use
- The optimum combination of these two ingredients, to have the desired properties and performance

Asphalt mix design is much more complex compared to the PCC mix design. In the PCC mix design, the major requirements are the compression strength and the freeze–thaw durability of the hardened concrete, as well as the workability of the fresh concrete. In asphalt mix design, the major considerations include the properties and gradation of aggregates, type of asphalt binder, optimum amount of asphalt binder, different volumetric properties (discussed later), moisture susceptibility, etc. The expected numbers of axle load repetitions and historical climate data are also considered during the asphalt mix design.

The *asphalt concrete* (AC) mix, shown in Figure 7.1, is broadly composed of asphalt binder and aggregates. The stiffness and performance of AC depend on aggregate types, aggregate gradation, binder grade, binder proportion, etc. Asphalt mixes are mainly used in constructing flexible (asphalt) pavement. Flexible pavements are the pavements in which the surface layer is predominantly asphalt concrete and bends just below the wheel load.

The following performances in the final mix are sought out by manipulating the variables of aggregate and asphalt binder.

a) **Deformation or Rutting Resistance.** AC deforms under traffic loading, as shown in Figure 7.2. The deformation of AC for a particular site increases with the increase in the proportion of round particles, fine particles, binder content, and inappropriate binder grade.

b) **Fatigue Resistance.** AC cracks when it is subjected to repeated loads over time, as shown in Figure 7.3. The fatigue cracking of AC, also called *alligator cracking*, is related to the asphalt content and hardened-mix stiffness. A higher asphalt content results in a lessened amount of fatigue cracking under a repeated load. However, a mix with a high asphalt content has a greater tendency to deform (rut) under loading.

c) **Low-Temperature Transverse Cracking Resistance.** AC cracks, as shown in Figure 7.4, when it is subjected to low ambient temperatures. Low temperature

FIGURE 7.1 Asphalt mixture at room temperature. *Photo taken at the Farmingdale State College of the State University of New York.*

FIGURE 7.2 Rutting (permanent deformation) in flexible pavements. *Photos taken in Pueblo, Colorado.*

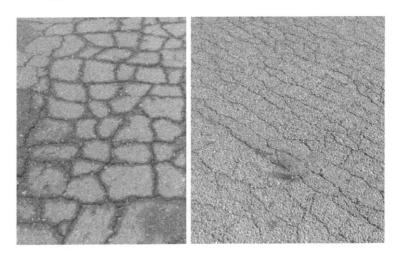

FIGURE 7.3 Fatigue (alligator) cracking in flexible pavements. *Photos taken in Pueblo, Colorado.*

cracking is primarily a function of the asphalt binder grade (low-temperature stiffness). Specifying the appropriate asphalt binder grade with the adequate low-temperature properties should prevent, or at least minimize, low temperature cracking.

d) **Durability.** AC should not oxidize excessively during its production and service life when exposed to air. If the mix has a high content of air voids (8% or more), the permeability of the mix increases. This high permeability allows air to access the asphalt binder easily and accelerates oxidation and volatilization. If the asphalt

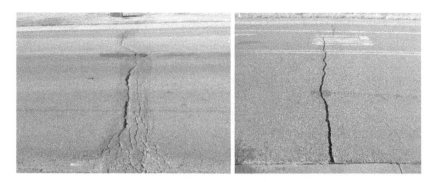

FIGURE 7.4 Transverse cracking in flexible pavements. *Photos taken in Pueblo, Colorado.*

FIGURE 7.5 A pothole after severe stripping in an asphalt pavement. *Photo taken in Pueblo, Colorado.*

coating thickness surrounding the aggregate is low, then the aggregate may become accessible to water through holes in the film, resulting in potholes, as shown in Figure 7.5.

e) **Skid resistance.** AC placed as a surface course should provide sufficient friction when it is in contact with tires. Smooth, rounded, or polish-susceptible aggregates are less skid resistant.

f) **Workability.** AC must be capable of being placed and compacted with reasonable effort. Flat, elongated or angular particles tend to interlock, which makes compaction difficult. An excess amount of fines in a mix causes tenderness. If the viscosity of the asphalt binder is too high at the mixing and laydown temperature, it becomes difficult to dump, spread, and compact.

The abovementioned objectives are to be satisfied during the mix designing process. In addition, selecting the proper asphalt grade, optimum asphalt content, and proposing a simple ratio of aggregate and asphalt binder must also be done. Understanding the weight–volume relationships of the asphalt mix is important for the correct selection of the proportions of the ingredients in order to satisfy the specification requirements. Let us now discuss the weight–volume relationships in an asphalt mix.

7.2   WEIGHT–VOLUME RELATIONSHIPS

Weight–volume relationships describe the weight–volume relationships of different components of a mix which are used while mix designing. Let us consider a small portion of asphalt mix first as shown in Figure 7.6. Aggregate particles have the major portion of solids, some internal *impermeable pores* which cannot be accessed by water or asphalt, *some permeable* pores accessible by water or asphalt. The impermeable pores cannot be accessed for water or asphalt and are considered the part of solid. Thus, the solid part of aggregate always includes the impermeable pores. While mixing some part of asphalt is absorbed by the permeable pores of asphalt which is called the *absorbed asphalt* and does not take part in binding aggregates. This is why the absorbed asphalt is also known as lost asphalt. The absorbed asphalt may fill the permeable pores or some part of permeable pores may remain unfilled. The other part of asphalt, called the *effective asphalt*, coats the surface of aggregate to create the asphalt film for binding other aggregates. The effective asphalt contributes to the binding action. As aggregates are irregular in shape, some air pockets are created between the coated aggregates throughout a compacted mixture and are known as the *air voids*. Some useful parameters of asphalt mix are introduced below:

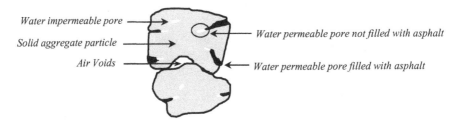

FIGURE 7.6   Absorbed asphalt in aggregate particles.

V_{mb} = Bulk volume of the mix (i.e., total volume of mix). It includes the solid aggregates, permeable pores, asphalt binder and the air pockets inside the mix.

V_b = Volume of asphalt binder. This includes the asphalt absorbed by the permeable pores which does not take part in binding aggregates and the effective asphalt coating the surface of aggregate to bind aggregates.

V_{ba} = Volume of absorbed asphalt binder in the permeable pores

V_{be} = Volume of effective asphalt binder ($V_b - V_{ba}$)

V_{sb} = Bulk volume of stone (aggregate). It includes the solid part, impermeable pores inside the solid that cannot be accessed by water or asphalt, and permeable pores.

V_{apa} = Apparent (solid) volume of aggregate including any impermeable pores. It does not include the permeable pores.

V_{eff} = Effective volume of aggregate (apparent/solid volume + the external voids – external voids filled with asphalt = $V_{sb} - V_{ba}$). More specifically, V_{eff} does not consider the portion of permeable pores that are filled by asphalt. In other words, V_{eff} includes the solid particle, any impermeable pores inside the solid, and any portion of permeable voids that are not filled by asphalt.

V_a = Total volume of air voids in mix.

P_{ba} = Absorbed asphalt content (percent or fraction).

M_{mb} = mass of the mix (i.e., total mass of mix).

M_{SSD} = Saturated surface dry (SSD) mass.

M_{sub} = Submerged mass in water.
M_b = Mass of the asphalt binder.
M_{be} = Mass of effective asphalt binder.
M_{ba} = Mass of absorbed asphalt binder.
M_{agg} = Mass of aggregate. Permeable and impermeable pores have no mass.
P_b = Asphalt content by weight of mix (percent or fraction), $P_b = M_b / M_{mb}$
P_s = Aggregate content by weight of mix (percent or fraction), $P_s = M_{agg} / M_{mb}$

where,

V = Volume
M = Mass
P = Proportion
a (first subscript) = air
a (second subscript) = absorbed
b (first subscript) = binder
b (second subscript) = bulk
s = stone
agg = aggregate
eff = effective
apa = apparent
sub = submerged

To be consistent, P_b, P_s, and P_{ba} are used as fractions in this text, i.e., $P_s + P_b = 1.0$.
Specific gravity (G) is the parameter connecting the weight and volume of a substance. In general, specific gravities of different substances in asphalt mixes are abbreviated as, G_{xy},

where,

First subscript, x:
    b = binder (asphalt)
    s = stone (i.e., aggregate)
    m = mixture
Second subscript, y:
    b = bulk
    e = effective
    a = apparent
    m = maximum

For example, G_{mm} = gravity, mixture, maximum = the maximum gravity of the mixture. Thus,

G_b = Specific gravity of asphalt binder
G_{sb} = Bulk specific gravity of the stone (aggregate)
G_{se} = Effective specific gravity of the stone (aggregate)
G_{mb} = Bulk specific gravity of the compacted mixture
G_{mm} = Maximum theoretical specific gravity of the mixture

AC is made up of three components: aggregate, asphalt binder, and air. These components are described by their volumes. Therefore, it is important to know how these three materials relate to one another volumetrically. The schematic in Figure 7.7 shows the phases of these three components. This sketch is used to define and correlate different parameters of mix design, which are discussed here.

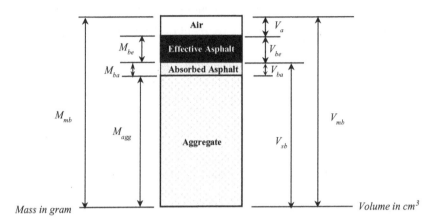

FIGURE 7.7 Volumetric relationships of key asphalt mix constituents.

Let us start with the different specific gravities used in the mix design. While studying the weight–volume relationship, Eq. 7.1 is very often used. This equation is also discussed in Chapter 2.

$$\text{Specific gravity}, G = \frac{M}{V \rho_w} = \frac{M \text{ gram}}{V \text{ cm}^3 \left(1.0 \frac{\text{gram}}{\text{cm}^3}\right)} = \frac{M \text{ gram}}{V \text{ cm}^3} \tag{7.1}$$

where,
G = Specific gravity
M = Mass of the substance (g)
V = Volume of the substance (cm³)
ρ_w = Density of water (g/cm³)

Note: No other unit is possible for Eq. 7.1. Some other terms defined from the volumetric relationships are discussed below.

7.2.1 Bulk Specific Gravity of the Compacted Asphalt Mixture (G_{MB})

The *bulk specific gravity of the compacted asphalt mixture*, G_{mb}, is the ratio of the mass in air of a unit volume of a compacted specimen (including both permeable and impermeable pores normal to the material) at a stated temperature to the mass in air (of equal density) of an equal volume of distilled water at that temperature. It can be expressed as shown in Eq. 7.2.

$$G_{mb} = \frac{\text{Total Mass}}{\text{Total Volume}} = \frac{M_{mb}}{V_{mb}} \tag{7.2}$$

In the laboratory, G_{mb} can be determined using Eq. 7.3. How the ratio of A and $(B - C)$ represents the bulk-specific gravity is discussed in Chapter 2.

$$G_{mb} = \frac{M_D}{M_{SSD} - M_{sub}} = \frac{A}{B - C} \tag{7.3}$$

where,

$A = M_D$ = Dry mass of the compacted specimen
$B = M_{SSD}$ = Saturated-surface dry (SSD) mass of the compacted specimen
$C = M_{sub}$ = Submerged mass of the compacted specimen

The G_{mb} test is conducted following the AASHTO T 166 standard. In this test, the specimen is dried overnight at $52 \pm 3°C$ ($125 \pm 5°F$) to a constant mass. The specimen is then cooled to room temperature at $25 \pm 5°C$ ($77 \pm 9°F$), and the dry mass is recorded as A. The specimen is then immersed in water at $25 \pm 1°C$ ($77 \pm 1.8°F$) for 4 ± 1 minutes, and the submerged mass is recorded as C. The specimen is then removed from the water and is damp-dried by blotting with a damp towel as quickly as possible (not to exceed 5 seconds), and the surface-dry mass is taken as B.

7.2.2 THEORETICAL MAXIMUM SPECIFIC GRAVITY OF BITUMINOUS PAVING MIXTURES (G_{MM})

The *theoretical maximum specific gravity of bituminous paving mixtures*, G_{mm}, is the ratio of the mass of a given volume of voidless ($V_a = 0$) asphalt loose mixture at a stated temperature (usually 25°C) to the mass of an equal volume of gas-free distilled water at the same temperature. It is also called the *Rice specific gravity*, named after James Rice. Theoretically, G_{mm} can be written as shown in Eqs. 7.4–7.9.

$$G_{mm} = \frac{\text{Mass of (Aggregate + Asphalt)}}{\text{Volume of } \left(\text{Solid Aggregate + Permeable Voids Not Filled with Asphalt + Total Asphalt}\right)} \tag{7.4}$$

$$G_{mm} = \frac{\text{Total Mass}}{\text{Effective Volume of Aggregate + Volume of Total Asphalt}} \tag{7.5}$$

Note: Apparent/solid volume of aggregate always includes the impermeable pores inside it. Therefore,

$$G_{mm} = \frac{M_{mb}}{V_{eff} + V_b} \tag{7.6}$$

$$G_{mm} = \frac{M_{mb}}{\dfrac{M_{agg}}{G_{se}} + \dfrac{M_b}{G_b}} \tag{7.7}$$

As $P_s = M_{agg} / M_{mb}$

and

$P_b = M_b/M_{mb}$, then:

$$G_{mm} = \frac{M_{mb}}{\dfrac{M_{mb}P_s}{G_{se}} + \dfrac{M_{mb}P_b}{G_b}} \tag{7.8}$$

Finally,

$$G_{mm} = \frac{1}{\dfrac{P_s}{G_{se}} + \dfrac{P_b}{G_b}} \tag{7.9}$$

The G_{mm} test is conducted following the AASHTO T 209 test standard. A weighed oven-dry (oven-dry is denoted as A) loose mixture is placed in a vacuum vessel. A sufficient amount of water at a temperature of $25 \pm 0.5°C$ ($77 \pm 0.9°F$) is added to completely submerge the specimen. A vacuum is applied for 15 ± 2 minutes to gradually reduce the residual pressure in the vacuum vessel to 3.7 kPa (0.54 psi), as shown in Figure 7.8. At the end of the vacuum period, the vacuum is gradually released. The volume of the specimen of paving mixture is obtained either by immersing the vacuum container with the specimen into a water bath and weighing, or by filling the vacuum container level full of water and weighing it in air. The G_{mm} is then calculated using Eq. 7.10.

$$G_{mm} = \frac{M_{D(loose)}}{M_{D(loose)} - M_{sub(loose)}} = \frac{A}{A - (E - D)} \tag{7.10}$$

where,

A = Dry mass of the loose asphalt coated specimens
$(E - D)$ = Submerged mass of the loose specimen
E = Submerged mass of the loose specimen and the vacuum bowl
D = Submerged mass of the vacuum bowl
$M_{D(loose)}$ = Dry mass of the loose asphalt coated specimens
$M_{sub(loose)}$ = Submerged mass of the loose asphalt coated specimens

At the time of weighing, the water temperature is measured as well. From the mass and volume measurements, the maximum specific gravity, or the Rice density, at $25°C$ ($77°F$)

Pycnometer Filled   Dry Loose Mixture   Immerse Loose Mixture   Vacuum-Suction of
with Water   The Immersed Mixture

FIGURE 7.8 Different phases of theoretical maximum specific gravity testing. *Photos taken at the Farmingdale State College of the State University of New York.*

is calculated. If the temperature employed is different from 25°C (77°F), an appropriate correction is applied.

7.2.3 VOID IN TOTAL MIX (VTM)

The *void in total mix* (VTM) is the total volume of the small pockets of air (V_a) between the coated aggregate particles throughout a compacted paving mixture, expressed as a percent of the bulk volume of the compacted paving mixture (V_{mb}). For typical mixes, air void contents below 3% result in an unstable mixture, while air voids above about 8% result in a water-permeable mixture. By definition, VTM can be expressed as shown in Eqs. 7.11–7.21:

$$VTM = \frac{V_a}{V_{mb}} \times 100 \tag{7.11}$$

Now,

$$V_a = V_{mb} - V_{be} - V_{sb} \tag{7.12}$$

Therefore,

$$VTM = \left(\frac{V_{mb} - V_{be} - V_{sb}}{V_{mb}} \right) \times 100 \tag{7.13}$$

$$VTM = \left(1 - \frac{V_{be} + V_{sb}}{V_{mb}} \right) \times 100 \tag{7.14}$$

$$VTM = \left(1 - \frac{\dfrac{1}{V_{mb}}}{\dfrac{1}{V_{be} + V_{sb}}} \right) \times 100 \tag{7.15}$$

$$VTM = \left(1 - \frac{\dfrac{M_{mb}}{V_{mb}}}{\dfrac{M_{mb}}{V_{be} + V_{sb}}} \right) \times 100 \tag{7.16}$$

Now,

$$V_{be} + V_{sb} = V_{be} + \left(V_{eff} + V_{ba} \right) \tag{7.17}$$

$$= V_{eff} + \left(V_{be} + V_{ba} \right) \tag{7.18}$$

$$= V_{eff} + V_b \tag{7.19}$$

Therefore,

$$VTM = \left(1 - \frac{\dfrac{M_{mb}}{V_{mb}}}{\dfrac{M_{mb}}{V_{eff} + V_b}} \right) \times 100 \tag{7.20}$$

Therefore,

$$VTM = \left(1 - \frac{G_{mb}}{G_{mm}} \right) \times 100 \tag{7.21}$$

7.2.4 Effective Specific Gravity (G_{se})

The *effective specific gravity*, G_{se}, is the specific gravity of aggregate considering the volume of apparent (solid) aggregate plus the external voids and minus the external voids filled with asphalt, as shown in Figure 7.6. By definition, G_{se} can be expressed as shown in Eqs. 7.22–7.28:

$$G_{se} = \frac{M_{agg}}{V_{eff}} \tag{7.22}$$

$$G_{se} = \frac{M_{mb} - P_b M_{mb}}{V_{mb} - V_b - V_a} \tag{7.23}$$

$$G_{se} = \frac{M_{mb} - P_b M_{mb}}{\dfrac{M_{mb}}{G_{mm}} - \dfrac{M_b}{G_b} - 0} \quad \left[G_{mm} = \frac{M_{mb}}{V_{mb}} \text{ if there is no air void, i.e., } V_a = 0 \right] \tag{7.24}$$

$$G_{se} = \frac{M_{mb} - M_{mb} P_b}{\dfrac{M_{mb}}{G_{mm}} - \dfrac{P_b M_{mb}}{G_b}} \quad \left[\text{As } P_b = \frac{M_b}{V_{mb}} \right] \tag{7.25}$$

$$G_{se} = \frac{G_{mb} \left(1 - P_b \right)}{G_{mb} \left(\dfrac{1}{G_{mm}} - \dfrac{P_b}{G_b} \right)} \tag{7.26}$$

$$G_{se} = \frac{P_s}{\dfrac{1}{G_{mm}} - \dfrac{P_b}{G_b}} \tag{7.27}$$

where,

$$P_s + P_b = 1 \tag{7.28}$$

Eqs. 7.24–7.27 are valid only if there are no air voids, i.e., $V_a = 0$.

7.2.5 Voids in the Mineral Aggregate (VMA)

The volume void space among the aggregate particles of a mixture, which includes the air voids and the effective asphalt content, is known as the *voids in mineral aggregate* (VMA).

Once a batch of aggregate is compacted, some voids occur which cannot be closed by compacting, as the aggregates are irregular in shape. Once the asphalt binder is mixed with a batch of aggregate, a part of the asphalt is absorbed by the aggregates, and the remaining part of asphalt coats the aggregates' surface. The coated asphalt binder is the *effective asphalt binder*, as it takes part in binding the aggregates. The effective asphalt binder also occupies some empty spaces among the aggregates. The VMA includes the empty spaces among the aggregates, including the empty space occupied by the asphalt binder. The VMA is expressed as a percent of the total volume, and can be calculated using Eqs. 7.29 through 7.36.

$$VMA = \left(\frac{V_a + V_{be}}{V_{mb}} \right) \times 100 \tag{7.29}$$

$$VMA = \left(\frac{V_{mb} - V_{sb}}{V_{mb}} \right) \times 100 \tag{7.30}$$

$$VMA = \left(\frac{V_{mb}}{V_{mb}} - \frac{V_{sb}}{V_{mb}} \right) \times 100 \tag{7.31}$$

$$VMA = \left(1 - \frac{\dfrac{M_{agg}}{G_{sb}}}{\dfrac{M_{mb}}{G_{mb}}} \right) \times 100 \tag{7.32}$$

$$VMA = \left(1 - \frac{G_{mb} M_{agg}}{G_{sb} M_{mb}} \right) \times 100 \tag{7.33}$$

$$VMA = \left(1 - \frac{G_{mb} \left(M_{mb} - M_{mb} P_b \right)}{G_{sb} M_{mb}} \right) \times 100 \quad \left[\text{As, } M_{agg} = M_{mb} - M_{mb} P_b \right] \tag{7.34}$$

$$VMA = \left(1 - \frac{G_{mb} \left(1 - P_b \right)}{G_{sb}} \right) \times 100 \tag{7.35}$$

$$VMA = 100 - \left(\frac{G_{mb} P_s}{G_{sb}} \right) \times 100 \tag{7.36}$$

Too low a VMA means there is not enough room in the mixture to add sufficient asphalt binder to coat the aggregates. Mixes with low amounts of VMA are more sensitive to small changes in binder content. Excessive VMA causes instability of the mix(es) when under loading.

7.2.6 VOIDS FILLED WITH ASPHALT (VFA)

The portion of the voids in the mineral aggregate which contain asphalt binder is known as the *voids filled with asphalt* (VFA). It can be described as the percent of VMA that is filled with asphalt. VFA is inversely related to air voids; as the VFA increases, the amount of air voids decreases. VFA can be calculated using Eqs. 7.37 or 7.38.

$$VFA = \left(\frac{V_{be}}{V_a + V_{be}} \right) \times 100 \tag{7.37}$$

$$VFA = \frac{VMA - VTM}{VMA} \times 100 \tag{7.38}$$

7.2.7 Absorbed Asphalt Content (P_{BA})

The asphalt binder in the AC that has been absorbed into the pore structure of the aggregate is called the *absorbed asphalt*. It is the portion of the asphalt binder in the AC that is not accounted for by the effective asphalt content. The absorbed portion of the binder does not take part in binding aggregate, and is very often termed the *lost binder*. Thus, careful selection of aggregate is important. The less the absorptive power of aggregate, the better the aggregate is. The percentage of the absorbed asphalt can be calculated using Eqs. 7.39 and 7.43.

$$P_{ba} = \frac{M_{ba}}{M_{agg}} \times 100 \tag{7.39}$$

$$P_{ba} = \frac{V_{ba} G_b}{M_{agg}} \times 100 \tag{7.40}$$

The *volume of absorbed asphalt, V_{ba}*, is the difference between the bulk volume of the aggregate and the effective volume, i.e., $V_{ba} = V_{sb} - V_{eff}$.
    Thus,

$$P_{ba} = \frac{\left(V_{sb} - V_{eff} \right) G_b}{M_{agg}} \times 100 \tag{7.41}$$

$$P_{ba} = \frac{\left(\dfrac{M_{agg}}{G_{sb}} - \dfrac{M_{agg}}{G_{se}} \right) G_b}{M_{agg}} \times 100 \tag{7.42}$$

Therefore,

$$P_{ba} = \left(\frac{G_{se} - G_{sb}}{G_{sb} G_{se}} \right) G_b \times 100 \tag{7.43}$$

7.2.8 Effective Asphalt Content (P_{BE})

The total asphalt binder content of the AC, less the portion of asphalt binder that is lost by the absorption into the aggregate, is called *the effective asphalt content, P_{be}*. This portion of the binder is coated on the aggregate surface and takes part in binding the aggregates. Effective asphalt also fills some part of the VMA. Too small an amount of effective binder produces weak bonding among aggregates, causing the mix to crack easily. Too large an amount of effective asphalt binder makes the mix tender (aggregates slide upon compacting) and causes the mix to deform greatly. P_{be} can be calculated using Eq. 7.44.

$$P_{be} = \left(P_b - P_{ba}P_s\right) \times 100 \tag{7.44}$$

7.2.9 DUST-TO-ASPHALT RATIO

Some amount of dust or fines is required in the asphalt mix. Recalling that smaller particles have a larger surface area, too much dust requires more asphalt binder to coat all surfaces of particles. Again, too much asphalt binder allows for greater deformation of the mix upon the traffic loading. If the binder content is limited, then the excessive dust dries out the mix, reducing film thickness and durability. On the other hand, insufficient dust allows for excessive asphalt films, resulting in a tender, unstable mix. The *dust-to-asphalt ratio (F/A)* is calculated using Eq. 7.45.

$$\frac{F}{A} = \frac{\%\text{ passing No.200}}{\%\,P_{be}} \tag{7.45}$$

Now, practice the following worked-out examples for better understanding. Note that P_b, P_{ba}, and P_s are to be used in fractions in the equations.

Example 7.1 Weight–Volume Relationship

A compacted AC mixture has the following properties.

- Bulk specific gravity of the mixture, $G_{mb} = 2.425$
- Theoretical maximum specific gravity, $G_{mm} = 2.521$
- Asphalt binder specific gravity, $G_b = 1.015$
- Asphalt content, $P_b = 5.0\%$ (by mass of the total mix)
- Percent passing No. 200 sieve $= 5.3$

The percent of each aggregate and the bulk specific gravity (G_{sb}) for each is listed Table 7.1. Based on the information given for this problem:

a) Calculate the bulk specific gravity of the combined aggregate
b) Calculate the effective specific gravity of the aggregate

TABLE 7.1
Bulk Specific Gravity Data for Example 7.1

Aggregate	% of Total Aggregate (P)	G_{sb} or G
A	50%	2.695
B	25%	2.711
C	25%	2.721

c) Calculate the percent of absorbed asphalt for the mixture
d) Calculate the percent of effective asphalt for the mixture
e) Calculate the percent of voids in total mix for the mixture
f) Calculate the percent of voids in mineral aggregate for the mixture
g) Calculate the percent of voids filled with asphalt for the mixture
h) Calculate the dust-to-asphalt ratio

Solution

Note: This example is a basic mathematical problem to determine different parameters required for a mix design process. This example is intended to provide knowledge in determining these parameters.

Given,

Asphalt content, $P_b = 5.0\% = 0.05$
Aggregate content, $P_s = 100 - 5\% = 95\% = 0.95$
Proportion of Aggregate A, $P_A = 50\% = 0.50$
Proportion of Aggregate B, $P_B = 25\% = 0.25$
Proportion of Aggregate C, $P_C = 25\% = 0.25$
Bulk specific gravity of Aggregate A, $G_A = 2.695$
Bulk specific gravity of Aggregate B, $G_B = 2.711$
Bulk specific gravity of Aggregate C, $G_C = 2.721$

a) Bulk Specific Gravity of the Combined Aggregate, G_{sb}

$$G_{sb} = \cfrac{1}{\left(\cfrac{P_A}{G_A} + \cfrac{P_B}{G_B} + \cfrac{P_C}{G_C}\right)} = \cfrac{1}{\left(\cfrac{0.50}{2.695} + \cfrac{0.25}{2.711} + \cfrac{0.25}{2.721}\right)} = 2.705$$

Specific gravity is commonly expressed with three significant figures following the decimal, in the asphalt community.

b) Effective Specific Gravity of Aggregate, G_{se}

$$G_{se} = \cfrac{P_s}{\cfrac{1}{G_{mm}} - \cfrac{P_b}{G_b}} = \cfrac{0.95}{\left(\cfrac{1}{2.521} - \cfrac{0.05}{1.015}\right)} = 2.735$$

[Note that P_b and P_s are to be used in fractions in the equation.]

c) Percent Absorbed Asphalt Binder, P_{ba}

$$P_{ba} = \left(\frac{G_{se} - G_{sb}}{G_{sb}G_{se}}\right)G_b \times 100 = \left(\frac{2.735 - 2.705}{(2.705)(2.735)}\right)(1.015) \times 100 = 0.0041 = 0.41\%$$

d) Effective Asphalt Content (P_{be})

$$P_{be} = (P_b - P_{ba}P_s) \times 100 = (0.05 - (0.0041)(0.95)) \times 100 = 4.61\%$$

[Note that P_b, P_{ba}, and P_s are to be used in fractions in the equation.]

e) Void in Total Mix (VTM)

$$VTM = \left(1 - \frac{G_{mb}}{G_{mm}}\right) \times 100 = \left(1 - \frac{2.425}{2.521}\right) \times 100 = 3.81\%$$

f) Voids in the Mineral Aggregate (VMA)

$$VMA = 100 - \left(\frac{G_{mb}P_s}{G_{sb}}\right) \times 100 = 100 - \left(\frac{2.425 \times 0.95}{2.705}\right) \times 100 = 14.83\%$$

[Note that P_s is to be used in fraction in the equation.]

g) Voids Filled with Asphalt (VFA)

$$VFA = \frac{VMA - VTM}{VMA} \times 100 = \frac{14.83 - 3.81}{14.83} \times 100 = 74.31\%$$

h) Dust-to-Asphalt ratio

$$\frac{F}{A} = \frac{\% \text{ passing No. } 200}{\%P_{be}} = \frac{5.3}{4.61} = 1.15$$

Answers:

a) $G_{sb} = 2.705$
b) $G_{se} = 2.735$
c) $P_{ba} = 0.41\%$
d) $P_{be} = 4.61\%$
e) $VTM = 3.81\%$
f) $VMA = 14.83\%$
g) $VFA = 74.31\%$
h) $F/A = 1.15$

This example shows how different properties of a mix, such as VMA, VTM, VFA, etc., can be determined if basic laboratory test parameters are known. Note that the absorbed binder in this example is very small (0.4%). In fact, the absorbed binder of a real mix is similar. If the absorbed binder is neglected, the calculation procedure is shown in the following example.

Example 7.2 Weight–Volume Relationship

A compacted asphalt concrete specimen contains 5% asphalt binder (by weight of the total mix) with a specific gravity of 1.023. The aggregate used in the mix has a specific gravity of 2.755. The bulk density of the specimen is 2.441 Mg/m³. If the absorption of the aggregate is negligible, determine the VMA, VTM, and VFA.

Solution

Given,

Asphalt content, $P_b = 5.0\% = 0.05$
Aggregate content, $P_s = 100 - 5\% = 95\% = 0.95$

Specific gravity of binder, $G_b = 1.023$

Bulk specific gravity of mix, $G_{mb} = \dfrac{\rho}{\rho_w} = \dfrac{2.441 \times 1000 \text{ kg/m}^3}{1000 \text{ kg/m}^3} = 2.441$

Specific gravity of aggregate, $G_{sb} = 2.755$ [Specific gravity means bulk specific gravity; 'bulk' is very often omitted.]

Absorbed asphalt, $P_{ba} = 0\%$

a) VMA

$$VMA = 100 - \left(\frac{G_{mb}P_s}{G_{sb}}\right) \times 100 = 100 - \left(\frac{2.441 \times 0.95}{2.755}\right) \times 100 = 15.82\%$$

[Note that P_s is to be used in fraction in the equation.]

b) VTM

$$P_{ba} = \left(\frac{G_{se} - G_{sb}}{G_{sb}G_{se}}\right)G_b \times 100$$

$$0 = \left(\frac{G_{se} - G_{sb}}{G_{sb}G_{se}}\right)G_b \times 100$$

Therefore, $G_{se} = G_{sb} = 2.755$

$$G_{mm} = \frac{1}{\dfrac{P_s}{G_{se}} + \dfrac{P_b}{G_b}} = \frac{1}{\dfrac{0.95}{2.755} + \dfrac{0.05}{1.023}} = 2.540$$

[Note that P_s and P_b are to be used in fractions in the equation.]

$$VTM = \left(1 - \frac{G_{mb}}{G_{mm}}\right) \times 100 = \left(1 - \frac{2.441}{2.540}\right) \times 100 = 3.89\%$$

c) VFA

$$VFA = \frac{VMA - VTM}{VMA} \times 100 = \frac{15.82 - 3.89}{15.82} \times 100 = 75.41\%$$

Answers:

a) *VMA = 15.82%*
b) *VTM = 3.89%*
c) *VFA = 75.41%*

Example 7.3 Weight–Volume Relationship

An aggregate blend is composed of 60% coarse aggregate by weight with a specific gravity of 2.755, 35% fine aggregate with a specific gravity of 2.710, and 5% filler particles with a specific gravity of 2.748. The compacted specimen contains 5% (by weight of the total mix) asphalt binder with a specific gravity of 1.088. The mixture has a bulk density of 144.4 pcf. If the absorption of the aggregate is negligible, determine the VMA, VTM, and VFA.

Solution

Note: The absorption of the aggregate is negligible in this example as well. However, the data is given in a different way.

Given,

Asphalt content, $P_b = 5.0\% = 0.05$
Aggregate content, $P_s = 100 - 5\% = 95\% = 0.95$
Proportion of Coarse Aggregate, $P_A = 60\% = 0.60$
Proportion of Fine Aggregate, $P_B = 35\% = 0.35$
Proportion of Filler, $P_C = 5\% = 0.05$
Bulk specific gravity of Coarse Aggregate, $G_A = 2.755$
Bulk specific gravity of Fine Aggregate, $G_B = 2.710$
Bulk specific gravity of Filler, $G_C = 2.748$
Specific gravity of binder, $G_b = 1.088$

The bulk specific gravity of the aggregate blend,

$$G_{sb} = \frac{1}{\dfrac{P_A}{G_A} + \dfrac{P_B}{G_B} + \dfrac{P_C}{G_C}} = \frac{1}{\dfrac{0.60}{2.755} + \dfrac{0.35}{2.71} + \dfrac{0.05}{2.748}} = 2.739$$

The bulk specific gravity of the mix, $G_{mb} = \dfrac{\rho}{\rho_w} = \dfrac{144.4 \text{ pcf}}{62.4 \text{ pcf}} = 2.314$

a) VMA

$$VMA = 100 - \left(\frac{G_{mb}P_s}{G_{sb}}\right) \times 100 = 100 - \left(\frac{2.314 \times 0.95}{2.739}\right) \times 100 = 19.74\%$$

Note that P_s is to be used in fraction in the equation.

b) VTM

$$VTM = \left(1 - \frac{G_{mb}}{G_{mm}}\right) \times 100$$

G_{mb} is known. Let us find G_{mm}.

$$P_{ba} = \left(\frac{G_{se} - G_{sb}}{G_{sb}G_{se}}\right) G_b \times 100$$

$$0 = \left(\frac{G_{se} - G_{sb}}{G_{sb}G_{se}} \right) G_b \times 100$$

Therefore, $G_{se} = G_{sb} = 2.739$

$$G_{mm} = \frac{1}{\dfrac{P_s}{G_{se}} + \dfrac{P_b}{G_b}} = \frac{1}{\dfrac{0.95}{2.739} + \dfrac{0.05}{1.088}} = 2.546$$

[Note that P_s and P_b are to be used in fractions in the equation.]

$$VTM = \left(1 - \frac{G_{mb}}{G_{mm}} \right) \times 100 = \left(1 - \frac{2.314}{2.546} \right) \times 100 = 9.11\%$$

c) VFA

$$VFA = \frac{VMA - VTM}{VMA} \times 100 = \frac{19.74 - 9.11}{19.74} \times 100 = 53.85\%$$

Answers:

VMA = 19.74%
VTM = 9.11%
VFA = 53.85%

7.3 HISTORY OF ASPHALT MIX DESIGN

Superpave is an abbreviation for *superior performing flexible pavements*. Superpave is currently the most widely used method of asphalt mix design in North America. It was developed in the early 1990s, as a part of the Strategic Highway Research Program (SHRP). The superpave method draws upon history and incorporates additional information. There are many mix design methods available in the literature, such as:

a) Hubbard Field Mix Design
b) Hveem Mix Design
c) Marshall Mix Design
d) Superpave Mix Design

7.3.1 HUBBARD FIELD MIX DESIGN

In the mid-1920s, Charles Hubbard and Frederick Field from the Asphalt Institute developed a method of mix design, called the *Hubbard Field Method of Design*. The Hubbard Field method was commonly used among state highway departments in the 1920s and 1930s, and was continued by some states up to the 1960s.

7.3.2 HVEEM MIX DESIGN

Francis Hveem, a Resident Engineer for the California Division of Highways had proposed a method to determine the asphalt content based on surface area, and developed this method for the California Department of Highways in 1927. Hveem's mix design concept was that the optimum amount of asphalt binder is needed to have a good film thickness on aggregates after satisfying the aggregates' absorption. Also, to carry load, the aggregates had to have a bonding strength or sliding resistance, and a minimum tensile strength to resist turning movement. The sliding resistance was measured by the Hveem stabilometer, and the tensile strength was measured by the cohesiometer. Hveem found that stability and cohesion were influenced by the aggregate characteristics and binder content. For durability, Hveem developed a swell test and moisture vapor sensitivity test to measure the reaction of the mix to water. The swell test used liquid water, and the vapor sensitivity test used moisture vapor. Then, the effect on the Hveem stability after conditioning was measured. Hveem found that thicker asphalt films had a greater resistance to moisture.

7.3.3 Marshall Mix Design

Bruce Marshall, from the Mississippi Department of Highways, developed the *Marshall Mix Design* in the late 1930s. The Marshall Mix Design is an improvement of the Hubbard–Field Mix Design method. In the 1950s, the Asphalt Institute adopted the Marshall standard. As a result, ASTM and AASHTO adopted the test methods for the Marshall Mix Design, but the properties specified within them were established by further research. The Marshall Mix Design procedure served as the primary means of designing dense mixtures, until the mid-1990s, when the Superpave procedure was introduced. The Marshall Mix Design method consists of six basic steps:

Step 1. Aggregate selection
Step 2. Asphalt binder selection
Step 3. Specimen preparation
Step 4. Stability determination, using the Marshall Stabilometer
Step 5. Density and voids calculations
Step 6. Optimum asphalt binder content selection

Step 1. Aggregate Selection
The Marshall Mix Design developed an aggregate evaluation and selection procedure, which includes three sub-steps:

1) Determine aggregate physical properties by various tests, such as:
   - Toughness and abrasion
   - Durability and soundness
   - Cleanliness and deleterious materials
   - Particle shape and surface texture
2) If the aggregate is acceptable according to the previous step, additional tests are done to fully characterize the aggregates. These tests determine:
   - Gradation and size
   - Specific gravity and absorption
3) Perform blending calculations to achieve the mix design aggregate gradation, if aggregates from more than one source or stockpile are used to obtain the final aggregate gradation.

Step 2. Asphalt Binder Evaluation

In the Marshall method, the asphalt binder is selected based on local experience.

Step 3. Specimen Preparation

Several trial aggregate–asphalt binder blends are prepared, each with a different asphalt binder content. By evaluating each trial blend's performance, an optimum asphalt binder content can be selected. The trial blends must contain a range of asphalt contents, both above and below the optimum asphalt content. Estimation of optimum asphalt content is done by experience. To prepare a specimen, the mix is heated to the anticipated compaction temperature and compacted with a *Marshall hammer*, a device that applies pressure to a specimen through a tamper foot (Figure 7.9). Some hammers are automatic while some are hand operated. Key parameters of the compactor are:

- Specimen size = 102 mm (4 in.) diameter, 64 mm (2.5 in.) height
- Tamper foot = Flat and circular, with a diameter of 98.4 mm (3.875 in.)
- Compaction pressure = 457.2 mm (18 in.) free fall drop distance of a hammer assembly, with a 4,536 g (10 lb) sliding weight.
- Number of blows = 35, 50, or 75 on each side, as listed in Table 7.2.
- Simulation method = The tamper foot strikes the specimen on the top, covering almost the entire specimen's top area. After the specified number of blows, the specimen is turned over and the procedure is repeated.

FIGURE 7.9 Compacting an asphalt specimen using the Marshall hammer. *Photo taken at the Farmingdale State College of the State University of New York.*

TABLE 7.2
Marshall Design Criteria

	Traffic Level					
	Light (<0.01 Million)		Medium (0.01–1.0 Million)		Heavy (>1.0 Million)	
	Min.	Max.	Min.	Max.	Min.	Max.
Compaction (number of blows on each side)	35		50		75	
Stability (minimum)	3,340 N		5,340 N		8,010 N	
	(750 lb)		(1,200 lb)		(1,800 lb)	
Flow (0.25 mm (0.01 in.))	8	18	8	16	8	14
Percent Air Voids	3	5	3	5	3	5
VFA (%)	70	80	65	78	65	75

Adapted from the Asphalt Mix Design Methods, MS-2, Asphalt Institute, 7th edition, Lexington, KY.

Step 4. The Marshall Stability and Flow Test

The performance of the trial mix is evaluated by the *Marshall Stability and Flow Test*. The prepared trial specimen is placed on the Marshall stabilometer, and a load is applied diagonally on the specimen, at a loading rate of 2 in./minute (50.8 mm/minute), shown in Figure 7.10. The load is increased until it reaches the peak, and the maximum load is recorded. During the loading, an attached dial gauge measures the specimen's deformation. The maximum deformation, called the *flow value*, is recorded in 0.01 in. (0.25 mm) increments at the same time the maximum load is recorded. The thickness measured from the test should ideally be 2.5 in. (63.5 mm). If not, the test results are multiplied by the Marshall stability adjustment factors, listed in Table 7.3.

Step 5. Density and Voids Analysis

All mix design methods use density and voids to determine basic AC physical characteristics. Two different measures of densities are typically taken:

- Bulk specific gravity (G_{mb})
- Theoretical maximum specific gravity (G_{mm})

These densities are then used to calculate the volumetric parameters of the AC. Measured void expressions are usually:

- Air voids (V_a), sometimes expressed as voids in the total mix (VTM)
- Voids in the mineral aggregate (VMA), VMA requirements are listed in Table 7.4
- Voids filled with asphalt (VFA)

Step 6. Selection of Optimum Asphalt Binder Content

The optimum asphalt binder content is finally selected based on the combined results of the Marshall stability and flow, density, and void analysis. The optimum asphalt binder content can be determined by the following procedures:

1) Plot the following graphs:
   - VTM vs. Asphalt Binder Content – Percent air voids should decrease with increase in asphalt binder content.

FIGURE 7.10   Marshall stability test equipment in a laboratory. *Photo taken at the Farmingdale State College of the State University of New York.*

- G_{mb} vs. Asphalt Binder Content – Density generally increases with increasing asphalt content, reaches a maximum, and then decreases.
- Marshall Stability vs. Asphalt Binder Content – This should follow one of two trends:
    - Stability increases with increasing asphalt binder content, reaches a peak, and then decreases.
    - Stability decreases with increasing asphalt binder content, and does not have a peak.
- Flow vs. Asphalt Binder Content
- VMA vs. Asphalt Binder Content – Percent VMA decreases with increasing asphalt binder content, reaches a minimum, and then increases.
- VFA vs. Asphalt Binder Content – Percent VFA increases with increasing asphalt binder content.

2) Determine the asphalt binder content that corresponds to the desired air void content (typically 4%). This is the optimum asphalt binder content.
3) Determine properties at this optimum asphalt binder content by referring to the plots. Compare each of these values against the specification values. If all are within the specifications listed in Table 7.2 and Table 7.4, then the preceding optimum asphalt binder content is satisfactory. If any of these properties is outside the specification range, then the mixture must be redesigned.

TABLE 7.3
Marshall Stability Adjustment Factors

Thickness of Specimen (mm)	Adjustment Factor	Thickness of Specimen (mm)	Adjustment Factor
50.8	1.47	65.1	0.96
52.4	1.39	66.7	0.93
54.0	1.32	68.3	0.89
55.6	1.25	69.8	0.86
57.2	1.19	71.4	0.83
58.7	1.14	73.0	0.81
60.3	1.09	74.6	0.78
61.9	1.04	76.2	0.76
63.5	1.00	–	–

Adapted from the Asphalt Mix Design Methods, MS-2, Asphalt Institute, 7th edition, Lexington, KY.

TABLE 7.4
Minimum VMA Requirements

	Min. VMA (%)		
	Design Air Voids (%)		
NMAS*	3.0	4.0	5.0
2.36 mm (No. 8)	19	20	21
4.75 mm (No. 4)	16	17	18
9.5 mm (3/8 in.)	14	15	16
12.5 mm (1/2 in.)	13	14	15
19.0 mm (3/4 in.)	12	13	14
25.0 mm (1.0 in.)	11	12	13

* NMAS – One size larger than the first sieve to retain more than 10%.

Adapted from the Asphalt Mix Design Methods, MS-2, Asphalt Institute, 7th edition, Lexington, KY.

Example 7.4 Marshall Mix Design

An asphalt concrete mixture must be designed using the Marshall Mix Design method. A PG 64-22 binder with a specific gravity of 1.031 was used. The mixture contains a 19-mm nominal maximum aggregate size with a bulk specific gravity of 2.696. The effective specific gravity of the aggregates used is 2.666. Trial mixes were made, with the average results shown in Table 7.5.

Determine the design asphalt content using the Asphalt Institute design criteria for medium traffic. Assume a design air void of the mix is 4.0%.

TABLE 7.5

Test Data for Example 7.4

Asphalt Content (%)	Bulk Specific Gravity, G_{mb}	Stability (kN)	Flow, 0.25 mm
4.0	2.360	6.3	9
4.5	2.378	6.7	10
5.0	2.395	5.4	12
5.5	2.405	5.1	15
6.0	2.415	4.7	22

Solution

Note: This example shows how the asphalt mix design is performed following the Marshall Mix Design procedure. This design procedure has six steps. The first four steps are physical work, such as selection of aggregate, selection of binder, specimen preparation, and testing using the Marshall Stabilometer. The last two steps (steps 5 and 6) are computations to determine the optimum binder content. It can be seen that the data from the first 4 steps are given. We are required to follow the fifth and sixth steps.

Given,

$$G_b = 1.031$$
$$G_{se} = 2.666$$
$$G_{sb} = 2.696$$

Step 5. Density and Voids Analysis

The following equations are used to complete Table 7.6 (columns 5 through 8).

$$G_{mm} = \frac{1}{\dfrac{P_s}{G_{se}} + \dfrac{P_b}{G_b}} = \frac{1}{\dfrac{1-P_b}{2.666} + \dfrac{P_b}{1.031}}$$

$$VTM = \left(1 - \frac{G_{mb}}{G_{mm}}\right) \times 100$$

TABLE 7.6

Data Analysis for Example 7.4

1	2	3	4	5	6	7	8
P_b (%)	G_{mb}	Stability (kN)	Flow, 0.25 mm	G_{mm}	VTM (%)	VMA (%)	VFA (%)
4.0	2.360	6.3	9	2.507	5.9	16	63.3
4.5	2.378	6.7	10	2.488	4.4	15.8	72.0
5.0	2.395	5.4	12	2.470	3.0	15.6	80.5
5.5	2.405	5.1	15	2.452	1.9	15.7	87.8
6.0	2.415	4.7	22	2.434	0.8	15.8	95.1

$$VMA = 100 - \left(\frac{G_{mb}P_s}{G_{sb}}\right) \times 100 = 100 - \left(\frac{G_{mb}(1-P_b)}{2.696}\right) \times 100$$

$$VFA = \frac{VMA - VTM}{VMA} \times 100$$

Note that P_b and P_s are to be used in fractions in the equations. For example, $P_b = 4.0\%$ is to be used as 0.04 and so on.

Step 6. Selection of Optimum Asphalt Binder Content

From Figure 7.11a, $P_b = 4.6\%$ at VTM = 4%

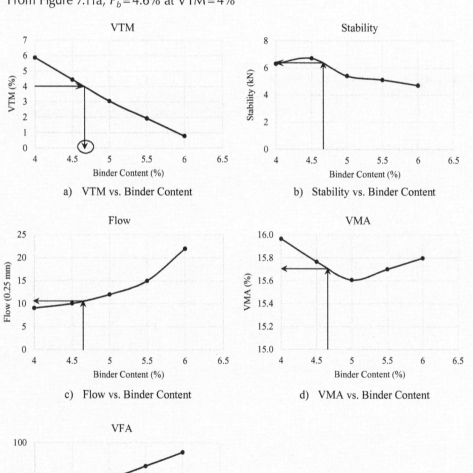

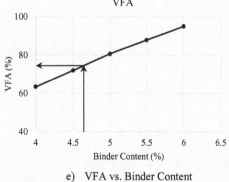

FIGURE 7.11 Data analysis for Example 7.4.

From Figure 7.11b, at $P_b = 4.6\%$, stability = 6.4 kN [okay, criterion is 5.34 kN minimum]
From Figure 7.11c, at $P_b = 4.6\%$, Flow = 10.5 [okay, criterion is 8–16]
From Figure 7.11d, at $P_b = 4.6\%$, $G_{mb} = 2.39$
From Figure 7.11e, at $P_b = 4.6\%$, VMA = 15.75% [okay, criterion is 13% minimum]
From Figure 7.11f, at $P_b = 4.6\%$, VFA = 73% [okay, criterion is 65–78%]
Therefore, the design binder is 4.6%.

This example is the common practice to follow when asphalt mix design is performed using the Marshall procedure. The design air void is assumed 4%, which is very common as well. Having too many air voids decreases strength and durability; not enough air voids causes shear-rutting upon loading. More specifically, when an asphalt mix has negligible air voids, it cannot deform under loading as there is no void space available. Then, the aggregate moves laterally outward below the tire loading.

Example 7.5 Marshall Mix Design

An asphalt concrete mixture is to be designed using the Marshall Mix Design method. A PG 64-22 binder with a specific gravity of 1.019 was used. The dense mixture contains a 25-mm nominal maximum aggregate size with a bulk specific gravity of 2.626. The theoretical maximum specific gravity of the mix (G_{mm}) is 2.514 at 4% asphalt content. Trial mixes were made, with the average results shown in Table 7.7.

TABLE 7.7
Test Data for Example 7.5

Asphalt Content (%)	G_{mb}	Stability (lb)	Flow (0.25 mm)
3.5	2.305	1,600	7
4	2.375	1,750	8
4.5	2.401	1,960	10
5	2.395	1,550	16
5.5	2.386	1,180	23

Determine the design asphalt content using the Asphalt Institute design criteria for medium traffic. Assume a design air void of the mix is 5.0%. Also, make comments regarding what happens if

a) The design air void is 3%
b) The nominal maximum aggregate size is 12.5 mm with the design air void of 5.0%
c) The traffic level is heavy with the design air void of 5.0%

Solution

Note: This example is similar to the previous one. However, this one examines more variables, such as different air voids, aggregates sizes, and traffic volumes.

The first four steps of the mix design process are given. We are required to follow the fifth and sixth steps.

$G_b = 1.019$
$G_{se} =$ not given, we need to find it to proceed
$G_{sb} = 2.626$

Step 5. Density and Voids Analysis
The following equations are used to complete Table 7.8. (columns 5 through 8).

TABLE 7.8

Data Analysis for Example 7.5

1	2	3	4	5	6	7	8	9
P_b	G_{mb}	Stability (lb)	Flow, 0.01 in.	G_{se}	G_{mm}	VTM	VMA	VFA
3.5	2.305	1,600	7	2.678	2.534	9.0	15.3	41.0
4	2.375	1,750	8	2.678	2.514	5.5	13.2	58.0
4.5	2.401	1,960	10	2.678	2.495	3.8	12.7	70.2
5	2.395	1,550	16	2.678	2.476	3.3	13.4	75.4
5.5	2.386	1,180	23	2.678	2.458	2.9	14.1	79.3

$$G_{mm} = \frac{1}{\dfrac{P_s}{G_{se}} + \dfrac{P_b}{G_b}}$$

$$VTM = \left(1 - \frac{G_{mb}}{G_{mm}}\right) \times 100$$

$$VMA = 100 - \left(\frac{G_{mb}P_s}{G_{sb}}\right) \times 100$$

$$VFA = \frac{VMA - VTM}{VMA} \times 100$$

Note that P_b and P_s are to be used in fractions in the equations. For example, $P_b = 4.0\%$ is to be used as 0.04 and so on. All the parameters are known except G_{se}.

$$G_{se} = \frac{P_s}{\dfrac{1}{G_{mm}} - \dfrac{P_b}{G_b}} = \frac{1 - P_b}{\dfrac{1}{G_{mm}} - \dfrac{P_b}{G_b}} = \frac{1 - 0.04}{\left(\dfrac{1}{2.514} - \dfrac{0.04}{1.019}\right)} = 2.678$$

Step 6. Selection of Optimum Asphalt Binder Content
   From Figure 7.12a, $P_b = 4.15\%$ at VTM = 5%
   From Figure 7.12b, at $P_b = 4.15\%$, stability = 1,800 lb [okay, criterion is 1,200 lb minimum]
   From Figure 7.12c, at $P_b = 4.15\%$, Flow = 8.5 [okay, criterion is 8–16]
   From Figure 7.12d, at $P_b = 4.15\%$, VMA = 13% [okay, criterion is 13% minimum]
   From Figure 7.12e, at $P_b = 4.15\%$, VFA = 62% [not okay, criterion is 65–78%]

Therefore, the mix design is not good.

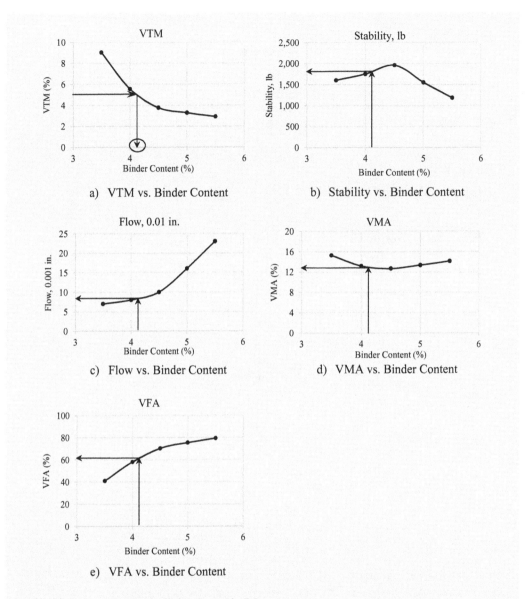

FIGURE 7.12   Data analysis for Example 7.5.

If the design air void is 3%:

From Figure 7.12a, $P_b = 5.4\%$ at VTM = 3%
From Figure 7.12b, at $P_b = 5.4\%$, stability = 1,200 lb [okay, criterion is 1,200 lb minimum]
From Figure 7.12c, at $P_b = 5.4\%$, Flow = 21.5 [not okay, criterion is 8–16]
From Figure 7.12d, at $P_b = 5.4\%$, VMA = 14.5% [okay, criterion is 11% minimum]
From Figure 7.12e, at $P_b = 5.4\%$, VFA = 78% [okay, criterion is 65–78%]

Therefore, the mix design is not good.

If the nominal maximum aggregate size is 12.5 mm:

From Figure 7.12a, $P_b = 4.15\%$ at VTM = 5%
From Figure 7.12b, at $P_b = 4.15\%$, stability = 1,800 lb [okay, criterion is 1,200 lb minimum]

From Figure 7.12c, at $P_b = 4.15\%$, Flow = 8.5 [okay, criterion is 8–16]
From Figure 7.12d, at $P_b = 4.15\%$, VMA = 13% [not okay, criterion is 15% minimum]
From Figure 7.12e, at $P_b = 4.15\%$, VFA = 62% [not okay, criterion is 65–78%]

Both VMA and VFA criteria are not satisfied.

If the traffic level is heavy:

From Figure 7.12a, $P_b = 4.15\%$ at VTM = 5%
From Figure 7.12b, at $P_b = 4.15\%$, stability = 1,800 lb [okay, criterion is 1,800 lb minimum]
From Figure 7.12c, at $P_b = 4.15\%$, Flow = 8.5 [okay, criterion is 8–14]
From Figure 7.12d, at $P_b = 4.15\%$, VMA = 13% [not okay, criterion is 15% minimum]
From Figure 7.12e, at $P_b = 4.15\%$, VFA = 62% [not okay, criterion is 65–75%]

7.3.4 Superpave Mix Design

The Marshall Mix Design procedure has some limitations, such as:

- Does not predict pavement performance
- Test load is applied perpendicular to compaction axis while testing the mix
- Does not orient aggregate particles in the way traffic does
- There is no specification for selecting asphalt binder
- Very few tests are performed on aggregates

The Superpave mix design procedure was developed to overcome these limitations. Superpave mix design was part of the Strategic Highway Research Program (SHRP) from 1987 to 1993. Its intent was to improve the performance and durability of roads in the United States. The objective of the Superpave program was to develop a performance-based asphalt binder specification, a performance-based asphalt mixture specification, and a mix design system. Superpave mixes have been widely used over the last few years, and are replacing the Marshall method, which was used for asphalt concrete mixture design for almost half a century. The Superpave mix design has the following process:

Step 1. Selection of aggregates
Step 2. Selection of asphalt binder
Step 3. Selection of design aggregate structure
Step 4. Determination of design binder content
Step 5. Evaluation of moisture susceptibility

Step 1. Selection of Aggregate
In the Superpave mix design, aggregate is selected in a way such that it satisfies both the source and consensus requirements. Source properties, commonly defined by the owner/contractors include the following methods of testing:

- Los Angeles Abrasion (AASHTO T 96) – Maximum allowable loss values typically range from approximately 35–45%.
- Soundness (AASHTO T 104) – Maximum allowable loss values typically range from approximately 10–20% for five cycles.

- Deleterious Materials (AASHTO T 112) – A wide range of the maximum permissible percentages of clay lumps and friable particles is evident. Values range from as little as 0.2% to as high as 10%, depending on the exact composition of the contaminant.

The national Superpave specification includes the following methods of testing for consensus properties:

- Coarse aggregate angularity (AASHTO MP 2 and ASTM D 5821)
- Fine aggregate angularity (AASHTO TP 33)
- Flat and elongated particles (ASTM D 4791)
- Sand equivalency (ASTM D 2419)

The consensus requirements, which depend on traffic level and pavement depth, are listed in Table 7.9.

TABLE 7.9

Superpave Aggregate Consensus Property Requirements

| Design ESAL[1] (million) | Coarse Aggregate Angularity (%), minimum[2] | | Uncompacted Void Content of Fine Aggregate (%), minimum | | Sand Equivalent (%), minimum | Flat and Elongated[3] (%), minimum |
| | Depth from Surface | | Depth from Surface | | | |
	≤100 mm	>100 mm	≤100 mm	>100 mm		
<0.3	55/–	–/–	–	–	40	–
0.3 to <3	75/–	50/–	40	40	40	10
3 to <10	85/80[2]	60/–	45	40	45	
10 to <30	95/90	80/75	45	40	45	
≥30	100/100	100/100	45	45	50	

(1) Design ESALs are the anticipated project traffic level expected on the design lane over a 20-year period. Regardless of the actual design life of the roadway, determine ESALs for 20 years and choose the appropriate N_{des} level.

(2) 85/80 denotes that 85% of the coarse aggregate has one fractured face and 80% has two or more fractured faces.

(3) Criterion based upon a 5:1 maximum-to-minimum ratio.

Adapted from Superpave Fundamentals Reference Manual, NHI Course #131053, Federal Highway Administration (FHWA), 2017, Washington DC.

Step 2. Selection of Asphalt Binder

The performance grade (PG) binder required for a project is based on environmental data, traffic level, and traffic speed. The environmental data is obtained by converting the historic air temperatures to pavement temperatures. The average 7-day maximum pavement temperature (T_{max}) and the minimum pavement temperature (T_{min}) define the binder laboratory test temperatures. A factor of safety can be incorporated into the PG system based on temperature reliability. The 50% reliability temperatures represent the average of the weather data. The 98% reliability temperatures are determined based on the standard deviations of the low ($\sigma_{Low\ Temp}$) and high ($\sigma_{High\ Temp}$) temperature data. From statistics, 98%

reliability is about two standard deviations from the average value, as presented in Eqs. 7.46 and 7.47.

$$T_{\text{max at }98\%} = T_{\text{max at }50\%} + 2\sigma_{\text{High Temp}} \qquad (7.46)$$

$$T_{\text{min at }98\%} = T_{\text{min at }50\%} - 2\sigma_{\text{Low Temp}} \qquad (7.47)$$

Traffic level and speed are also considered in adjusting the PG binder for high temperature, as listed in Table 7.10. No adjustment is used for low temperature.

TABLE 7.10

Numbers of Increase in High-Temperature Grade for Traffic Load and Speed

Design ESALs (millions)	Traffic Speed (miles per hour)		
	Less than 12	12–42	Above 42
Less than 0.3	–	–	–
0.3 to less than 3	2	1	–
3 to less than 10	2	1	1

1 grade equivalent is 6°C.
Adapted from Superpave Fundamentals Reference Manual, NHI Course #131053, Federal Highway Administration (FHWA), 2017, Washington DC.

The mixing and compaction temperatures of the mix are selected such that the binder has sufficiently less viscosity to mix and compact properly. The temperatures are selected such that the binder's Brookfield or Rotational viscosity ranges as follows:

Range for mixing temperature: $150 - 190$ centiStokes

Range for compaction temperature: $250 - 310$ centiStokes

The rotational viscometer measures the viscosity in centipoises (cP), and the values are reported in Pascal-seconds (Pa-s). The conversion from centipoises to Pascal-seconds is as follows (Eq. 7.48):

$$1 \text{ Pa-s} = 1,000 \text{ centipoises} \qquad (7.48)$$

Step 3. Selection of Design Aggregate Structure

Trial blends are established mathematically by combining the gradations of the individual materials into a single blend in order to select the design aggregate structure. Details of the blending procedure are discussed in Chapter 2. The blend is then compared to the specification requirements for the appropriate sieves, as listed in Table 7.11. The gradation control is based on four control sieves: the maximum sieve, the nominal maximum sieve, the 2.36 mm sieve, and the 0.075 mm sieve.

TABLE 7.11

Aggregate Grading Requirements for Superpave

Sieve Size, mm	Nominal Maximum Aggregate Size – Control Points (% Passing)											
	37.5 mm		25.0 mm		19.0 mm		12.5 mm		9.5 mm		4.75 mm	
	Min.	Max.	Min.	Max.	Min.	Max.	Min.	Max.	Min.	Max.	Min.	Max.
50	100	–	–	–	–	–	–	–	–	–	–	–
37.5	90	100	100	–	–	–	–	–	–	–	–	–
25.0	–	90	90	100	100	–	–	–	–	.	–	–
19.0	–	–	–	90	90	100	100	–	–	–	–	–
12.5	–	–	–	.	–	90	90	100	100		100	–
9.5	–	–	–	–	–	.	—	90	90	100	95	100
4.75	–	–	–	–	–	–	–	–	–	90	90	100
2.36	15	41	19	45	23	49	28	58	32	67	–	–
1.18	–	–	–	–	–	–	–	–	–	–	30	55
0.075	0	6	1	7	2	8	2	10	2	10	6	13

Adapted from Superpave Fundamentals Reference Manual, NHI Course #131053, Federal Highway Administration (FHWA), 2017, Washington DC.

Trial blends are evaluated by compacting specimens and determining the volumetric properties of each trial blend. The compaction is conducted using the *Superpave Gyratory Compactor* (Figure 7.13), with a gyration angle of 1.16° with respect to the horizontal direction, at a constant vertical pressure of 600 kPa (87 psi). The number of gyrations used for compaction is determined based on the traffic volume expected, with three critical stages of compaction: the initial (N_{ini}), design (N_{des}), and maximum (N_{max}) number of gyrations, as listed in Table 7.12. The initial compaction stage, N_{ini}, is used to identify the tenderness of a mix, which is undesirable. A tender mix has less stability during construction, and displaces rather than densifies, when the roller arrives. The causes of a tender mix are excessive moisture, excessive binder, excessive rounded particles, excessive mix temperature, insufficient fines, etc. The design compaction stage, N_{des}, denotes the anticipated level of compaction occurring during construction. The maximum compaction stage, N_{max}, corresponds to the ultimate compaction that occurred after years of traffic operation.

For determining the design aggregate structure, specimens are compacted with N_{des} gyrations and the volumetric properties are determined, as listed in Table 7.12. The following tasks are performed:

Task 1: Determine the G_{mm} of the loose mix and G_{mb} of the compacted specimen, using the laboratory test

Task 2: Calculate VTM, $VTM = \left(1 - \dfrac{G_{mb}}{G_{mm}}\right) \times 100$

FIGURE 7.13 Superpave gyratory compactor ejecting the compacted specimen. *Photo taken at the Farmingdale State College of the State University of New York.*

VTM should be 4.0%, otherwise adjust other properties.

Task 3: Calculate VMA, $VMA = 100 - \left(\dfrac{G_{mb} P_s}{G_{sb}} \right) \times 100$

Task 4: Calculate VFA, $VFA = \left(\dfrac{VMA - VTM}{VMA} \right) \times 100$

Task 5: Calculate % G_{mm} at N_{ini}, $\%G_{mm,N_{ini}} = \%G_{mm,N_{des}} \left(\dfrac{h_{des}}{h_{ini}} \right) = \left(100 - VTM \right) \left(\dfrac{h_{des}}{h_{ini}} \right)$

Task 5: Calculate dust to asphalt Ratio, $\dfrac{F}{A} = \dfrac{\% \text{ passing No.} 200}{\% P_{be}}$

where, Effective binder, $P_{be} = \left(P_b - P_{ba} P_s \right) \times 100$

Absorbed binder, $P_{ba} = \left(\dfrac{G_{se} - G_{sb}}{G_{sb} G_{se}} \right) G_b \times 100$

TABLE 7.12

Superpave Gyratory Compaction Effort

Design ESALs[1] (million)	Compaction Parameters			Typical Roadway Application[2]
	N_{ini}	N_{des}	N_{max}	
<0.3	6	50	75	Applications include roadways with very light traffic volumes, such as local roads, county roads, and city streets, where truck traffic is prohibited or at a very minimal level. Traffic on these roadways would be considered local in nature, and not regional, intrastate, or interstate. Special purpose roadways serving recreational sites or areas may also be applicable to this level.
0.3 to <3	7	75	115	Applications include many collector roads or access streets. Medium-trafficked city streets and the majority of county roadways may be applicable to this level.
3 to <30	8	100	160	Applications include many two-lane, multilane, divided, and partially or completely controlled access roadways. Among these are medium- to highly trafficked city streets, many state routes, US highways, and some rural interstates.
≥30	9	125	205	Applications include the vast majority of the US Interstate system, both rural and urban in nature. Special applications, such as truck-weighing stations or truck-climbing lanes on two-lane roadways, may also be applicable to this level.

(1) Design ESALs are the anticipated project traffic level expected on the design lane over a 20-year period. Regardless of the actual design life of the roadway, determine the design ESALs for 20 years and choose the appropriate N_{des} level.

(2) Typical Roadway Applications as defined by A Policy on Geometric Design of Highway and Streets, 1994, AASHTO.

Adapted from Superpave Fundamentals Reference Manual, NHI Course #131053, Federal Highway Administration (FHWA), 2017, Washington DC.

Effective specific gravity, $G_{se} = \dfrac{P_s}{\dfrac{1}{G_{mm}} - \dfrac{P_b}{G_b}}$

As mentioned in Task 2, the volumetric calculations must be made at 4% VTM. If the VTM is not 4%, the results of the volumetric calculations are estimated at 4% as follows (Eqs. 7.49–7.54):

$$P_{b,est} = P_b - 0.4(4 - VTM) \tag{7.49}$$

$$VMA_{est} = VMA + C(4 - VTM) \tag{7.50}$$

where, $C = 0.1$ for VTM < 4% and $C = 0.2$ for VTM ≥ 4%.

$$VFA_{est} = 100\left(\frac{VMA_{est} - 4.0}{VMA_{est}}\right) \tag{7.51}$$

$$\%G_{mm,N_{ini},est} = \%G_{mm,N_{ini}} - \left(4 - VTM\right) \tag{7.52}$$

$$P_{be,est} = P_{b,est} - P_s G_b \left(\frac{G_{se} - G_{sb}}{G_{sb}G_{se}}\right) \tag{7.53}$$

$$\left(\frac{F}{A}\right)_{est} = \frac{\% \text{ passing No.}200}{\left(\% P_{be}\right)_{est}} \tag{7.54}$$

The calculated volumetric properties are compared to the Superpave Mix Design criteria listed in Table 7.13. If more than one blend of aggregate satisfies the criteria, the most economical blend or the designer's choice should be selected. If none of the blends satisfy the requirement, another blend of aggregate should be considered.

TABLE 7.13
Superpave Volumetric Mixture Design Requirements

Design ESALs[1] (million)	Required Density (% of Theoretical Maximum Specific Gravity)			Voids in the Mineral Aggregate (%), minimum					Voids Filled with Asphalt (%)	Dust-to-Binder Ratio
	N_{ini}	N_{des}	N_{max}	Nominal Maximum Aggregate Size, mm						
				37.5	25.0	19.0	12.5	9.5		
<0.3	≤91.5	96.0	≤98.0	11.0	12.0	13.0	14.0	15.0	70–80[3,4]	0.6–1.2
0.3 to <3	≤90.5								65–78[4]	
3 to <10	≤89.0								65–75[2,4]	
10 to <30										
≥30										

(1) Design ESALs are the anticipated project traffic level expected on the design lane over a 20-year period. Regardless of the actual design life of the roadway, determine the design ESALs for 20 years and choose the appropriate N_{des} level.

(2) For 9.5-mm nominal maximum size mixtures, the specified VFA range shall be 73–76% for design traffic levels $3 million ESALs.

(3) For 25.0-mm nominal maximum size mixtures, the specified lower limit of the VFA shall be 67% for design traffic levels <0.3 million ESALs.

(4) For 37.5-mm nominal maximum size mixtures, the specified lower limit of the VFA shall be 64% for all design traffic levels.

Adapted from Superpave Fundamentals Reference Manual, NHI Course #131053, Federal Highway Administration (FHWA), 2017, Washington DC.

Step 4. Determination of Design Binder Content

Design binder content corresponds to the binder content, which produces 4% air void in the mix. Once the design aggregate structure is selected, specimens are compacted at varying asphalt binder contents. The mixture properties are then evaluated to determine a design asphalt binder content. Superpave requires a minimum of two specimens compacted at each of the following asphalt contents (FHWA recommends three specimens compacted at each asphalt binder content):

a) Two specimens at the estimated binder content
b) Two specimens 0.5% less than the estimated binder content
c) Two specimens 0.5% more than the estimated binder content
d) Two specimens 1.0% more than the estimated binder content

The volumetric properties are calculated at the design number of gyrations (N_{des}) for each trial of asphalt binder content. Optimum asphalt binder content can be determined by the following procedures:

a) Plot the following graphs:
   • VTM vs. Asphalt Binder Content – Percent air voids should decrease with increasing asphalt binder content.
   • VMA vs. Asphalt Binder Content – Percent VMA should decrease with increasing asphalt binder content, reach a minimum, and then increase.
   • VFA vs. Asphalt Binder Content – Percent VFA increases with increasing asphalt binder content.
   • G_{mm} vs. Asphalt Binder Content – Density generally increases with the asphalt content, reaches the maximum, and then decreases. Peak density usually occurs at a high asphalt binder content rather than peak stability.
   • F/A Ratio vs. Asphalt Binder Content.
b) Determine the asphalt binder content that corresponds to the desired air void content (typically 4%). This is the optimum asphalt binder content.
c) Determine properties at this optimum asphalt binder content by referring to the plots. Compare each of these values against the specification values and, if all are within specification, then the preceding optimum asphalt binder content is satisfactory. Otherwise, if any of these properties is outside the specification range, the mixture should be redesigned.

Step 5. Evaluation of Moisture Susceptibility (AASHTO T 283)
   The last step in the volumetric mix design process is to evaluate the moisture sensitivity of the design mixture. This step is accomplished by performing AASHTO T 283 on the design aggregate blend at the design asphalt binder content. Specimens are compacted to approximately 7.0% (±1.0%) air voids. One batch of three specimens is considered the control/unconditioned subset. The other subset of three specimens is the conditioned subset. The conditioned subset is subjected to partial vacuum saturation, followed by an optional freeze cycle, and then followed by a 24-hour heating cycle at 60°C. All specimens are tested to determine their indirect tensile strengths.
   A deformation rate of 2-in. (50 mm) per minute is applied diametrically until failure occurs, as shown in Figure 7.14, and the peak load is recorded. Then, the indirect tensile strength is calculated using Eq. 7.55.

$$ITS = \frac{2P}{\pi Dt} \tag{7.55}$$

FIGURE 7.14 Indirect tensile strength test setup. *Photo taken at the University of New Mexico.*

where, P, D, and t are the measured peak load, specimen diameter, and specimen thickness, respectively.

The moisture sensitivity is determined as a ratio of the tensile strengths of the conditioned subset to the tensile strengths of the control subset, as shown in Eq. 7.56.

$$\text{Tensile strength ratio} = \frac{\text{Tensile strength after conditioning}}{\text{Tensile strength before conditioning}} \qquad (7.56)$$

The minimum criterion for the tensile strength ratio is 0.8 or 80%.

Example 7.6 Selection of Asphalt Binder (Step 2 of Superpave Mix Design)

The 7-day maximum pavement temperature of a pavement site is 52°C, with a standard deviation of 4°C, and the single-day minimum pavement temperature of that pavement site is 12°C, with a standard deviation of 5°C. Determine the recommended asphalt binder at 50% reliability and at 98% reliability.

Solution

At 50% reliability:

$$T_{\text{max at 50\%}} = 52°C$$

$$T_{\text{min at 50\%}} = -12°C$$

The recommended binder is PG 52-16 at 50% reliability.
   At 98% reliability:

$$T_{max\,at\,98\%} = T_{max\,at\,50\%} + 2\sigma_{HighTemp} = 52 + 2(4) = 60°C$$

$$T_{min\,at\,98\%} = T_{min\,at\,50\%} - 2\sigma_{LowTemp} = -12 - 2(5) = -22°C$$

The recommended binder is PG 64-22 at 98% reliability.
   Answers:

   The recommended binder is PG 52-16 at 50% reliability.
   The recommended binder is PG 64-22 at 98% reliability.

Example 7.7 Selection of Asphalt Binder (Step 2 of Superpave Mix Design)

The 7-day maximum pavement temperature of a pavement site is 52°C, with a standard deviation of 4°C, and the single-day minimum pavement temperature of that pavement site is 12°C, with a standard deviation of 5°C. If the design traffic is 5 million and the design speed is 50 mph, determine the binder grade to be selected at 98% reliability.

Solution

$$T_{max\,at\,98\%} = T_{max\,at\,50\%} + 2\sigma_{HighTemp} = 52 + 2(4) = 60°C$$

$$T_{min\,at\,98\%} = T_{min\,at\,50\%} - 2\sigma_{LowTemp} = -12 - 2(5) = -22°C$$

The recommended binder at 98% reliability is PG 64-22 without considering the traffic and speed.
   For 5 million traffic and 50 mph speed, the increase in the high-temperature grade, from Table 7.10, is 1. Therefore, the final binder is PG 70-22.
   Answers:
   The final binder is PG 70-22.

Example 7.8 Superpave Mix Design (Step 3 of Superpave Mix Design)

Using the Superpave design aggregate structure, select the best aggregate blend for a 9-million ESAL and a 25-mm nominal aggregate size from the test data shown in Table 7.14.

TABLE 7.14

Test Data for Example 7.8

Data	Blend 1	Blend 2	Blend 3
G_{mb}	2.438	2.397	2.462
G_{mm}	2.612	2.598	2.613
G_b	1.023	1.023	1.023
P_b	6.5%	4.9%	5.6%
P_s	93.5%	95.1%	94.4%
Fines	7.5%	12.3%	3.5%
G_{sb}	2.629	2.632	2.615
h_{ini}	131	125	129
h_{des}	112	115	119

Solution

Conduct a volumetric analysis to select an aggregate blend (Table 7.15). Note that P_b and P_s are to be used in fractions in the equations. For example, $P_b = 4.0\%$ is to be used as 0.04 and so on.

TABLE 7.15

Data Analysis for Example 7.8

Computed	Equation	Blend 1	Blend 2	Blend 3
VTM (%)	$VTM = \left(1 - \dfrac{G_{mb}}{G_{mm}}\right) \times 100$	6.7	7.7	5.8
VMA (%)	$VMA = 100 - \left(\dfrac{G_{mb}P_s}{G_{sb}}\right) \times 100$	13.3	13.4	11.1
VFA (%)	$VFA = \left(\dfrac{VMA - VTM}{VMA}\right) \times 100$	49.9	42.2	48.0
% G_{mm} at N_{ini}	$\%G_{mm,N_{ini}} = (100 - VTM)\left(\dfrac{h_{des}}{h_{ini}}\right)$	79.8	84.9	86.9
G_{se}	$G_{se} = \dfrac{P_s}{\dfrac{1}{G_{mm}} - \dfrac{P_b}{G_b}}$	2.928	2.822	2.878
P_{ba} (%)	$P_{ba} = \left(\dfrac{G_{se} - G_{sb}}{G_{sb}G_{se}}\right)G_b \times 100$	4.0	2.6	3.6
P_{be} (%)	$P_{be} = (P_b - P_{ba}P_s) \times 100$	2.8	2.4	2.2
F/A	$\dfrac{F}{A} = \dfrac{\% \text{ passing No.200}}{\% P_{be}}$	2.7	5.1	1.6

The volumetric calculations must be made at 4% VTM. Since the VTM here is not 4%, the results of the volumetric calculations are to be estimated at 4% as listed in Table 7.16:

None of the blends satisfies all the requirements. Therefore, no blend is satisfactory.

However, Blend 1 is the best among them. It satisfies all except the F/A ratio; it is very close to the recommended range of the F/A ratio.

Answer: No blend is satisfactory.

TABLE 7.16
Data Analysis for Example 7.8

At 4% Air Void	Equation	Blend 1	2	3	Criteria	OK
$P_{b,est}$	$P_{b,est} = P_b - 0.4(4 - VTM)$	7.6	6.4	6.3	–	–
VMA_{est}	$VMA_{est} = VMA + C(4 - VTM)$	12.8	12.6	10.8	Min. 12	1, 2
VFA_{est}	$VFA_{est} = 100\left(\dfrac{VMA_{est} - 4.0}{VMA_{est}}\right)$	68.7	68.4	62.9	65–75	1, 2
% G_{mm} at $N_{ini,est}$	$\%G_{mm,N_{ini},est} = \%G_{mm,N_{ini}} - (4 - VTM)$	82.5	88.6	88.7	Max. 89	All
$P_{be,est}$	$P_{be,est} = P_{b,est} - P_s G_b\left(\dfrac{G_{se} - G_{sb}}{G_{sb}G_{se}}\right)$	3.8	3.9	2.9	–	–
F/A,$_{est}$	$\left(\dfrac{F}{A}\right)_{est} = \dfrac{\% \text{ passing No.}200}{(\% P_{be})_{est}}$	1.9	3.1	1.2	0.6–1.2	3

Example 7.9 Superpave Mix Design (Step 4 of Superpave Mix Design)

Using the Superpave design aggregate structure, select the design binder content for a 20 million ESAL and a 12.5-mm nominal aggregate size from the test data shown in Table 7.17.

TABLE 7.17
Test Data for Example 7.9

Data	Binder Content (%) 4.2	4.9	5.3	5.9
G_{mb}	2.412	2.428	2.435	2.449
G_{mm}	2.614	2.591	2.531	2.63
G_b	1	1	1	1
$P_s(\%)$	95.8	95.1	94.7	94.1
% Fines	3	4	5	5.5
G_{sb}	2.695	2.695	2.695	2.695
h_{ini}	131	128	125	132
h_{des}	117	115	112	119

Solution

Conduct a volumetric analysis to select optimum binder content (Table 7.18). Note that P_b and P_s are to be used in fractions in the equations.

Volumetric Analysis to Charts:

From Figure 7.15a, $P_b = 6.3\%$ at VTM = 4%

From Figure 7.15b, VMA = 14.4% at $P_b = 6.3\%$ [okay, criterion is 14% minimum]

From Figure 7.15c, VFA = 72% at $P_b = 6.3\%$ [okay, criterion is 65–75%]

From Figure 7.15d, F/A Ratio = 1.1 at $P_b = 6.3\%$ [okay, criterion is 0.6–1.2]

From Figure 7.15d, G_{mm} at $N_{ini} = 86\%$ at $P_b = 6.3\%$ [okay, criterion is 89% maximum]

Therefore, design binder content is 6.3%

Answer: The design binder content is 6.3%

TABLE 7.18

Data Analysis for Example 7.9

Computed	Equation	Binder Content (%)			
		4.2	**4.9**	**5.3**	**5.9**
VTM (%)	$VTM = \left(1 - \dfrac{G_{mb}}{G_{mm}}\right) \times 100$	7.7	6.3	3.8	6.9
VMA (%)	$VMA = 100 - \left(\dfrac{G_{mb}P_s}{G_{sb}}\right) \times 100$	14.3	14.3	14.4	14.5
VFA (%)	$VFA = \left(\dfrac{VMA - VTM}{VMA}\right) \times 100$	45.8	56.1	73.7	52.5
% G_{mm} at N_{ini}	$\%G_{mm,N_{ini}} = (100 - VTM)\left(\dfrac{h_{des}}{h_{ini}}\right)$	82.4	84.2	86.2	83.9
G_{se}	$G_{se} = \dfrac{P_s}{\dfrac{1}{G_{mm}} - \dfrac{P_b}{G_b}}$	2.813	2.822	2.768	2.929
P_{ba} (%)	$P_{ba} = \left(\dfrac{G_{se} - G_{sb}}{G_{sb}G_{se}}\right)G_b \times 100$	1.56	1.67	0.98	2.97
P_{be} (%)	$P_{be} = (P_b - P_{ba}P_s) \times 100$	2.71	3.31	4.37	3.11
F/A	$\dfrac{F}{A} = \dfrac{\% \text{ passing No.} 200}{\% P_{be}}$	1.1	1.2	1.1	1.8

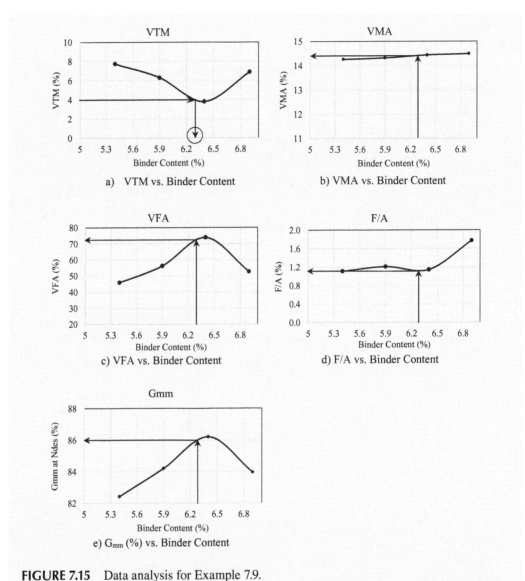

FIGURE 7.15   Data analysis for Example 7.9.

Example 7.10 Superpave Mix Design (Steps 3 and 4 of Superpave Mix Design)

Using the Superpave design aggregate structure, select the blend and the design binder content for a 9 million ESAL and a 19-mm nominal aggregate size from the test data shown in Table 7.19.

Once the design aggregate structure is selected, the following data (Table 7.20) at varying asphalt binder contents are obtained.

TABLE 7.19
Test Data for Example 7.10

Data	Blend		
	1	2	3
G_{mb}	2.457	2.441	2.477
G_{mm}	2.598	2.558	2.664
G_b	1.025	1.025	1.025
$P_b(\%)$	5.9	5.7	6.2
$P_s(\%)$	94.1	94.3	93.8
% Passing No. 200	4.5	4.5	4.5
G_{sb}	2.692	2.688	2.665
h_{ini}	125	131	125
h_{des}	115	118	115

TABLE 7.20
Test Data for Example 7.10

Data	Binder Content (%)			
	5.4	5.9	6.4	6.9
G_{mb}	2.351	2.441	2.455	2.469
G_{mm}	2.570	2.558	2.530	2.510
$G_b(\%)$	1.025	1.025	1.025	1.025
$P_s(\%)$	94.6	94.1	93.6	93.1
% Passing No. 200	4.5	4.5	4.5	4.5
G_{sb}	2.688	2.688	2.688	2.688
H_{ini}	125	131	126	130
H_{des}	115	118	114	112

Solution

Conduct a volumetric analysis to select an aggregate blend (Step 3 of Superpave mix design) (Table 7.21). Note that P_b and P_s are to be used in fractions in the equations.

The volumetric calculations must be made at 4% VTM. As the VTM is not 4% the results of the volumetric calculations are to be estimated at 4% as listed in Table 7.22.

Both 1 and 2 blends satisfy the requirements. Select Blend 2 as it fulfills the requirements better.

Volumetric Analysis to Design Binder Content (Step 4 of Superpave mix design) (Table 7.23):

TABLE 7.21
Data Analysis for Example 7.10

Computed	Equations	Blend 1	Blend 2	Blend 3
VTM (%)	$VTM = \left(1 - \dfrac{G_{mb}}{G_{mm}}\right) \times 100$	5.4	4.6	7.0
VMA (%)	$VMA = 100 - \left(\dfrac{G_{mb}P_s}{G_{sb}}\right) \times 100$	14.1	14.4	12.8
VFA (%)	$VFA = \left(\dfrac{VMA - VTM}{VMA}\right) \times 100$	61.5	68.2	45.2
% G_{mm} at N_{ini}	$\%G_{mm,N_{ini}} = (100 - VTM)\left(\dfrac{h_{des}}{h_{ini}}\right)$	87	86	85.5
G_{se}	$G_{se} = \dfrac{P_s}{\dfrac{1}{G_{mm}} - \dfrac{P_b}{G_b}}$	2.875	2.812	2.979
P_{ba} (%)	$P_{ba} = \left(\dfrac{G_{se} - G_{sb}}{G_{sb}G_{se}}\right)G_b \times 100$	2.42	1.68	4.05
P_{be} (%)	$P_{be} = (P_b - P_{ba}P_s) \times 100$	3.62	4.11	2.4
F/A	$\dfrac{F}{A} = \dfrac{\% \text{ passing No.200}}{\% \, P_{be}}$	1.2	1.1	1.9

TABLE 7.22
Data Analysis for Example 7.10

Estimated at 4% Air Void	Equation	Blend 1	Blend 2	Blend 3	Criteria	OK
$P_{b,est}$	$P_{b,est} = P_b - 0.4(4 - VTM)$	6.5	5.9	7.4	–	–
VMA$_{est}$	$VMA_{est} = VMA + C(4 - VTM)$	13.8	14.3	12.2	Min. 13	1, 2
VFA$_{est}$	$VFA_{est} = 100\left(\dfrac{VMA_{est} - 4.0}{VMA_{est}}\right)$	71.1	72.0	67.2	65–75	All
% G_{mm} at $N_{ini,est}$	$\%G_{mm,N_{ini},est} = \%G_{mm,N_{ini}} - (4 - VTM)$	88.4	86.5	88.6	Max. 89	All
$P_{be,est}$	$P_{be,est} = P_{b,est} - P_sG_b\left(\dfrac{G_{se} - G_{sb}}{G_{sb}G_{se}}\right)$	4.2	4.3	3.6	–	–
F/A$_{,est}$	$\left(\dfrac{F}{A}\right)_{est} = \dfrac{\% \text{ passing No.200}}{(\% \, P_{be})_{est}}$	1.1	1.0	1.2	0.6–1.2	1, 2

TABLE 7.23

Data Analysis for Example 7.10

Computed	Equation	Binder Content (%)			
		5.4	**5.9**	**6.4**	**6.9**
VTM (%)	$VTM = \left(1 - \dfrac{G_{mb}}{G_{mm}}\right) \times 100$	8.5	4.6	3.0	1.6
VMA (%)	$VMA = 100 - \left(\dfrac{G_{mb}P_s}{G_{sb}}\right) \times 100$	17.3	14.5	14.5	14.5
VFA (%)	$VFA = \left(\dfrac{VMA - VTM}{VMA}\right) \times 100$	50.6	68.6	79.6	88.7
% G_{mm} at N_{ini}	$\%G_{mm,N_{ini}} = \left(100 - VTM\right)\left(\dfrac{h_{des}}{h_{ini}}\right)$	84.2	86.0	87.8	84.7
G_{se}	$G_{se} = \dfrac{P_s}{\dfrac{1}{G_{mm}} - \dfrac{P_b}{G_b}}$	2.812	2.823	2.812	2.812
P_{ba} (%)	$P_{ba} = \left(\dfrac{G_{se} - G_{sb}}{G_{sb}G_{se}}\right)G_b \times 100$	1.68	1.82	1.69	1.68
P_{be} (%)	$P_{be} = \left(P_b - P_{ba}P_s\right) \times 100$	3.81	4.19	4.82	5.34
F/A	$\dfrac{F}{A} = \dfrac{\%\ \text{passing No.}200}{\%\ P_{be}}$	1.2	1.1	0.9	0.8

From Figure 7.16a, $P_b = 6.0\%$ at VTM = 4%

From Figure 7.16b, VMA = 14.5% at $P_b = 6.0\%$ [okay, criterion is 13% minimum]

From Figure 7.16c, VFA = 72% at $P_b = 6.0\%$ [okay, criterion is 65–75%]

From Figure 7.16d, G_{mm} at N_{ini} = 86.3% at $P_b = 6.0\%$ [okay, criterion is 89% maximum]

From Figure 7.16e, F/A Ratio = 0.95 at $P_b = 6.0\%$ [okay, criterion is 0.6–1.2]

Therefore, design binder content is 6.0%

Finally, the design aggregate blend is 2 and the design binder content is 6.0%.

Answer: The design aggregate blend is 2 and the design binder content is 6.0%.

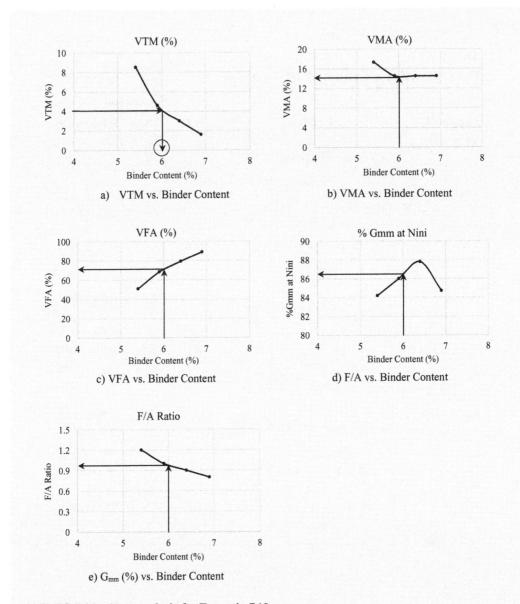

FIGURE 7.16   Data analysis for Example 7.10.

Example 7.11 Moisture-Susceptibility (Step 5 of Superpave mix design)

In an indirect tensile strength test, the peak load required for a specimen of 6 in. diameter and 2-in. thick asphalt concrete to fail is 3,500 lb. The peak load required for a moisture-conditioned asphalt concrete specimen of 6-in.diameter and 1.5-in.thickness to fail is 2,500 lb. Calculate the tensile strength ratio of the asphalt specimen.

Solution

Before conditioning:

$$ITS = \frac{2P}{\pi Dt} = \frac{2(3,500\ \text{lb})}{\pi(6.0\ \text{in.})(2.0\ \text{in.})} = 185.7\ \text{psi}$$

After conditioning:

$$ITS = \frac{2P}{\pi Dt} = \frac{2(2,500\ \text{lb})}{\pi(6.0\ \text{in.})(1.5\ \text{in.})} = 176.8\ \text{psi}$$

$$\text{Tensile strength ratio} = \frac{\text{Tensile strength after conditioning}}{\text{Tensile strength before conditioning}}$$

$$= \frac{176.8\ \text{psi}}{185.7\ \text{psi}} = 0.95 = 95\%$$

Answer: 95%

7.4   CHAPTER SUMMARY

The asphalt industry is newer and growing faster compared to the concrete industry. Many federal agencies are largely involved with asphalt, and its usages, characterizations, and specifications are rapidly improving day by day. This chapter describes the mixed-design procedures which are currently most commonly used. However, with continuing research and practice, this may change in the near future.

The asphalt mix design process is used in determining which aggregate and grade of asphalt binder to use, as well as what the optimum combination of these two ingredients is to obtain the desired performance of the asphalt. Some of the desired performance qualities include deformation resistance, fatigue cracking resistance, thermal cracking resistance, durability, serviceability, easiness of work, etc. In order to achieve all of these, a very careful volumetric analysis is required.

Bulk specific gravity is one of the most commonly used parameters. It is the ratio of mass per unit bulk volume of the compacted specimen to the mass of an equal volume of distilled water at that temperature. Theoretical maximum specific gravity is the mass per unit volume of a loose mix after expelling all the air from the mix to the mass of an equal volume of gas-free distilled water at that temperature. Both the bulk specific gravity and the theoretical maximum specific Gravity are used to calculate the air void content of the compacted specimen. VMA describes the voids among aggregate particles of a mixture, including the effective asphalt. The portion of the voids in the mineral aggregate which contain asphalt binder is known as the VFA. This does not include the asphalt that is absorbed within the aggregates. The volume of asphalt binder in the mix that has been absorbed into the pore structure of the aggregate, and does not take part in binding aggregates, is called the absorbed asphalt binder. The portion of binder that is coated on the aggregate surface, and does take part in binding aggregates, is called the effective asphalt binder. The dust-to-asphalt ratio is the ratio of the amount of dust or fines to the effective asphalt content of the mix. All these parameters have desirable ranges which provide the desired mix properties and performances.

The Hubbard and Hveem mix design methods are now obsolete. The Marshall mix design is still used in some parts of the world, but the newest method, the Superpave mix design, is replacing other methods as time progresses.

In the Marshall mix design, aggregates are selected based on laboratory tests, such as toughness and abrasion, durability and soundness, cleanliness and deleterious materials, and particle shape and surface texture. If the aggregates are found to be good in these tests, then additional tests are run to finalize the aggregate. These tests include the gradation, size, specific gravity, and absorption tests. Once aggregates are selected, the asphalt binder is selected based on local experience. Then, trial cylindrical specimens are prepared (4-in. diameter and 2.5-in. tall). Then, the Marshall stability and flow tests are conducted on the prepared specimens. If the specimens pass the Marshall stability and flow tests, then further volumetric analysis is conducted to determine the optimum asphalt binder content. More clearly, specimens of different asphalt binder contents are prepared and compared. The binder content which meets the desirable ranges of VTM, VMA, VFA, and the Marshall and flow is selected.

The Superpave mix design method is more robust than the Marshall method. When selecting aggregates, two types of properties are tested: source and consensus properties. Source properties include Los Angeles abrasion, soundness, and deleterious materials. Consensus properties include coarse aggregate angularity, fine aggregate angularity, flat and elongated particles, and sand equivalency. The asphalt binder is selected based on the performance grading binder, traffic level, and traffic speed. The binder must be good enough to reach the average 7-day maximum pavement temperature and the minimum single-day pavement temperature. Based on traffic load and speed, the binder grade may be increased if and as necessary. For example, for a traffic load of 3–10 million ESALs with a design speed of over 42 mph, the binder grade is increased by one (6°C). Aggregate blends (gradation of the blended aggregates) are evaluated by preparing specimens and comparing the volumetric properties (VMA, VFA, $\%G_{mm}$, effective binder content, and dust-to-asphalt ratio) of each trial blend, with the specification at 4% VTM. The optimum binder content is determined by using the asphalt content which produces the air void (VTM) of 4.0% and satisfies the volumetric properties (VMA, VFA, $\%G_{mm}$, and dust-to-asphalt ratio). Finally, the mix is evaluated for moisture susceptibility by conducting the indirect tensile strength test on freshly prepared specimens and moisture-conditioned specimens. The decrease in indirect tensile strength after the moisture conditioning should not exceed the threshold.

ORGANIZATIONS DEALING WITH ASPHALT MIX DESIGN

AI, AASHTO, and ASTM International are resources for asphalt mix design.

REFERENCES

Asphalt Institute. 2014. *Asphalt Mix Design Method, Manual Series No. 2 (MS-2)*, 7th Edition. Asphalt Institute, Lexington, KY.
FHWA. 2017. Superpave Fundamentals Reference Manual, NHI Course #131053, Federal Highway Administration (FHWA), Washington, DC.

FUNDAMENTALS OF ENGINEERING (FE) EXAM STYLE QUESTIONS

FE Problem 7.1

The main objective of the asphalt mix design is:

A. To determine an economical mix proportion
B. To determine the densest mix proportion possible
C. To determine the mix proportion for the best performance
D. All of the above

Solution: C

Asphalt mix design is performed to determine the proportion of constituents which produces the best performing mix. Economy is a concern, but not the main objective. The densest mix is not commonly sought out as too dense a mixture causes shear rutting (the material moves laterally upon applying a load).

FE Problem 7.2

Volume of dust is controlled in the asphalt mix design as:

A. Too much dust means too much surface area, meaning more binder is needed
B. Too less dust means too less surface area, leaving excessive film thickness
C. Both A and B
D. Neither A and B

Solution: C

Excessive dust dries out the mix, reducing film thickness and durability. Insufficient dust allows excessive asphalt films, resulting in a tender, unstable mix.

FE Problem 7.3

How is the load carried by the asphalt concrete layer?

A. Aggregate transfers the load to the film binder, and the film binder transfers the load to adjacent aggregate
B. Aggregate interlocks and the asphalt binder aids in interlocking
C. Asphalt layer works as a solid mass/slab
D. None of the above

Solution: B

In asphalt concrete pavement, aggregate interlocks and asphalt binder helps with the interlocking process. Thus, it supports the applied traffic loading. Angular aggregate is important for asphalt concrete. In PCC, cement and aggregate work is done with a solid mass, so aggregate interlocking is not a concern.

PRACTICE PROBLEMS

PROBLEM 7.1

The 7-day maximum pavement temperature of a pavement site is estimated to be 48°C, with a standard deviation of 5°C. The 1-day minimum temperature of that pavement site is −25°C, with a standard deviation of 4°C.

a) What PG asphalt binder is required for the pavement site at 50% reliability?
b) What PG asphalt binder is required for the pavement site at 98% reliability?
c) If the design traffic is 6 million ESALs and the design speed is 10 mph, then what is the binder grade to be selected at 98% reliability?

PROBLEM 7.2

The 7-day maximum pavement temperature of a pavement site is estimated to be 62°C, with a standard deviation of 5°C. The 1-day minimum temperature of that pavement site is −15°C, with a standard deviation of 4°C.

a) What PG asphalt binder is required for the pavement site at 50% reliability?
b) What PG asphalt binder is required for the pavement site at 98% reliability?
c) If the design traffic is 2 million ESALs and the design speed is 50 mph, then what is the binder grade to be selected at 98% reliability?

PROBLEM 7.3

A compacted AC mixture has the following properties:

Bulk specific gravity of the mixture, $G_{mb} = 2.325$
Theoretical maximum specific gravity, $G_{mm} = 2.571$
Asphalt binder specific gravity, $G_b = 1.015$
Asphalt content, $P_b = 5.0\%$ (by mass of the total mix)
Percent passing No. 200 = 7.3

The percent of each aggregate and the bulk specific gravity (G_{sb}) for each test data is shown in Table 7.24.

TABLE 7.24
Test Data for Problem 7.3

Aggregate	% of Total Aggregate	G_{sb}
A	40%	2.685
B	40%	2.611
C	20%	2.421

Based on the information given for this problem:

a) Calculate the bulk specific gravity of the combined aggregate
b) Calculate the effective specific gravity of the aggregate
c) Calculate the percent of absorbed asphalt for the mixture
d) Calculate the percent of effective asphalt for the mixture
e) Calculate the percent of voids in total mix for the mixture
f) Calculate the percent of voids in mineral aggregate for the mixture
g) Calculate the percent of voids filled with asphalt for the mixture
h) Calculate the dust-to-asphalt ratio

PROBLEM 7.4

A compacted AC mixture has the following properties:

Bulk specific gravity of the mixture, $G_{mb} = 2.525$
Theoretical maximum specific gravity, $G_{mm} = 2.671$
Asphalt binder specific gravity, $G_b = 1.023$
Asphalt content, $P_b = 6.0\%$ (by mass of total mix)
Percent passing No. 200 = 6.3

The percent of each aggregate and the bulk specific gravity (G_{sb}) for each test data is shown in Table 7.25.

TABLE 7.25
Test Data for Problem 7.4

Aggregate	% of Total Aggregate	G_{sb}
A	20%	2.585
B	50%	2.411
C	30%	2.721

Based on the information given for this problem:

a) Calculate the bulk specific gravity of the combined aggregate
b) Calculate the effective specific gravity of the aggregate
c) Calculate the percent of absorbed asphalt for the mixture
d) Calculate the percent of effective asphalt for the mixture
e) Calculate the percent of voids in total mix for the mixture
f) Calculate the percent of voids in mineral aggregate for the mixture
g) Calculate the percent of voids filled with asphalt for the mixture
h) Calculate the dust-to-asphalt ratio

Problem 7.5

An asphalt concrete mixture is to be designed using the Marshall Mix Design method. A PG 58-34 binder with a specific gravity of 1.00 is to be used. The mixture contains a 9.5-mm nominal maximum aggregate size with a bulk specific gravity of 2.721. The effective specific gravity of the aggregates used is 2.692. Trial mixes were made, with the average results shown in Table 7.26.

Determine the design asphalt content using the Asphalt Institute design criteria for heavy traffic. Assume a design air void of 4.0%.

TABLE 7.26
Test Data for Problem 7.5

Asphalt Content (%)	Bulk Specific Gravity, G_{mb}	Stability (kN)	Flow (0.25 mm)
4.0	2.320	8.3	9
4.5	2.348	8.7	10
5.0	2.365	9.4	12
5.5	2.398	9.1	15
6.0	2.402	8.7	22

Problem 7.6

An asphalt concrete mixture is to be designed using the Marshall Mix Design method. A PG 58-34 binder with a specific gravity of 1.00 is to be used. The mixture contains a 19-mm nominal maximum aggregate size with a bulk specific gravity of 2.696. The theoretical maximum specific gravity of the mix is 2.47 at 5% asphalt content. Trial mixes were made, with the average results shown in Table 7.27.

TABLE 7.27
Test Data for Problem 7.6

Asphalt Content (%)	Bulk Specific Gravity, G_{mb}	Stability (N)	Flow (0.25 mm)
4.0	2.303	7,076	9
4.5	2.386	8,411	10
5.0	2.412	7,565	12
5.5	2.419	5,963	15
6.0	2.421	4,183	22

Determine the design asphalt content using the Asphalt Institute design criteria for medium traffic. Assume a design air void of 4.0%.

PROBLEM 7.7

Select a blend using the Superpave design aggregate structure for a 5-million ESAL and a 19-mm nominal maximum aggregate size for the test data shown in Table 7.28.

TABLE 7.28
Test Data for Problem 7.7

Data	Blend		
	1	2	3
G_{mb}	2.451	2.341	2.377
G_{mm}	2.528	2.528	2.604
G_b	1.0	1.0	1.0
P_b	5.0	4.7	4.2
P_s	95.0	95.3	95.8
% Passing No. 200	4.5	4.5	4.5
G_{sb}	2.722	2.625	2.605
h_{ini}	126	132	129
h_{des}	116	117	117

PROBLEM 7.8

Determine the design binder content for a 5 million ESAL and a 19-mm nominal maximum aggregate size for the test data shown in Table 7.29.

TABLE 7.29
Test Data for Problem 7.8

Data	Binder Content (%)			
	5.0	5.5	6.0	6.5
G_{mb}	2.451	2.491	2.455	2.469
G_{mm}	2.570	2.558	2.530	2.510
G_b	1.023	1.023	1.023	1.023
P_s	95.0	94.5	94.0	93.5
% Passing No. 200	4.5	4.5	4.5	4.5
G_{sb}	2.688	2.688	2.688	2.688
h_{ini}	125	131	126	130
h_{des}	115	118	114	112

8 Asphalt Mixtures

This chapter discusses the types of hardened asphalt concretes, such as hot-mix, warm-mix, and cold-mix asphalt. Some recycled mixtures are also discussed with their potential usages. The standard laboratory characterization methods commonly used by different private and government agencies are also presented.

8.1 BACKGROUND

The previous chapter discussed the asphalt mix design details. Once the mix design has been completed, the mixture is produced by mixing aggregate, asphalt binder, and any optional modifier(s). The mixing temperature is dependent on the viscosity of the asphalt binder used. After the compaction, the mixture is allowed to cool down for several hours. Once the compacted mix has completely cooled down, it can work as a solid, stiff, and hard material to support loading. Some asphalt concrete samples are shown in Figure 8.1.

8.2 TYPES OF ASPHALT MIXTURES

Asphalt mixtures can be broadly categorized into three types:

a) Hot-mix asphalt (HMA)
b) Warm-mix asphalt (WMA)
c) Cold-mix asphalt (CMA)

8.2.1 HOT-MIX ASPHALT (HMA)

Typically, asphalt mixtures are commonly known as the *hot-mix asphalt* (HMA), although this is one kind of AC mixture (not all AC mixtures are HMA). HMA is an AC mixture which is produced by heating and mixing the aggregates and asphalt binder at a certain level of temperature, commonly used as 285–325°F. The level of temperature is determined based on the viscosity of the asphalt binder as discussed in the previous chapter. There are several types of HMAs:

Dense-Graded Mixture. The term 'dense' depicts that a dense-graded mix is a well-graded HMA mix. This is the most common mix used in regular paving projects. A dense-graded mix (Figure 8.2) has a very low permeability when properly designed and compacted. It is suitable for all types of asphalt pavements and for all traffic conditions. It also works well for structural, friction, leveling, and patching needs.

Open-Graded Friction Course (OGFC). Open-graded mixes (Figures 8.2 and 8.3) use only crushed stone (or gravel) and a small percentage of manufactured sands. Typically, a minimum of 15% of air voids is specified, with no maximum. This mix type is used for surface courses only, as it allows water to drain very quickly and thus, reduces the tire splash during the rainfall. It also results in smoother surfaces than the dense-graded HMA, and reduces the thermal cracking in asphalt pavement (Islam et al. 2018b).

a) Laboratory Compacted   b) Field Cored

FIGURE 8.1   Asphalt concrete samples. *Photos taken at the University of New Mexico.*

Open Graded Mix   Dense Graded Mix

FIGURE 8.2   Comparison of open-graded and dense-graded AC. *Photo taken in Albuquerque, New Mexico.*

FIGURE 8.3   OGFC specimens. *Photo taken in Albuquerque, New Mexico.*

Asphalt Treated Permeable Bases (ATPB). This is essentially an aggregate base layer with some asphalt binder and a high permeability. It is used as a drainage layer only, under the dense-graded HMA or PCC for drainage.

Sand Asphalt Mix. Sand asphalt mix is a dense-graded mix with a nominal maximum aggregate size of less than # inches (9.5 mm).

Stone Matrix Asphalt (SMA). *Stone matrix asphalt* (SMA) uses mostly coarse and granular stone to create gap-graded and stone-on-stone contacts within a mixture (Figure 8.4). Sometimes, this mix is called the stone mastic asphalt. This HMA was originally developed to maximize rutting resistance and durability. Since aggregates do not deform as much as asphalt binder under loading, this stone-on-stone contact greatly reduces the rutting. SMA is generally more expensive than a typical dense-graded HMA because it requires more durable aggregates, higher asphalt content, modified asphalt binder, and fibers.

FIGURE 8.4   Stone mix asphalt.

8.2.2   WARM MIX ASPHALT (WMA)

Without heating aggregate and binder like the HMA, asphalt mixtures can be produced at lower temperatures with proper mixing of water, water-based additives, water-bearing mineral additives, chemical additives, waxes, and organic additives, or a combination of such technologies. All of these technologies reduce the viscosity (the thickness) of the asphalt binder so that asphalt aggregates can be coated at lower temperatures. Such a produced mixture is known as the *warm-mix asphalt* (WMA). This technology allows the production and compaction of asphalt pavement material using a lower temperature (200–275°F). Producing WMA offers many benefits, such as low fuel consumption, minimized production of greenhouse gases, superior compaction on the road, the ability to haul paving mixes for longer distances, and extending the paving season.

8.2.3   COLD MIX ASPHALT (CMA)

Cold mix asphalt (CMA) concrete is produced by emulsifying the asphalt in water with soap or cement prior to mixing with the aggregate. While in its emulsified state, the asphalt is less

viscous, and the mixture is easy to work and compact. In current practice, the old, deteriorated pavement is milled off, crushed to desirable size inside a large truck, mixed with the
emulsion, and can be paved in place within an hour. Sometimes, the milled material is carried to the plant for better gradation and mixing of the emulsion. Cold mix asphalt is being
used even in highway pavements nowadays. Much research shows that proper usages of cold
mix asphalt are very competitive with those of the traditional HMA (Islam et al. 2018a). A
batch of field-cored, cold-mix specimens and some loose mixes are shown in Figure 8.5.

(a) Cold Mix Asphalt Specimens

(b) Cold Mix Loose Asphalt held by the Author

FIGURE 8.5   Cold mix asphalt. *Photos taken in Pueblo, Colorado.*

8.3   RECYCLED ASPHALT MATERIALS

8.3.1   RECYCLED ASPHALT PAVEMENT (RAP)

Recycled (or reclaimed) asphalt pavement (RAP) is the milling of asphalt surface layer
from an old pavement containing aggregates and asphalt binder. The RAP is obtained by
milling the old pavement (Figure 8.6), which is screened afterward and mixed with the
new aggregates. Nowadays, RAP is being used up to 40%, by weight, of all mixtures in the

FIGURE 8.6   Collection of RAP from an interstate highway. *Photo taken in Albuquerque, New Mexico.*

United States (Islam et al. 2014). As RAP has some amount of aged-stiff binder, the binder grade of the combined binder stiffens. Therefore, a revised mix design is sought out if a mix has a considerable amount of RAP.

The amount of RAP used in asphalt mixtures was roughly 66.7 million tons in 2011, which saved approximately 3.6 million tons of virgin asphalt binder (about 12% of the total binder used in 2011). Nearly 87 million tons of RAP were milled from existing pavements in 2011, of which 81 million tons were recycled. About 74 million tons (92%) of the recycled RAP were used in new asphalt pavement construction (Hansen and Copeland 2013).

8.3.2 RECYCLED ASPHALT SHINGLE (RAS)

Recycled (or reclaimed) asphalt shingles (RAS) are collected from roof tear-offs, processed, and re-used for pavement. Shingles can contain between 20% and 36% of asphalt, by weight. This asphalt can be used to bind aggregates like the conventional asphalt, or other similar materials. Shingle waste can be collected from two sources. The first source is the manufacturer waste shingles, directly from manufacturers of asphalt shingles. This waste is highly appreciated, as the composition of the material is fairly well known. The other type of waste shingle is known as tear-off from re-roofing or roof removal projects. Literature shows that RAS is beneficial for pavement in terms of increased rutting resistance, reduced cracking, and less compaction effort required (Roque et al. 2018). RAS is also economical and environmentally friendlier, as it saves virgin aggregate, virgin asphalt binder, and reduces landfill demands.

8.3.3 RUBBERIZED ASPHALT CONCRETE (RAC)

Rubberized asphalt concrete (RAC), also known as asphalt rubber or just rubberized asphalt, is a noise-reducing pavement material that consists of regular asphalt concrete mixed with crumb rubber from the recycled tires. About 2.4 million tires are recycled as asphalt rubber annually, and this recycling trend is growing (Dower et al. 1985). RAC is obtained by blending the ground-up recycled tires with asphalt to produce a binder, which is then mixed with the conventional aggregates. RAC is a cost-effective, durable, safe, quiet, and environmentally friendly alternative to the traditional road paving materials. The performance of RAC is very similar to the conventional asphalt materials, but it requires production machinery.

8.3.4 RECYCLED ASPHALT PAVEMENT (RAP) IN BASE AND SUBGRADE

The use of RAP in aggregate base or subbase of pavement has also become popular recently (Tarefder and Islam 2015, Islam et al. 2014, Hasan et al. 2018). There are different ways RAP can be used in base and subbase layers. One approach is the plant processing, where milled materials are transported to a central plant, crushed, and screened. The better-quality RAP is used with the new asphalt mix production. The inferior RAP is mixed with the virgin base or subgrade materials. This improves the base and subgrade strength, and reduces waste by preventing the RAP from being dumped. An example of a RAP-mixed base course used in an interstate highway in New Mexico is shown in Figure 8.7. About 50% of RAP is mixed with virgin aggregates to produce this base layer. In this interstate highway, RAP is also mixed with the subgrade to produce an improved subgrade.

FIGURE 8.7 RAP-mixed base course in an interstate pavement. *Photo taken in Albuquerque, New Mexico.*

8.4 CHARACTERIZATIONS OF ASPHALT MIXTURES

Characterization of asphalt mixtures is required when designing pavement. Some properties essential for pavement design are discussed here.

8.4.1 DYNAMIC MODULUS

The *dynamic modulus* ($|E^*|$) is the primary material property required for pavement design. It dictates the potentiality of deformation upon loading. The dynamic modulus test is conducted by applying sinusoidal loads at different, uniaxial frequencies on a cylindrical asphalt concrete specimen. The dynamic modulus is defined mathematically as the ratio of peak dynamic stress (σ_o) to the peak recoverable axial strain (ε_o), which is presented by Eq. 8.1:

$$\left| E^* \right| = \frac{\text{Peak Stress}}{\text{Peak Strain}} = \frac{\sigma_0}{\varepsilon_0} \tag{8.1}$$

where,

$|E^*|$ = Dynamic modulus

σ_o = Peak dynamic stress, applied by the test equipment

ε_o = Peak recoverable axial strain, measured upon loading

Thus, the dynamic modulus is defined mathematically as the ratio of σ_o and ε_o. The dynamic modulus is similar to Young's modulus, with the difference that the load applied is dynamic (sinusoidal), and the resulting strain is also dynamic. In testing Young's modulus, a monotonic increasing load is applied (say, 10 lb, 20 lb, 30 lb…etc.) to the specimen.

To prepare the test specimen, cylindrical specimens of 6-in. (150-mm) diameter and about 7-in. (170-mm) height are compacted. The compacted specimens are then cored and sawed to the diameter of 4-in. (100 mm) and height of 6-in. (150-mm), as shown in Figure 8.8.

a) Cored Specimen   b) Edge Cut Specimen

FIGURE 8.8   Specimen preparation for dynamic modulus testing. *Photos taken at the University of New Mexico.*

The dynamic modulus test is conducted according to the AASHTO T 342 test protocol. The test setup is shown in Figure 8.9. The *linear variable displacement transducer* (LVDT) attached to the specimen measures the axial deformation of the specimen upon loading. Strain is calculated by dividing this deformation by the gage length (the length of the LVDT). Stress is calculated by dividing the applied load by the cross-sectional area of the specimen.

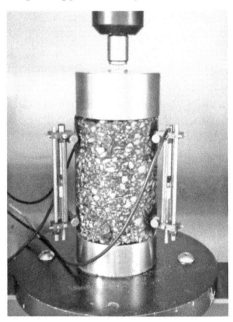

FIGURE 8.9   A dynamic modulus test specimen. *Photo taken at the University of New Mexico.*

Example 8.1 Dynamic Modulus

In a dynamic modulus test of an asphalt concrete specimen, a sinusoidal stress of 360 sin (ωt) is applied and a sinusoidal strain of 0.001 sin ($\omega t + 80$) is obtained, where the stress is in psi, ω is the angular frequency, and t is the time in seconds. Calculate the dynamic modulus of the asphalt specimen.

Solution

Given,

Peak stress, $\sigma_o = 360$ psi

Peak stain, $\varepsilon_o = 0.001$

From Eq. 8.1: Dynamic modulus,

$$|E^*| = \frac{\text{Peak Stress}}{\text{Peak Strain}} = \frac{\sigma_0}{\varepsilon_0} = \frac{360 \text{ psi}}{0.001} = 360,000 \text{ psi}$$

Answer: 360 ksi

8.4.2 INDIRECT TENSILE STRENGTH

Indirect tensile strength (ITS) is required to determine the transverse cracking (also called thermal cracking) resistance of asphalt pavement. Due to decreases in temperature, asphalt concrete contracts, causing tensile stress to develop in the asphalt mixture. Once the developed tensile stress exceeds the tensile capacity of asphalt mixture, thermal cracks develop. It is impossible to test an asphalt specimen in direct tension, like a steel rod. Thus, the tensile test is conducted indirectly with ease, known as the indirect tensile strength test. ITS is a measure of resistance capacity to low-temperature cracking of asphalt pavement (Islam et al. 2015b). In this test, a thin cylindrical specimen is compressed diametrically so that indirect tension is developed along its diameter.

To prepare a specimen, cylindrical specimens are cut into thin circular specimens having 4-in. (100-mm) diameter and about 2-in. (50-mm) thickness. A uniform compressive load of 2.0 in. (50 mm) per minute is applied diametrically until failure occurs, as shown in Figure 8.10. The peak load is recorded, and the ITS is calculated as (Eq. 8.2):

$$ITS = \frac{2P}{\pi Dt} \tag{8.2}$$

where,

P = Peak force required to crack the specimen diagonally

D = Diameter of the specimen

t = Thickness of the specimen

Prepared Specimens   Test Setup   Few Failed Specimens

FIGURE 8.10   Indirect tensile strength testing. *Photos taken at the University of New Mexico.*

This test is similar to the PCC splitting test, with the exception of the specimen size. A larger specimen is used when testing PCC.

Example 8.2 Indirect Tensile Strength Test

In an ITS test, the peak load required to fail a specimen of 5.5-in. diameter and 2.2-in. thickness is 2,900 lb. Calculate the ITS of the specimen.

Solution

Given,

   Peak load, $P = 2,900$ lb
   Specimen diameter, $D = 5.5$ in.
   Specimen thickness, $t = 2.2$ in.
   From Eq. 8.2: Indirect tensile strength,

$$ITS = \frac{2P}{\pi Dt} = \frac{2(2,900\text{ lb})}{\pi(5.5\text{ in.})(2.2\text{ in.})} = 152.6 \text{ psi}$$

Answer: 153 psi

8.4.3 FATIGUE ENDURANCE LIMIT (FEL)

The *fatigue endurance limit* (FEL) is an input parameter in asphalt pavement design, although it is very often neglected because of its highly expensive and time-consuming test procedure. Fatigue damage occurs in asphalt pavement due to the developed repeated tensile strain at the bottom of AC under the applied, repeated traffic loading (Islam 2015). If the developed strain at the bottom of AC under the applied, repeated traffic loading is below the threshold value (called, FEL), no fatigue damage occurs. More specifically, if the developed strain at the bottom of AC under the applied repeated traffic loading is very small, the damage produced at each cycle of loading is also very small. This minimal damage can be 'healed,' as asphalt concrete has some self-restoring capability, and some rest period exists between two consecutive traffic loadings.

The beam fatigue test is conducted following the AASHTO T 321 test standard. Beam slabs of 18 in.×6 in.×3 in. (450 mm×150 mm×75 mm) are prepared as shown in Figure 8.11.

a) Compaction of Mixture   b) Cutting the Slabs   c) Prepared Specimens

FIGURE 8.11 Beam specimen preparation. *Photos taken at the University of New Mexico.*

Once cooled down, each slab is cut into two beams of 15 in.×2.5 in.×2 in. (380 mm×63 mm×50 mm) using a laboratory saw.

The prepared beam specimen is clamped similarly to two roller supports at the ends in the loading frame, as shown in Figure 8.12. The middle two clamps apply the load to bend the beam. Strain-controlled loading is applied using a sinusoidal waveform at a frequency of 5 or 10 Hz at a fixed temperature of 20°C, and the resulting stress waveforms are measured. This test is conducted at different applied-strain levels and different fatigue lives are obtained.

Then, a curve (shown in Figure 8.13) is drawn considering the applied strain (ε) and the fatigue life (N). From the regression model, the strain value (ε) is calculated for which the fatigue life (N) is 50 million. This strain value is referred to as the FEL. The reason for choosing 50 million is that it is considered that, if a specimen can withstand 50 million cycles of load in the laboratory, it is able to withstand unlimited cycles of loading in real pavement.

FIGURE 8.12   Beam fatigue testing by the author. *Photos taken at the University of New Mexico.*

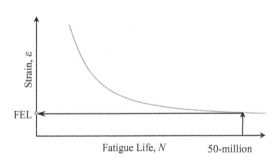

FIGURE 8.13   Determining the FEL from the ε–N curve.

Example 8.3 Fatigue Endurance Limit

In a fatigue test of asphalt specimen, the data listed in Table 8.1 are obtained:

Determine the FEL if the perpetual pavement can withstand 50 million cycles of load.

TABLE 8.1
Different Fatigue Lives due to
Different Strain Levels

Applied Strain (Micro-Strain)	Fatigue Life
600	10,000
400	30,000
300	600,000
200	4,000,000
100	21,000,000
50	100,000,000

Solution

From the applied-strain fatigue life curve shown in Figure 8.14, the FEL is about 70 micro-strain (corresponding to the 50-million fatigue life).

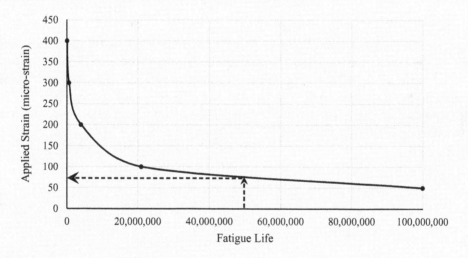

FIGURE 8.14 Determining the FEL for Example 8.3.

Answer: 70 $\mu\varepsilon$, or 70 $\times$ 10^{-6}

Note: Graph reading most often produces erroneous results. Thus, the answer may vary from person to person. In practice, a best-fit curve and its regression equation are generated. The regression equation can provide more accurate output.

8.4.4 Creep Compliance

Creep compliance is defined as the time-dependent strain divided by the applied stress. This test is conducted following the AASHTO T 322 test protocol on a compacted cylindrical

FIGURE 8.15   Creep compliance test setup. *Photo taken at the University of New Mexico.*

AC specimen with a diameter of 6 in. (150 mm) and a thickness of 1.5–2 in. (38–50 mm), shown in Figure 8.15. A static load is applied along a diametric axis of the temperature-controlled specimen for 100 seconds. During the loading period, the vertical and horizontal deformations are measured on the two parallel faces of the specimen using extensometers. Using these displacements (or strains) and the applied stress, the creep compliance can be calculated using Eq. 8.3:

$$D(t) = \frac{\varepsilon(t)}{\sigma_o} \tag{8.3}$$

where,

$D(t)$ = Creep compliance, 1/psi (or 1/pa)
σ_o = Applied stress, psi (or Pa)
$\varepsilon(t)$ = Time-dependent strain
t = Time of loading

8.4.5   POISSON'S RATIO

Poisson's ratio (μ) is the ratio of the proportional decrease in a lateral measurement to the proportional increase in length in a specimen of material that is elastically stretched. It is an important property required to determine the lateral deformation of a material upon a load. Often, a flat value of 0.35 is assumed, although the μ of AC varies with the loading frequency and temperature (Islam et al. 2015a). The μ of new asphalt concrete can also be determined by laboratory testing (no standard test method is currently available), or by using Eq. 8.4 (AASHTO 2015).

$$\mu = 0.15 + \frac{0.35}{1 + e^{a+b(E^*)}} \tag{8.4}$$

where,

Regression constant, $a = -1.63$

Regression constant, $b = 3.84 \times 10^{-6}$

$E^* = $ Dynamic modulus, psi

Example 8.4 Poisson's Ratio of AC

The dynamic modulus of an AC is 500 ksi. Calculate Poisson's ratio of the AC.

Solution

Known:

$a = -1.63$

$b = 3.84 \times 10^{-6}$

$E^* = 500{,}000$ psi

From Eq. 8.4: Poisson's ratio,

$$\mu = 0.15 + \frac{0.35}{1 + e^{a+b(E^*)}}$$

$$= 0.15 + \frac{0.35}{1 + e^{-1.63+3.84\times10^{-6}(500{,}000)}} = 0.299$$

Answer: 0.30

8.5   ASPHALT MODIFIERS

Asphalt cement modification has been practiced for over 50 years to attain special properties of AC, but has received added attention in the past decade or so. There are numerous binder additives available on the market today, as listed in Table 8.2. The benefits of modified asphalt cement can only be realized by a judicious selection of the modifier(s); not all modifiers are appropriate for all applications. In general, asphalt cement should be modified to achieve the following types of improvements:

- Lower stiffness (or viscosity) at the high temperatures associated with construction. This facilitates pumping of the liquid asphalt binder, as well as mixing and compaction of AC.
- Higher stiffness at high service temperatures. This reduces rutting and shoving.
- Lower stiffness and faster relaxation properties at low service temperatures. This reduces thermal cracking.
- Increased adhesion between the asphalt binder and the aggregate in the presence of moisture. This reduces the likelihood of stripping.

TABLE 8.2

Asphalt Cement and HMA Modifiers

Type	General Purpose	Examples
Filler	• Fill voids and reduce optimum asphalt content • Meet aggregate gradation specifications • Increase stability • Improve the asphalt cement-aggregate bond	• Mineral filler • Crushed fines • Lime • Portland cement • Fly ash • Carbon black
Extender	Substituted for a portion of asphalt cement (20–35%) to decrease the amount of asphalt cement	• Sulfur • Lignin
Rubber	• Increase HMA stiffness at high temperatures • Increase HMA elasticity at medium temperatures to resist fatigue cracking • Decrease HMA stiffness at low temperatures to resist thermal cracking	• Natural latex • Synthetic latex (e.g., polychloroprene latex) • Block copolymer (e.g., styrene-butadiene-styrene (SBS)) • Reclaimed rubber (e.g., crumb rubber from old tires)
Plastic		• Polyethylene/polypropylene • Ethylene acrylate copolymer • Ethyl-vinyl-acetate (EVA) • Polyvinyl chloride (PVC) • Ethylene propylene or polyolefin
Rubber-Plastic Combinations		Blends of rubber and plastic
Fiber	• Improve tensile strength of HMA mixtures • Improve cohesion of HMA mixtures • Permit higher asphalt content without significant increase in drain-down	Natural: Asbestos and rock wool Manufactured: • Polypropylene • Polyester • Fiberglass • Mineral • Cellulose
Oxidant	Increase HMA stiffness after the HMA is placed	Manganese salts
Antioxidant	Increase the durability by retarding the oxidation	• Lead compounds • Carbon • Calcium salts
Hydrocarbon	• Restore aged asphalt cements to current specifications • Increase HMA stiffness	• Recycling and rejuvenating oils • Hard and natural asphalts
Antistripping Agents	Minimize stripping of asphalt cement from aggregates	• Amines • Lime
Waste Materials	Replace aggregate or asphalt volume with a cheaper waste product	• Roofing shingles • Recycled tires • Glass

8.6   CHAPTER SUMMARY

Asphalt mixtures are broadly characterized into three types: HMA, WMA, and CMA. HMA is produced by heating the aggregates and binder at elevated temperature. HMA can be a dense-graded mixture, open-graded friction course, sand asphalt mix, and stone matrix asphalt. Another kind of HMA is ATPBs. WMA is similar to HMA but is heated at a temperature lower than that of HMA. The temperature can be lowered with proper mixing of water, water-based additives, water-bearing mineral additives, chemical additives, waxes, and organic additives, or a combination of technologies. CMA is produced at normal temperature by emulsifying the asphalt in water with soap or cement prior to mixing with the aggregate.

Recycled asphalt materials such as RAP, RAS, and RAC are used widely nowadays due to the increased awareness of the environment. RAP is the milled material of old pavement, and can be reused in new asphalt mixtures or in unbound aggregate base layers. RAS is collected from roof tear-offs, processed, and reused in new pavement. RAC is the mixture of regular asphalt concrete and crumb rubber from recycled tires.

The dynamic modulus is determined by applying uniaxial, cyclic load on a cylindrical specimen. The ratio of applied peak-stress and the peak-strain is called the dynamic modulus. The indirect tensile strength is determined by compressing a thin cylindrical specimen diametrically so that indirect tension is developed along its diameter. FEL is the applied strain level below which no fatigue damage occurs, or withstands at least 50 million cycles of loading. Creep compliance is defined as the time-dependent strain divided by the applied constant stress. There is still no standard test method to determine Poisson's ratio of asphalt mixes. Either an empirical equation or the default value of 0.35 is most commonly used for pavement design.

Several modifiers are used in asphalt concrete to produce desirable properties of asphalt concrete. For example, fillers such as crushed fines, lime, cement, and fly ash can be used to increase stability, improve the asphalt cement-aggregate bond, and reduce optimum binder content. Similarly, some other modifiers, such as extender, rubber, plastic, fiber, oxidant, antioxidant, hydrocarbon, etc., are used very often.

ORGANIZATIONS DEALING WITH ASPHALT CONCRETE

The National Asphalt Pavement Association (NAPA) is the trade association that exclusively represents the interests of the asphalt pavement material producer and paving contractor. NAPA provides technical, educational, and marketing materials and information to its members, and supplies technical information to users and specifiers of paving materials.

Location: Lanham, MD
Website: http://www.asphaltpavement.org

AI, AASHTO, TRB, and ASTM International are also resources for asphalt mixtures.

REFERENCES

AASHTO. 2015. *Mechanistic-Empirical Pavement Design Guide: A Manual of Practice*, 2nd edition. American Association of State Highway and Transportation Officials, Washington, DC.

Dower, R. C., Rand S. D. and Scodari, P. F. 1985. The scrap tire problem: a preliminary economic study, Grant No. CR-811897-01. US Environmental Protection Agency, Washington, D.C.

Hansen, K. and A. Copeland. 2013. *2nd Annual Asphalt Pavement Industry Survey on Reclaimed Asphalt Pavement, Reclaimed Asphalt Shingles, and Warm-Mix Asphalt Usage*: 2009–2011. Information Series 138. National Asphalt Pavement Association, Lanham, MD.

Hasan, M. M., Islam, M. R. and Tarefder, R. A. 2018. "Characterization of subgrade soil mixed with recycled asphalt pavement (RAP)." *Journal of Traffic and Transportation Engineering*, 5(3): 207–214.

Islam, M. R. 2015. *Thermal Fatigue Damage of Asphalt Pavements*, Doctoral Dissertation, Department of Civil Engineering, University of New Mexico, Albuquerque, NM.

Islam, M. R., Faisal, H. and Tarefder, R. A. 2015a. "Determining temperature and time dependent Poisson's ratio of asphalt concrete using indirect tension test." *Journal of Fuel*, 146: 119–124.

Islam, M. R., Hossain, M. I. and Tarefder, R. A. 2015b. "A study of asphalt aging using Indirect Tensile Strength test." *Construction and Building Materials*, 95: 218–223.

Islam, M. R., Kalevela, S. A. and Rivera, J. 2018a. *Dynamic Modulus of Cold-in-Place Recycling Asphalt Materials*. Final Report, 2018–10, Colorado Department of Transportation (CDOT), Denver, CO, pp. 1–158.

Islam, M. R., Mannan, U. A., Rahman, A. and Tarefder, R. A. 2014. "Effects of reclaimed asphalt pavement on hot-mix asphalt." *ASTM Journal of Advances in Civil Engineering Materials*, 3(1): 291–307.

Islam, M. R., Rahman, A. and Tarefder, R. A. 2018b. "Open graded friction course in resisting transverse cracking in asphalt pavement." *ASCE Journal of Cold Regions Engineering*, 32(2), 04018006-1-7.

Roque, R., Yan, Y. and Lopp, G. 2018. *Impact of Recycled Asphalt Shingles (RAS) on Asphalt Binder Performance*. Final Report BDV31-977-36, Florida Department of Transportation, Tallahassee, FL.

Tarefder, R. A. and Islam, M. R. 2015. *Study and Evaluation of Materials Response in Hot Mix Asphalt Based on Field Instrumentation*. Final Report, Project ID. NM11MSC-03, Research Bureau, New Mexico Department of Transportation (NMDOT), pp. 1–195.

FUNDAMENTALS OF ENGINEERING (FE) EXAM STYLE QUESTIONS

FE PROBLEM 8.1

Asphalt is a:

A. Pure elastic material
B. Pure viscous material
C. Viscoelastic material
D. Visco-plastic material

Solution: C

The behavior of asphalt has both elastic and viscous components. Thus, it is considered a viscoelastic material.

FE PROBLEM 8.2

Sand asphalt mix is a mixture of:

A. Sand and asphalt
B. Stone and asphalt
C. Coarse aggregate, asphalt, and some sand
D. Fines and asphalt

Solution: A

Sand asphalt mix is a dense-graded mix with the NMAS less than 9.5 mm.

FE PROBLEM 8.3

Creep compliance is defined as the:

- A. Time-dependent strain over the applied stress
- B. Strain over the time-dependent applied stress
- C. Strain divided by the applied stress
- D. Time-dependent strain over the time-dependent applied stress

Solution: A

Creep compliance is defined as the time-dependent strain divided by the applied, constant stress. In this test, a constant stress is applied, and the resulting time-dependent strain is measured with time.

PRACTICE PROBLEMS

PROBLEM 8.1

In an ITS test, the peak load required to fail a specimen having 6-in. diameter and 2-in. thickness is 3,500 lb. Calculate the tensile strength of the material.

PROBLEM 8.2

In a fatigue test on asphalt specimens, the data shown in Figure 8.16 are obtained:
   Determine the FEL of the asphalt specimens, if the perpetual pavement withstands 50 million cycles of load.

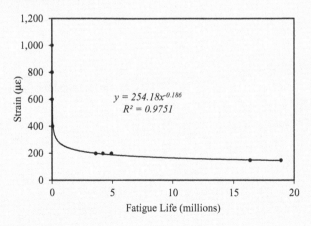

FIGURE 8.16   Fatigue test result data.

PROBLEM 8.3

If the dynamic modulus of an AC specimen is 13,000 MPa, calculate Poisson's ratio of it.

PROBLEM 8.4

Poisson's ratio of an asphalt specimen is calculated to be 0.39. Calculate how much a 4-in. diameter and 6-in. tall cylindrical asphalt specimen deforms upon applying a load of 800 lb.

9 Steels

Steel is one of the most preferred and commonly used materials in civil and construction engineering. This chapter discusses the compositions and types of steel products, as well as their standard shapes, grades, and associated reinforcing bars and fasteners. The standard laboratory characterization methods commonly used, as well as the 'weldability' calculations, are presented.

9.1 BACKGROUND

Steel is a mixture of several metals, making it an *alloy*. However, the majority of steel's composition is made up of iron. Still, steel itself is harder and stronger than iron. Steels are often iron alloys with between 0.02% and 1.7% of carbon, by weight. Steel is different from wrought iron, which has little (0.02–0.08%) or no carbon, and cast iron, which has a high (2–4%) amount of carbon. Changing the amount of carbon (or other atoms) added to steel changes the properties of steel. Some properties of steel are:

- Hardness – the surface does not deform easily with loading
- Ductility – shows significant deformation before failing
- Strength – can withstand high amounts of load
- Steel with more carbon is harder and stronger than pure iron, but it also breaks more easily (brittle)

There are enormous numbers of structures constructed using steel; one is shown in Figure 9.1. Many items manufactured from iron in the past are now being manufactured of steel. Structural steel is widely used in the United States for the construction of different types of building structures, from low-rise buildings to high-rise buildings, bridges, reinforcing bars, floors, arches, railways, trains, cables, machines, vehicles, and so on.

Steel offers the most advantages when a high strength-to-weight ratio is required. Some advantages of structural steel as a civil engineering material include the following:

- Very high strength-to-weight ratio
- Uniform and homogeneous (the properties of steel can be measured very easily)
- Similar tension and compression properties
- High ductility, providing adequate warning of any impending collapse
- Ready made and can be installed within hours, whereas concrete may take weeks or months
- Easily constructed (mainly machine operated, not dependent on weather or climate)
- Easily connectable
- Installation of steel member does not require curing, and strength is readily available
- Can be dismantled and reassembled easily
- Fully recyclable with a high scrap value
- Termite- and rot-proof (unlike wood structures)

FIGURE 9.1   A small steel structure for a school building. *Photo taken in Pueblo, Colorado.*

Some disadvantages of steel as a civil engineering material include:

- Softens due to rises in temperature (needs to be protected from fire). This disadvantage is mainly present in industrial building
- Brittle failure can occur at low temperature (recall the failure of the *Titanic* on April 14, 1912)
- Corrodes with the presence of air and moisture (coating with zinc-rich paint may be required). This disadvantage can be overcome by using a corrosion-resistant steel
- Due to its high strength, sizes required are often very thin (prone to buckling). Special design provision is available to overcome this issue

Approximately 8 million tons of reinforcing steel (rebar) is manufactured per year in the United States, using the scrap steel in efficient manufacturing operations. It is estimated that the industry impacts over 75,000 people in steel transportation and placement (CRSI 2019).

9.2   PRODUCTION OF STEEL

9.2.1   Basic Oxygen Furnace (BOF)

The *basic oxygen furnace* (BOF) process is the current primary method of producing steel from raw iron ores. Iron ores extracted from the naturally occurring ores consist of a common feature of being rich in iron oxides, often in the form of magnetite (Fe_3O_4) or hematite (Fe_2O_3). However, a wide range of complex chemical compounds may present in the iron ores. A mixture of raw materials, such as the iron ore, carbon in the form of coke, and limestone, is fed into the top of a furnace. Hot air of 1,650–2,370°F (900–1,300°C) is blasted through this mixture from the bottom of the furnace. The materials take 6–8 hours to descend to the bottom of the furnace, during which time they are transformed into molten iron and molten slag. Both the molten iron and the molten slag are then separated at intervals and allowed to cool down. The cooled iron forms into ingots of pig iron. The molten slag is separated as by-products, which are used as aggregates. This molten iron has a high carbon content (4–5%) and some impurities, such as silica. The molten iron is further processed to convert it into usable steel. This molten iron is fed into a furnace converter, which is a large vessel with an opening at the top. The vessel can be rotated to either receive the materials or discharge the processed products. It is first loaded with some scrap steels, which act as coolant to control the high temperatures produced by the subsequent exothermic reactions. Molten materials from the blast furnace (about 3–4 times the amount of the scrap steels) are then poured into this furnace converter. Then, the blasting oxygen is blown through a lance that is lowered into the molten materials. No further heating is required because the reaction of the oxygen with the impurities of carbon, silicon, manganese, and phosphorus is exothermic (Domone and Illston 2010). Carbon monoxide is given off, and the other acidic oxides are separated from the metal by adding calcium oxide to the furnace, thus producing a slag. The slag becomes a solid waste and can be used as aggregate. The other molten part is shaped by being poured into a mold and cooled down. The product is further finished and shaped as necessary for desired engineering use.

9.2.2 Electric Arc Process

Nowadays, about 30% of steel is manufactured using the basic oxygen furnace (BOF) process (Aghayere and Vigil 2014). The reason for this is that steel is a fully recyclable material. Most of the steel used in civil engineering in the United States is produced from some sort of recycled steel, and about 95% of the steel used in structural shapes is produced from recycled steel. When manufacturing new structural steel from the recycled steel, steel scraps are fed into an electric arc furnace (Figure 9.2) of about 3,000°F (1,650°C) (Mckee and Hursely 2007). The scraps are melted down to a liquid state, and can then be shaped as desired. The resulting byproduct (slag) floats to the surface. The floats are separated, cut into pieces, and used as aggregate in road construction. The carbon content is continuously monitored during this heating process. Other chemicals, such as copper, vanadium, nickel, molybdenum, manganese, silicon, etc., can be added to produce the desired chemical composition of the molten steel. After the quality control process, the liquid steel is poured into a mold to having the shape of the desired section. The red-hot rough shapes are cut to manageable lengths and are finished by being passed through machines. At the end, the cold finished sections are further sized for easy transportation.

FIGURE 9.2 Steel production from recycled steel using the electric arc process. *Photo taken in Pueblo, Colorado.*

9.3 TYPES OF STEELS

Metals are broadly classified into ferrous and non-ferrous metals, as shown in Figure 9.3. Steel is a ferrous metal. It is broadly classified as carbon steel (low-alloy) and alloy steel (high-alloy).

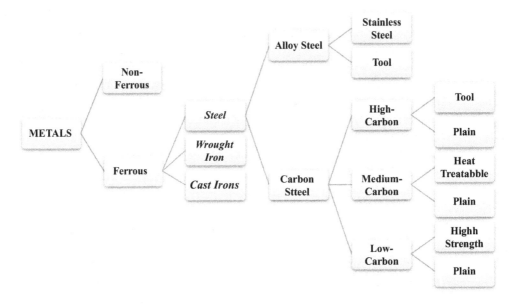

FIGURE 9.3 Types of metals with focus on steel.

9.3.1 CARBON STEEL

Carbon steel is mainly composed of two elements, iron and carbon. Generally, little to no alloy elements are added in steel. Carbon steel, in addition to carbon, generally contains silicon, manganese, sulfur, and phosphorus. The carbon content of commonly used structural steel varies from about 0.12% to about 2.0% by weight, with the iron content of as high as 95%, manganese of up to 1.65%, and silicon of up to 0.6% (Tamboli 1997). The higher the carbon content, the higher the yield stress and ultimate strength, but the lower the ductility and weldability. Higher carbon steels are also more brittle in nature.

Depending on the amount of carbon content, steel can be divided into three categories:

a) Low-carbon steel (less than 0.25%)
b) Medium-carbon steel (0.25–0.60%)
c) High-carbon steel (more than 0. 60%)

Depending on the amount of phosphorus and sulfur content, steel can be divided into two categories:

a) Ordinary carbon steel to carbon steel (high phosphorus and sulfur)
b) High-quality carbon steel (low phosphorus and sulfur)

The presence of phosphorus and sulfur in steel can cause brittleness.

With superior performance, easy processing, and low cost, the usage of carbon steel is dominant. However, the performance of carbon steel is lacking due to its:

- Low strength
- Low hardenability
- Temperature susceptibility
- Tempering poor stability

In order to overcome some of the pitfalls of carbon steel, the performance of the carbon steel is improved by adding some alloying elements in the iron, thus forming alloy steel

9.3.2 ALLOY STEEL

Alloy steel is a type of steel that contains certain other elements apart from iron and carbon. Alloy steel contains primarily silicon and manganese as the alloying elements, as well as some other minor elements such as chromium, nickel, molybdenum, vanadium, titanium, copper, tungsten, aluminum, cobalt, niobium, zirconium, etc., and some nonmetallic elements such as boron and nitrogen. The quantity of these metals in alloy steel is primarily dependent upon the use of such steel. Normally, alloy steel is made with the intention of having certain desired physical characteristics in the steel. Alloy steels are divided into *low-alloy steels* and *high-alloy steels*. When the percentage of added elements goes past 8% by weight, the steel is referred to as a high-alloy steel. In cases where added elements remain below 8% by weight of the steel, it is a low-alloy steel. Low-alloy steels are used more often than the high-alloy steel in the industry. In general, the addition of one or more of such elements to steel makes it harder and more durable. Such steel is also resistant to corrosion and tougher than typical steel. To alter the properties of steel, heat treatment must be conducted when elements are added to it.

The carbon content needs to be reduced in order to keep the alloy steel weldable. As such, carbon content is lowered at 0.1% to 0.3%, and alloying elements are also decreased in proportion. These alloys of steel are known as high-strength, low-alloy steels. Stainless steel is also an alloy steel with a minimum of 10% of chromium by weight.

9.3.2.1 Stainless Steel

Stainless steels generally contain between 10% and 20% chromium as the main alloying element and are valued for high corrosion resistance. With over 11% chromium, stainless steel is about 200 times more resistant to corrosion than the mild steel. These steels can be divided into three groups based on their crystalline structure:

a) *Austenitic* – Austenitic steels are non-magnetic and non-heat-treatable, and generally contain 18% chromium, 8% nickel, and less than 0.8% carbon. Austenitic steels form the largest portion of the global stainless steel market, and are often used in food processing equipment, kitchen utensils, and piping.

b) *Ferritic* – Ferritic steels contain trace amounts of nickel, 12–17% chromium, and less than 0.1% carbon, along with other alloying elements, such as molybdenum, aluminum or titanium. These magnetic steels cannot be hardened by heat treatment, but can be strengthened by cold working.

c) *Martensitic* – Martensitic steels contain 11–17% chromium, less than 0.4% nickel, and up to 1.2% carbon. These magnetic and heat-treatable steels are used in knives and cutting tools, as well as dental and surgical equipment.

9.3.2.2 Tool Steel

Tool steels contain tungsten, molybdenum, cobalt, and vanadium in varying quantities to increase heat resistance and durability, making them ideal for cutting and drilling equipment. The main difference between tool steel and stainless steel is the higher amount of carbon in tool steel, which increases brittleness. Because stainless steel has nickel in its composition, it resists corrosion better than tool steel.

9.3.3 WROUGHT IRON AND CAST IRON

Wrought iron is a relatively pure iron alloy with a very low carbon content (0.02–0.08%). Iron ore is heated with charcoal or coal, and released oxygen combines with carbon. A spongy mass of bloom is produced that contains traces of charcoal and slag. Impurities and slag are driven off by hammering the hot bloom. Wrought iron has been produced since the beginning of the Iron Age, and was used in construction until in the 19th century (example, Eiffel Tower in France). With the advancements in steel production, applications of wrought iron are now limited, mainly for decorative purposes.

On the other hand, cast iron has a high carbon content (2–4%). Iron is heated at high temperatures to produce molten iron. After removing the impurities using limestone, the molten iron is cast into desired shapes. Cast iron is good in compression and hardness. However, it is weak in tension, brittle, and not malleable. In the past, cast iron was used in compressive structures, such as arches and columns. Now, it too has been replaced by steel for civil engineering, but still is used for cookware, piping, etc.

9.4   HEAT TREATMENT OF STEEL

Heat treatment is the process of altering the properties of steel by applying heat to different levels. The amount of heating and its response to steel depends on the composition of the steel. Several popular methods are introduced below:

9.4.1 QUENCHING

Quenching, also known as hardening, is the process of heating steel above the transformation range of about degrees Fahrenheit (910°C), and then rapidly immersing (quenching) it in cool brine, water, or oil. As the surface cools quickly, the hardness increases dramatically at the surface (not the interior part) of the steel. The disadvantage of quenching is that the surface becomes more brittle and susceptible to cracking.

9.4.2 TEMPERING

Quenched steel is reheated by submerging into oil or nitrate salts for about 2 hours and then cooled down normally. Slow cooling in the air relieves internal stresses and decreases hardness. However, this process causes an increase in toughness and ductility.

9.4.3 ANNEALING

Steel is heated above its transformation range, which varies from 1,341–2,098°F (727–1,148°C), depending on the carbon content, and is maintained at that temperature until all steel transforms into austenite or austenite-cementite. It is then cooled down about 68°F (20°C) per hour initially, until it reaches about 1,256°F (680°C), and is then dried normally, in air. Slow cooling to room temperature improves ductility, removes internal stresses, toughness and softens the steel. To relieve internal stress, the steel is heated to about 1,112–1,202°F (600–650°C), and held at this temperature for about 1 hour. Afterward, the steel is cooled down slowly in air. This process is called the *stress-relief annealing.*

9.5 STRUCTURAL STEEL SHAPES

9.5.1 HOT-ROLLED STEEL

A rolling process at temperatures over 1,000°F is used to create *hot-rolled steel.* Steel products in Massachusetts that have been processed in this manner have a blue-gray finish that feels rough to the touch. Hot-rolled steel reconfigures itself during the cooling process, giving the finished product looser tolerances compared to the original material and cold-rolled steel products. Hot-rolled steel is very malleable, allowing it to be forced into a variety of different shapes. This makes hot-rolled steel an excellent choice for the manufacturing of structural components, such as I-beams or simple cross-sections, like rail tracks. It is also used to produce sheet metal. Shapes commonly used are mentioned below:

- Wide-flange shapes (W-, HP-, and M-shapes)
- I-beams (S-shapes)
- Channel shapes (C-and MC-shapes)
- Angle shapes (L-shapes)
- Hollow structural shapes (HSS)
- Pipes

Some commonly used shapes by the American Institute of Steel Construction (AISC) are shown in Figure 9.4. *Wide-flanged shapes (W-shapes)* have the inner and outer flange

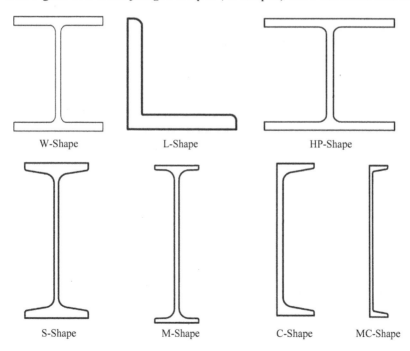

FIGURE 9.4   Some AISC standard shapes (hot-rolled).

surfaces parallel. *Miscellaneous shapes (M-shapes)* are similar to the W-shape, but their depth is limited to 12.5 in. and the flange width is limited to 5-in. *S-shapes* are also similar to W-shapes, except that the inside flange surfaces usually have a slope of 2:12, with the larger flange thickness closest to the web of the beam. *HP-shapes* are similar to W-shapes, and the nominal depth of these sections is usually approximately equal to the flange width, with the flange and web thicknesses approximately equal. To precisely describe a specific type of shape, for example, a W 18×40 is a W-shape with the nominal depth of 18 in. and self-weight of 40 lb per feet of length. The notation is similar for M, S, and HP sections. *Channels* are *C-shaped* members with the inside faces of the channel flanges sloped. C-shapes are American Standard channels, while *MC-shapes* are miscellaneous channels. C 15×50 is a C-shape with a nominal depth of 15 in. and self-weight of 50 lb per feet of length. *Angles* are *L-shaped* members which have equal or unequal length legs, looking similar to the letter 'L.' L 4×3×1/2 is an L-shape with the long leg of 4 in., short leg of 3 in., and thickness of ½ in. *Structural tees* are made by cutting a W-shape, M-shape, or S-shape in half. For example, if a W 18×40 section is cut in half, the resulting shape is WT 9×20, where the nominal depth is 9 in. and the self-weight of each piece is 20 lb per linear ft. *Hollow structural section (HSS)* members are rectangular, square, or round tubular members that have a hollow inside. Examples of designations are listed below:

- HSS 8×8×1/2 means a rectangular, hollow, structural steel with outside wall dimensions of 8 in. in one direction and 8 in. in the orthogonal direction, and a wall thickness of 1/2 in., except at the rounded corners.
- HSS 8×4×1/2 means a rectangular hollow structural steel with outside wall dimensions of 8 in. in one direction and 4 in. in the orthogonal direction, and a wall thickness of 1/2 in., except at the rounded corners.
- HSS 4×¼ means a round hollow structural steel with an outside wall diameter of 4 in. and a uniform wall thickness of 1/4 in.

Structural pipes are round tubes similar to HSS members that are available in three strength categories: *standard* (STD), *extra strong* (X-strong), and *double-extra strong* (XX-strong).

Steel pipes are designated with the letter P, followed by the nominal diameter, and then the letter X for extra strong or XX for double-extra strong, as applicable. For example, the designation P4 represents a nominal 4 in. standard pipe, P4X represents a 4-in. extra strong pipe, and P4XX represents a 4 in. double-extra strong pipe.

Sometimes, plate girders and plates are welded to the top or bottom flanges of W-sections to support heavy loads which the listed, standard steel sections are inadequate. These types of sections are called *built sections*.

9.5.2 COLD-ROLLED STEEL

Cold-rolled steel is processed at temperatures close to normal. Cold-rolling causes an increase in the strength of the produced steel through the use of strain hardening by about 20%. This process produces steel dimensionally more precisely than the hot-rolled process. This is because steel is already closer to the finished dimension, as it has already passed through the cooling process. Cold-rolled steel is limited to a few shapes, such as round, square, flat and variations of those types of shapes, as shown in Figure 9.5. Other shapes can be cold-rolled if the cross section is uniform and the transverse dimension is small. A

FIGURE 9.5 Cold-rolled steel shapes.

series of shaping operations such as sizing, breaking down, semi-roughing, semi-finishing, roughing, and finishing are used to produce cold-rolled shapes. Final shapes produced by the cold-rolled process include bars, strips, rods, and sheets, which are usually smaller than the same products available through hot-rolling. The smaller products are also much more tolerant than the larger, hot-rolled products.

9.6 STRUCTURAL STEEL GRADES

Steels used in civil and construction engineering in the United States use standard alloys identified and specified by the ASTM International. These steels have an alloy identification beginning with A, followed by two, three, or four numbers, such as A36, A500, A572, etc. ASTM specifications for various structural shapes, their type, yield strength (F_y), ultimate strength (F_u), preferred specification, etc. are presented in Table 9.1.

The preferred material specification, ASTM designation, F_y, F_u, etc. for structural plates and bars are listed in Table 9.2. Plates and bars are flat stock members. In the past, flat stock with widths equal to or less than 8 in. were known as bars, and widths greater than 8 in. were known as plates. Now, any flat stock members are referred to as plates. For example, a PL $4.5 \times 1/2$ implies a 4.5 in. wide by 1/2-in. thick plate. Plate widths are usually specified in 1/2-in. increments, and thicknesses are specified in 1/8-in. increments. The practical minimum width is 3 in. to satisfy the required clearance and the minimum thickness is 1/4 in.

9.7 REINFORCING BAR

Rebar (short for reinforcing bar) is a steel bar or mesh of steel wires used as a tension device in reinforced concrete and reinforced masonry structures to strengthen and hold the concrete in tension. It is a common hot-rolled steel bar that is widely used in the construction industry, especially for concrete reinforcement. Concrete is very strong in compression and weak in tension. To compensate for this weakness, reinforcement bars are cast into the concrete to carry the tensile loads. Common steel or concrete reinforcement bars are supplied with heavy ridges (ribbed) to assist in binding the reinforcement to the concrete mechanically, which is referred to as the *deformed bar*. Coating is used on the bars if they are to be used in close proximity to water, so as to avoid corrosion, as shown in Figure 9.6.

9.7.1 BAR IDENTIFICATION

Rebar provides some information, listed physically on the bar itself, to such things as producing mill, size, type of steel, etc. This identification system helps in providing useful information about the manufacturing and potential usages of each bar. Each individual reinforcing bar is manufactured with a series of individual markings (Figure 9.7):

TABLE 9.1
ASTM Specifications for Various Structural Shapes

Steel Type	ASTM Designation	Grade	F_y (ksi)	F_u (ksi)	W	M	S	HP	C	MC	L	HSS □	HSS ○	Pipe
Carbon	A36	–	36	58–80										
	A53	B	35	60										
	A500	B	42	58										
			46	58										
		C	46	62										
			50	62										
	A501	A	36	58										
		B	50	70										
	A529	50	50	65–100										
		55	55	70–100										
	A709	36	36	58–80										
	A1043	36	36–52	58										
		50	50–65	65										
	A1085	50	50–65	65										
High-Strength Low-Alloy	A572	42	42	60										
		50	50	65										
		55	55	70										
		60	60	75										
		65	65	80										
	A618	Ia, Ib, & II	50	70										
		III	50	65										
	A709	50	50	65										
		50 S	50–65	65										
		50W	50	70										
	A913	50	50	65										
		60	60	75										
		65	65	80										
		70	70	90										
	A992	–	50	65										
Corrosion Resistant High-Strength Low-Alloy	A588	–	50	70										
	A847	–	50	70										
	A1065	50W	50	70										

	= Preferred material specification.
	= Other applicable material specification, the availability should be confirmed prior to specification.
	= Material specification does not apply.

Reference: Steel Construction Manual, 15th Edition, American Institute of Steel Construction (AISC), 2017, Chicago, IL. Table 2–4, Used with permission.

TABLE 9.2
ASTM Specifications for Plates and Bars

Steel Type	ASTM Designation	Grade	F_y ksi	F_u ksi	Plates and Bars, in.									
					≤0.75	0.76–1.25	1.26–1.5	1.51–2.0	2.1–2.5	2.6–4.0	4.1–5.0	5.1–6.0	6.1–8.0	≥8.0
Carbon	A36	–	32	58–80										
			36	58–80										
	A283	C	30	55–75										
		D	33	60–80										
	A529	50	50	70–100										
		55	55	70–100										
	A709	36	36	58–80										
High-Strength Low-Alloy	A572	42	42	60										
		50	50	65										
		55	55	70										
		60	60	75										
		65	65	80										
	A709	50	50	65										
	A1043	36	36–52	58										
		50	50–65	65										
Corrosion Resistant High-Strength Low-Alloy	A242		42	63										
			46	67										
			50	70										
	A588		42	63										
			46	67										
			50	70										
Quenched and Tempered Alloy	A514		90	100–130										
			100	110–130										
Corrosion Resistant Quenched and Tempered Low-Alloy	A709	50W	50	70										
		HPS 50W	50	70										
		HPS 70W	70	85–110										
		HPS	90	100–130										
		100W	100	110–130										

	= Preferred material specification.
	= Other applicable material specification, the availability should be confirmed prior to specification.
	= Material specification does not apply.

Reference: Steel Construction Manual, 15th Edition, American Institute of Steel Construction (AISC), 2017, Chicago, IL. Table 2–5, Used with permission.

FIGURE 9.6   Steel reinforcement in a bridge prepared to be cast. *Photo by Armando Perez, taken in Pueblo, Colorado.*

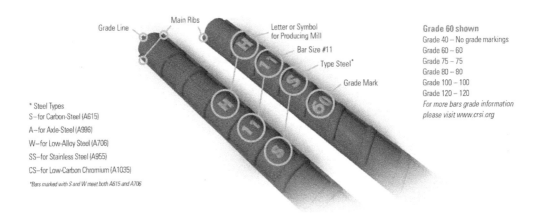

FIGURE 9.7   ASTM bar marking sequence. *Courtesy of Concrete Reinforcing Steel Institute (CRSI), Schaumburg, IL. Used with permission.*

- The first letter or symbol identifies the producing mill
- The next marking is the bar size, commonly in number (#) or metric number
- The third marking designates the type of reinforcing steel: 'S' for carbon-steel (ASTM A 615) or 'W' for low-alloy steel (ASTM A 706), etc.
- Finally, there is a grade marking designated as a number (60, 75, 80, 100, or 120) or by the addition of one line (60), two lines (75), three lines (80, 100), or four lines (120), that must be at least five deformations long

9.7.2   BAR SIZES

Different standard-sized bars are available (#3 to #11, #14, and #18) as listed in Table 9.3. The size designations up through #8 are the number of eighths of an inch in the diameter of a plain round bar having the same weight per foot as the deformed bar. For example, a #5 bar would have the same mass per foot as a plain bar 5/8 in. in diameter. The metric size is the same dimension expressed to the nearest millimeter. The sizes of the large bars are

TABLE 9.3
Standard Reinforcing Bars with Their Diameters and Weights

ASTM Bar Number	Diameter (in.)	Metric Number	Weight (lb/ft)
3	0.375	10	0.376
4	0.500	13	0.668
5	0.625	16	1.043
6	0.75	19	1.502
7	0.875	22	2.044
8	1.00	25	2.670
9	1.128	29	3.400
10	1.270	32	4.303
11	1.410	36	5.313
14	1.693	43	7.650
18	2.257	57	13.60

based on the square rebar formerly made (CRSI 2019). #9 has the same weight per foot and cross-sectional area as a 1-in.^{2} bar, #10 as 1.125-in.^{2}, #11 as 1.25-in.^{2}, #14 as 1.5-in.^{2}, and #18 as a 2-in.^{2} bar.

9.7.3 CORROSION PROTECTION

Steel readily corrodes when exposed to air and water, unless it is coated. The cost of the corrosion of infrastructure in the United States alone is about \$22.6 billion each year (Mamlouk and Zaniewski 2014). However, when steel is placed in concrete it develops a passive oxide film, due to the high pH of the concrete. This passive film prevents further corrosion of the steel for over 100 years. If steel corrodes in concrete, it causes cracking or spalling of the concrete. Corrosion of reinforcing steel may occur if the pH of the concrete is decreased, either from chemical attack or from the reaction of the concrete with carbon dioxide in the atmosphere. It may also occur if sufficient chloride ions reach the bar. These are typically introduced into the concrete, from either deicing salts or sea water. There are many ways to reduce the risk of corrosion-related distress in concrete. The first layer of defense is the concrete itself, which should be dense with minimal cracks. There are a few types of rebar with improved corrosion resistance, such as:

- Stainless steel bars
- Galvanized steel bars
- Epoxy-coated reinforced bars

9.8 FASTENER

A *fastener* is a hardware device that mechanically joins or affixes two or more objects together, shown in Figure 9.8. In general, fasteners are used to create non-permanent joints; that is, joints that can be removed or dismantled without damaging the joining components. There are three major types of steel materials used for fasteners used in industries: stainless steel, carbon steel, and alloy steel. They include the following types of fasteners:

FIGURE 9.8   Steel fasteners. *Courtesy of Randhir Metal and Alloys Pvt Ltd.*

- Conventional nuts and bolts
- Anchor rods
- Twist-off tension control mechanisms
- Washers
- Threaded rods
- Forged steel hardware

The preferred material specification, ASTM designation, F_y, F_u, etc. for bolts, nuts, washers, anchor rods, and threaded rods are listed in Table 9.4.

9.9   WIRE FABRIC

Wire fabric is a prefabricated mesh-type material available in rolls or sheets to be used in slabs, pavements, parking lots, footpaths, etc. The common practice is to tie the wire fabric with reinforcing bars tightly, and then fresh concrete is poured, as shown in Figure 9.9. Wires used in the wire fabrics can be *plain wires* or *deformed wires*. Plain or deformed wires are differentiated by letter 'W' for plain wire and 'D' for deformed wire. The letters, W or D, are followed by numbers representing the cross-sectional area of the wire in hundreds of a square in. For example, D1 is the deformed wire of 0.01-in.^{2} cross-sectional area. W2 is the plain wire of 0.02-in.^{2} cross-sectional area.

Welded wire fabric is designated by two numbers and two sets of letter-numbers, say, 4 X 6 or W2.0 X W2.0. The first number, 4, is the spacing in inches of the longitudinal wires. The second number, 6, is the spacing in inches of the transverse wires. The letter of the first letter-number combination, W, means plain wire in the longitudinal wire. The letter of the second letter-number combination, W, means plain wire in the transverse wire. The number of the first letter-number combination, 2, means the cross-sectional area of the wire in hundreds of a square in. in the longitudinal wire. The number of the second letter-number combination, 2, means the cross-sectional area of the wire in hundreds of a square in. in the transverse wire.

ASTM A 82 covers the specification for plain steel wire and ASTM A 185 covers the specification of welded fabric with A 82 plain wire. ASTM A 496 covers the specification for deformed steel wire and ASTM A 497 covers the specification of welded fabric with A 496 plain wire.

TABLE 9.4
ASTM Specifications for Various Structural Fasteners

ASTM Designation		F_y, ksi	F_u, ksi	Diameter Range (in.)	Bolts High Strength Conventional	Bolts High Strength Twist-Off-Type Tension Control	Bolts Common	Nuts	Washers Hardened	Washers Plain	Washers Direct Tension Indicator	Threaded Rods	Anchor Rods Hooked	Anchor Rods Headed	Anchor Rods Threaded and Nutted
F3125	A325	–	120	0.5 to 1.5											
	F1852	–	120	0.5 to 1.25											
	A490	–	150	0.5 to 1.5											
	F2280	–	150	0.5 to 1.25											
F3111		–	150	0.5 to 1.5											
F3043		–	150	0.5 to 1.5											
A194 Gr. 2H		–	–	0.25 to 4											
A563		–	–	0.25 to 4											
F436		–	–	0.25 to 4											
F844		–	–	Any											
F959		–	–	0.5 to 1.5											
A36		36	58–80	to 10											
A193 Gr. B7		75	100	Over 4 to 7											
		95	115	Over 2.5 to 4											
		105	125	≤2.5											
A307 Gr. A		–	60	0.25 to 4											
A354	Gr. BC	109	125	0.25 to 2.5											
		99	115	Over 2.5 to 4											
	Gr. BD	115	140	2.5 to 4											
		130	150	0.25 to 2.5											
A449		58	90	Over 1.5 to 3											
		81	105	Over 1 to 1.5											
		92	120	Over 0.25 to 1											
A572	Gr. 42	42	60	To 6											
	Gr. 50	50	65	To 4											
	Gr. 55	55	70	To 2											
	Gr. 60	60	75	To 3.5											
	Gr. 65	65	80	To 1.25											
A588		42	63	Over 5 to 8											
		46	67	Over 4 to 5											
		50	70	≤4											
F1554	Gr. 36	36	58–80	0.25 to 4											
	Gr. 55	55	75–95	0.25 to 4											
	Gr. 105	105	125–150	0.25 to 3											

	= Preferred material specification.
	= Other applicable material specification, the availability should be confirmed prior to specification.
	= Material specification does not apply.

Reference: Steel Construction Manual, 15th Edition, American Institute of Steel Construction (AISC), 2017, Chicago, IL. Table 2.6, Used with permission.

FIGURE 9.9   Wire fabrics in a construction site. *Photo by Shelby Nesselhauf, taken in Pueblo, Colorado.*

9.10   CHARACTERIZATIONS OF STEEL

9.10.1   TENSION TEST

Tensile testing in materials is conducted to determine the modulus, yield point, ultimate stress, failure stress, etc. following the ASTM A 370 or ASTM E 8 test standard. A specimen is clamped and pulled axially, as shown in Figure 9.10. The force and deformation are recorded. The test can be force controlled or displacement controlled. The stress–strain diagram is plotted to determine the desired mechanical properties.

9.10.1.1   Engineering Stress or Simply Stress

Engineering axial stress, or simply stress (F), can be defined as follows (Eq. 9.1):

$$F = \frac{P}{A_o} \tag{9.1}$$

Side view   Attaching the Dial Gage   Failed Specimen

FIGURE 9.10   Tensile testing on a steel specimen by the author. *Photos taken at the Colorado State University–Pueblo.*

where,

P = Applied force on the tension member
A_o = Initial cross-sectional area of the tension member

Note that in strength of materials concepts, stress is commonly denoted by σ, in steel member design by F, and in wood and concrete member design by f. This is why, in this chapter, stress is denoted by F.

9.10.1.2 Engineering Strain or Simply Strain

Engineering axial strain, commonly known as strain, is the change in length over the initial length. It can be expressed as follows (Eq. 9.2):

$$\varepsilon_a = \frac{\Delta L}{L_o} \tag{9.2}$$

where,

ε_a = Engineering axial strain (or just strain)
ΔL = Change in length of member
L_o = Initial length of member

If the stress (F) and strain (ε_a) are within the proportional limit, then Young's Modulus (E) is expressed as follows (Eq. 9.3):

$$E = \frac{F}{\varepsilon_a} \tag{9.3}$$

The schematic stress–strain curve of mild steel is shown Figure 9.11. A similar curve was shown in chapter 1. Note that steel is a ductile material, as it shows a remarkable amount of elongation before failing.

All the definitions discussed in chapter 1 for a ductile material are valid for a steel material as well. Some important definitions are repeated here to emphasize the topic.

- Modulus of elasticity: The slope of axial stress and axial strain diagram up to the proportional limit (A).
- Proportional limit: The linear portion of the stress–strain diagram (OA region).

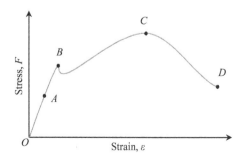

FIGURE 9.11 Stress-strain schematic curve of steel.

- Elastic limit: The stress level up to which no plastic strain occurs (B).
- Yield point: The stress level above which plastic strain occurs (B). It is also called yield stress or yield strength.
- Ultimate stress: The maximum stress a ductile material sustains before failing (C).
- Fracture stress: The stress at which a material fails (D).

If the yield point on the $F-\varepsilon$ curve is not distinct, then a straight line parallel to the initial line is drawn at the 0.2% strain point, as shown in Figure 9.12. The intersection of the straight line and the original $F-\varepsilon$ curve is considered the yield point (F_y).

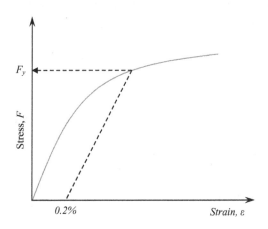

FIGURE 9.12   Determining yield stress if yield point is not distinct.

9.10.1.3   Percent Elongation
Percent elongation is the increase in length compared to the initial length, and is expressed as a percentage. It can be calculated as shown in Eq. 9.4.

$$\% \text{ elongation} = \left(\frac{\Delta L}{L_o} \right) \times 100 \tag{9.4}$$

9.10.1.4   Percent Reduction in Area
The reduction in area from initial area, A_i, to final area, A_f, can be calculated as shown in Eq. 9.5.

$$\% \text{ R}A = \left(\frac{A_i - A_f}{A_i} \right) \times 100 \tag{9.5}$$

9.10.1.5   True Stress and True Strain
True strain, commonly known simply as strain, is the differential change in length over the initial length. It can be expressed as follows (Eq. 9.6):

$$\varepsilon_T = \frac{dL}{L_o} \tag{9.6}$$

where,

ε_T = True strain = $\ln(1+\varepsilon)$
dL = Differential change in length of member
L_o = Initial length of member

9.10.1.6 Poisson's Ratio

Poisson's ratio can be defined as the ratio of lateral strain (ε_t) and longitudinal strain (ε_a). If tension force is applied axially on a body, its length increases and the lateral dimension decreases. If a compressive force is applied, the scenario is the opposite. Consequently, if lateral strain is positive, then longitudinal strain is negative, and vice-versa. Therefore, Poisson's ratio is always negative, and its value is between 0 and 0.5. Mathematically, Poisson's ratio can be expressed as (Eq. 9.7):

$$\mu = -\frac{\varepsilon_t}{\varepsilon_a} = -\frac{\dfrac{\Delta D}{D_o}}{\dfrac{\Delta L}{L_o}} = -\frac{\text{Lateral Strain}}{\text{Axial Strain}} \tag{9.7}$$

where,

L_o and D_o = The original length and lateral dimension, respectively
ΔL and ΔD = The change in length and lateral dimension, respectively

Note that the lateral strain can also be written in terms of radius or lateral dimension if the body is not cylindrical, as shown in Eq. 9.8.

$$\varepsilon_t = \frac{\Delta D}{D_o} = \frac{\Delta r}{r_o} = \frac{\Delta w}{w_o} = \frac{\Delta t}{t_o} \tag{9.8}$$

where,

r_o = Initial radius
Δr = Change in radius (final radius minus the original radius)
w_o = Initial width
Δw = Change in width (final width minus the original width)
t_o = Initial thickness
Δt = Change in thickness (final thickness minus the original thickness)

Example 9.1 Tension Test of Steel Rod

A 2-in.-long steel rod with a diameter of 0.5 in. is subjected to a tensile force until failure occurs. The test data are listed in Table 9.5.

Plot the stress–strain curve up to failure and determine the following parameters:

a) The proportional limit
b) The yield strength
c) The modulus of elasticity
d) The ultimate strength
e) The failure stress

TABLE 9.5
Load-Deformation
Data for Example 9.1

Deformation (in.)	Load (kip)
0	0
0.001	2.95
0.002	5.90
0.003	8.85
0.004	11.80
0.005	8.00
0.006	9.50
0.007	12.45
0.008	15.40
0.009	16.60
0.01	17.80
0.025	25.00
0.05	25.00
0.075	22.00
0.1	14.00

Solution:

Stress (F) and strain (ε) can be calculated as follows (Table 9.6):

From Eq. 9.1: $F = P/A_o$
From Eq. 9.2: $\varepsilon_a = \Delta L/L_o$

TABLE 9.6
Stress–Strain Data Analysis for Example 9.1

Deformation (in.)	Load (kip)	Strain, $\dfrac{\Delta L}{L_o}$	Stress $= \dfrac{\text{Load}}{\text{Area}}$ (ksi)
0	0	0	0
0.001	2.95	0.0005	15.02
0.002	5.90	0.0010	30.05
0.003	8.85	0.0015	45.07
0.004	11.80	0.0020	60.09
0.005	8.00	0.0025	40.74
0.006	9.50	0.0030	48.38
0.007	12.45	0.0035	63.41
0.008	15.40	0.0040	78.43
0.009	16.60	0.0045	84.54
0.01	17.80	0.0050	90.60
0.025	25.00	0.0125	127.32
0.05	25.00	0.0250	127.32
0.075	22.00	0.0375	112.04
0.1	14.00	0.0500	71.301

where,

Initial Cross-sectional Area, $A_o = \dfrac{\pi D_o^2}{4} = \dfrac{\pi (0.5 \text{ in.})^2}{4} = 0.1964 \text{ in.}^2$

Initial Length, $L_o = 2.0$ in.

The stress–strain curve up to failure is shown in Figure 9.13

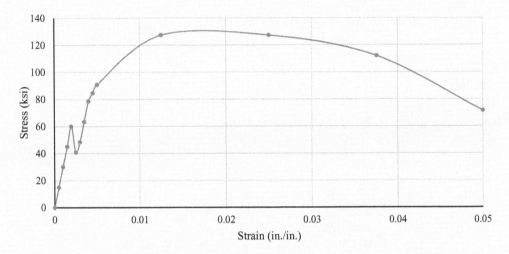

FIGURE 9.13 Stress–strain curve for Example 9.1.

a) The proportional limit
   From the stress–strain curve, the proportional limit is about 60 ksi. (Answer)
b) The yield strength
   From the stress–strain curve, the yield point is about 60 ksi. (Answer)
      Note: In this problem, the curve is linear up to the yield point. Thus, both the proportional limit and the yield point are equal. The proportional limit is mostly less than the yield point.
c) The modulus of elasticity

From Eq. 9.3: $E = \dfrac{\Delta F}{\Delta \varepsilon_a} = \dfrac{F_2 - F_1}{\varepsilon_2 - \varepsilon_1} = \dfrac{30.05 - 15.02}{0.001 - 0.0005} = 30{,}060 \text{ ksi} = 30{,}000 \text{ ksi}$

(Answer)

   The first two stress–strain data are used here to perform this calculation. Any two points within the proportional limit can be used to perform this calculation.
d) The ultimate strength
   From the stress–strain curve, the ultimate stress is about 128 ksi. (Answer)
e) The failure stress
   From the stress–strain curve, the failure stress is about 71 ksi. (Answer)

Example 9.2 Compression Test of Steel Pipe

A steel pipe has a length of 6 ft, outer diameter of 10 in., and wall thickness of 1.0 in. The modulus of elasticity of steel is 29,000 ksi with a Poisson's Ratio of 0.30. If the pipe is subjected to an axial compression of 100 kip, determine the following:

a) The axial stress on the pipe
b) The axial strain of the pipe
c) The shortening in length of the pipe
d) The final outer diameter
e) The increase in the wall thickness

Solution

Initial length, $L_o = 6$ ft
Initial outer radius, $r_o = 10$ in./2 = 5 in.
Initial inner radius, $r_i = 5$ in. – 1.0 in. = 4 in.
Initial Cross-sectional area, $A_o = \pi\left(r_o^2 - r_i^2\right) = \pi\left(5^2 - 4^2\right) = 9\pi$ in.^{2}

a) From Eq. 9.1: Axial stress, $F = \dfrac{P}{A_o} = \dfrac{100\ \text{kips}}{9\pi} = 3.54\ \text{ksi}$ (Answer)

b) From Eq. 9.3: Axial strain, $\varepsilon_a = \dfrac{F}{E} = \dfrac{3.54\ \text{ksi}}{29{,}000\ \text{ksi}} = 0.000122$ (Answer)

c) From Eq. 9.2: The shortening of the pipe, $\Delta L = \varepsilon_a L_o = 0.000122(6 \times 12) = 0.0088$ in. (Answer)

d) The final outer diameter
From Eq. 9.7: $\varepsilon_t = \mu\varepsilon_a = 0.30(0.000122) = 0.0000366$

Then, from Eq. 9.8, $\varepsilon_t = \dfrac{\Delta D}{D_o}$

Or, $0.0000366 = \dfrac{\Delta D}{10}$

Therefore, $\Delta D = 0.000366$ in.
Final outer diameter = Initial outer diameter + ΔD
Final outer diameter = 10 in. + 0.000366 in. = 10.000366 in. (Answer)

e) The increase in the wall thickness

From Eq. 9.8, $\Delta t = \varepsilon_t \left(t_o\right) = 0.0000366\ (1\,\text{in.}) = 0.0000366$ in. (Answer)

Note: A problem can be solved following different ways; the answers must be the same.

9.10.2 Torsion Test

Shear modulus, or the modulus of rigidity (G), is defined as the ratio of shear stress and shear strain up to the proportional limit. The shear modulus is a material property useful in calculating compliance of structural materials in torsion provided they follow Hooke's law,

FIGURE 9.14 Torsion test apparatus. *Photos taken at the Colorado State University–Pueblo.*

that is, the angle of twist is proportional to the applied torque. Examples of the use of shear modulus are in the design of rotating shafts and helical compression springs.

The ASTM E 143 test standard is followed to determine the shear modulus of materials. In this test, a cylindrical or tubular specimen is clamped, as shown in Figure 9.14, and torque is applied to cause uniform twisting within the gauge length. The applied torque and the corresponding angle of twist are measured throughout the test. Then, the shear stress–strain diagram is plotted. The initial slope of the shear stress–strain diagram is the shear modulus.

The applied shear stress (τ) is calculated using Eq. 9.9:

$$\tau = \frac{Tr}{J} \tag{9.9}$$

The applied shear strain (γ) is calculated using Eq. 9.10:

$$\gamma = \frac{\varphi r}{L} \tag{9.10}$$

Then, the shear modulus is calculated using Eq. 9.11:

$$G = \frac{\tau}{\gamma} = \frac{TL}{\varphi J} \tag{9.11}$$

Or,

$$\phi = \frac{TL}{JG} \tag{9.12}$$

where,

G = Shear modulus of the specimen

T = Torque

r = Radius of the of specimen

L = Gage length

J = Polar moment of inertia of the section about its center

ϕ = Angle of twist, in radian

If Poisson's ratio (μ) of the material is known, the shear modulus (G) and modulus of elasticity (E) can be correlated using Eq. 9.13:

$$G = \frac{E}{2(1+\mu)}$$ (9.13)

Example 9.3 Angle of Twist of Shaft

A 2 ft circular shaft of inner diameter of 12 in. and outer diameter of 14 in. is subjected to a torque of 10,000 kip.ft. The shear modulus of the shaft material is 10,000 ksi. Calculate the angle of twist of the shaft.

Solution

Given,

Applied Torque, $T = 10{,}000$ kip.ft $= 120{,}000$ kip.in.
Length, $L = 2$ ft $= 24$ in.
Shear Modulus, $G = 10{,}000$ ksi
Outer radius, $r_o = 14$ in./2 $= 7$ in.
Inner radius, $r_i = 12$ in./2 $= 6$ in.

Polar moment of inertia, $J = \dfrac{\pi}{2}\left(r_0^4 - r_i^4\right) = \dfrac{\pi}{2}\left(7^4 - 6^4\right) = 1{,}736$ in.^{4}

From Eq. 9.12: Angle of twist, $\phi = \dfrac{TL}{JG} = \dfrac{120{,}000 \text{ kip.in.}(24 \text{ in.})}{1{,}736 \text{ in.}^4(10{,}000 \text{ ksi})} = 0.166$ rad

Answer: 0.166 rad

Example 9.4 Torsion of Shaft

A drill rod with a diameter of 0.5 in. has a shear modulus of 12,000 ksi. The rod is subjected to a torque of T. If the allowable shear stress of the rod is 40,000 psi, calculate the minimum length of the rod so that one end of the rod rotates 10° with respect to the other.

Solution

Given,

Angle of twist, $\phi = 10° = 10\pi/180 = 0.175$ rad
Shear Modulus, $G = 12{,}000$ ksi $= 12{,}000{,}000$ psi
Radius of the rod, $r = 0.5$ in./2 $= 0.25$ in.
Allowable shear stress, $\tau = 40{,}000$ psi

From Eq. 9.11: $G = \dfrac{\tau}{\gamma} = \dfrac{TL}{\varphi J}$

Then, $L = \dfrac{G\varphi J}{T} = G\varphi\left(\dfrac{J}{T}\right) = ?$

From Eq. 9.9: $\tau = \dfrac{Tr}{J}$

Therefore, $\dfrac{J}{T} = \dfrac{r}{\tau} = \dfrac{0.25\,\text{in.}}{40,000\,\text{psi}} = 0.00000625$

Finally, $L = G\phi\left(\dfrac{J}{T}\right) = 12,000,000(0.175\,\text{rad})(0.00000625) = 13.1\,\text{in.}$

Answer: 13.1 in.

9.10.3 IMPACT TEST

The impact test (ASTM E 23) determines the amount of energy absorbed by a material during fracture. Two types of tests are possible:

a) Charpy (simple-beam) test – specimen is placed as a simply supported beam with a notch.
b) Izod (cantilever-beam) test – specimen is placed as a cantilever beam with a notch.

This absorbed energy is a measure of a given material's notch toughness. The test apparatus is mainly a pendulum of known mass and length, as shown in Figure 9.15. The pendulum is released without vibration, and the specimen is impacted by the striker. The energy

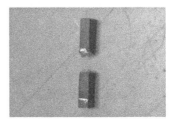

a) Impact Test Equipment

b) Notched and Failed Specimens

FIGURE 9.15 Impact testing (Charpy simple-beam) by the author. *Photos taken at the Colorado State University–Pueblo.*

transferred to the material can be inferred by comparing the difference in the height of the hammer before and after the fracture (energy absorbed by the fracture event).

In the Charpy test, the test specimen is thermally conditioned and positioned on the specimen supports against the anvils as a simply supported mini beam. The notch is made at the center of the specimen and is faced on the side opposite that which will be struck by the pendulum. The Charpy test can be conducted at the following temperatures:

- −40°C (conditioned in dry ice and isopropyl alcohol)
- −18°C (conditioned in dry ice, 30% isopropyl alcohol, and 70% water)
- 4°C (conditioned in dry ice and isopropyl alcohol)
- 40°C (conditioned in oven)

The Charpy test can be used to determine whether or not a material experiences a ductile-to-brittle transition as the temperature is decreased. In such a transition, at higher temperatures the impact energy is relatively large since the fracture is ductile. As the temperature is lowered, the impact energy drops over a narrow temperature range as the fracture becomes more brittle, as shown in Figure 9.16. The transition can also be observed from the fracture surfaces, which appear fibrous or dull for entirely ductile fracture, and granular and shiny for entirely brittle fracture. Over the course of the ductile-to-brittle transition, features distinctive of both types exist.

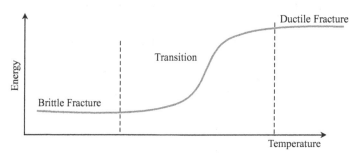

FIGURE 9.16 Brittle to ductile transition temperature.

While the transition may occur very suddenly at a temperature for pure materials, for many materials, the transition occurs over a range of temperatures. This causes difficulties when trying to define a single transition temperature, and no specific criterion has been established. If a material experiences a ductile-to-brittle transition, the temperature at which it occurs can be affected by the variables mentioned earlier, namely the strain rate, size and shape of the specimen, and the relative dimensions of the notch.

In the Izod test, the test specimen is positioned in the specimen-holding fixture as a cantilever beam, and the pendulum is released without vibration. The notching is faced on the same side as that which will be struck by the pendulum. Testing at temperatures other than room temperature is difficult because the specimen-holding fixture for the Izod specimens is often part of the base of the machine and cannot be readily cooled (or heated). Therefore, the Izod testing procedure is not recommended at other than room temperature.

9.10.4 HARDNESS TEST

The hardness test is an indentation hardness test that can provide useful information about metallic materials. This information may correlate to tensile strength, wear resistance, ductility, or other physical characteristics of metallic materials, and may be useful in quality control and selection of materials. The hardness test measures the resistance of a material to indentation/penetration. For a given indenter with fixed load, the greater the penetration, the lower the hardness of the material surface.

9.10.4.1 Brinell Hardness Test

In this test (ASTM E 10), the indenter (a tungsten sphere-ball of 1, 2.5, 5 or 10-mm diameter, as shown in Figure 9.17) is brought into contact with the test specimen in a direction perpendicular to the surface, and the test force is applied and held for a specified dwell time, and then removed. The test force is applied within 1–8 seconds and held for 10–15 seconds. The diameter of the indentation should be between 24% and 60% of the ball diameter.

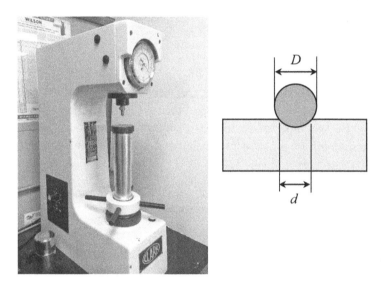

FIGURE 9.17   Rockwell hardness tester. *Photo taken at the Colorado State University–Pueblo.*

The Brinell hardness number (HBW) can be calculated using Eq. 9.14:

$$HBW = \frac{2F_{kgf}}{\pi D\left(D - \sqrt{D^2 - d^2}\right)}$$ (9.14)

where,
   F_{kgf} = Test force, kgf or N
   D = Diameter of the indenter, mm
   d = Measured mean diameter of the indentation, mm

The hardness is then empirically determined using the HBW.

Example 9.5 Brinell Hardness Number

A hardness test is conducted on a brass specimen using a tungsten sphere-ball of 10-mm diameter. Upon applying a force of 1,263 kgf, the indented diameter of 5 mm is measured. Calculate the HBW of the brass specimen.

Solution

Test force, $F_{kgf} = 1{,}263$ kgf
Diameter of the indenter, $D = 10$ mm
Measured mean diameter of the indentation, $d = 5$ mm

From Eq. 9.14: $\text{HBW} = \dfrac{2F_{kgf}}{\pi D\left(D - \sqrt{D^2 - d^2}\right)} = \dfrac{2(1{,}263)}{\pi(10)\left(10 - \sqrt{10^2 - 5^2}\right)} = 60$

Answer: 60

9.10.4.2   Rockwell Hardness Test

The Rockwell (ASTM E 18) indenters are either diamond spheroconical indenters or tungsten carbide balls of 1.588 mm (1/16 in.), 3.175 mm (1/8 in.), 6.35 mm (1/4 in.), or 12.70 mm (1/2 in.) in diameters. During the test, the force on the indenter is increased from a preliminary test force to a total test force, and then returned to the preliminary test force. The difference in the two indentation depth measurements while under the preliminary test force is measured as h. For scales using a diamond spheroconical indenter, the Rockwell hardness is derived as shown in Eq. 9.15:

$$\text{Rockwell Hardness} = 100 - \frac{h}{0.002} \qquad (9.15)$$

For scales using the ball indenter, the Rockwell hardness is derived as shown in Eq. 9.16:

$$\text{Rockwell Hardness} = 130 - \frac{h}{0.002} \qquad (9.16)$$

where h is in mm.

Example 9.6 Rockwell Hardness

In a Rockwell hardness test, upon applying a force of 200 N, the ball indenter indents 0.001 mm. When the force is increased to 20 kN, the indentation is 0.075 mm. Upon a decrease in the force to 200 N, the indentation is measured to be 0.001 mm again. Calculate the Rockwell hardness of the material.

Solution

Indentation depth, $h = (0.075 - 0.001)$ mm
For scales using the ball indenter, the Rockwell hardness is derived as

From Eq. 9.16: $\text{Rockwell Hardness} = 130 - \dfrac{h}{0.002} = 130 - \dfrac{(0.075 - 0.001)}{0.002} = 93$

Answer: 93

9.11   WELDABILITY OF STEEL

The ASTM A 6 specification prescribes the permissible maximum percentages of alloy elements, such as carbon, manganese, chromium, nickel, copper, molybdenum, vanadium, and so forth, in structural steel to ensure adequate weldability and resistance to corrosion and brittle fracture. In the specification, the percentage by weight of each of these chemical elements is combined to produce an equivalent percentage carbon content, which is called the *carbon equivalent* (CE). The carbon equivalent is useful in determining the weldability of older steels used in the repair or rehabilitation of existing or historical structures where the structural drawings and specifications are no longer available. It is also used for determining what, if any, special precautions are necessary for welding these steels in order to prevent brittle fractures. To ensure the good weldability already established above, the carbon equivalent, as calculated from Eq. 9.17, should be no greater than 0.5%. Precautionary measures for steels with higher carbon equivalents include preheating the steel and using low-hydrogen welding electrodes. Alternatively, bolted connections could be used in lieu of welding. The equivalent carbon content or carbon equivalent is given as shown in Eq. 9.17:

$$CE = C + \frac{Cu + Ni}{15} + \frac{Cr + Mo + V}{5} + \frac{Mn + Si}{6} \le 0.5 \qquad (9.17)$$

where,

C = Percentage of carbon content by weight
Cu = Percentage of copper content by weight
Ni = Percentage of nickel content by weight
Cr = Percentage of chromium content by weight
Mo = Percentage of molybdenum by weight
V = Percentage of vanadium content by weight
Mn = Percentage of manganese content by weight
Si = Percentage of silicon by weight

Example 9.7 Weldability of Steel

A steel floor girder in an existing building needs to be strengthened by welding a structural member to its bottom flange. The steel grade is unknown and to determine its weldability, material testing has revealed the following percentages by weight of the alloy chemicals in the girder:

C = 0.20%
Cr = 0.15%
Cu = 0.25%
Mn= 0.45%
Mo= 0.12%
Ni = 0.30%
V = 0.12%
Si = 0.20%

Calculate the carbon equivalent (CE) and determine the weldability of the steel.

Solution

From Eq. 9.16:

$$CE = C + \frac{Cu + Ni}{15} + \frac{Cr + Mo + V}{5} + \frac{Mn + Si}{6}$$

$$= 0.20 + \frac{0.25 + 0.30}{15} + \frac{0.15 + 0.12 + 0.12}{5} + \frac{0.45 + 0.20}{6}$$

$$= 0.42\% < 0.5\%$$

Answers: CE = 0.42%, and the steel is weldable.

9.12   CHAPTER SUMMARY

Steel is a widely used material in civil engineering. It is a mixture of several metals, though the majority of its composition is iron. Various amounts of carbon are added to attain certain properties such as strength, ductility, etc. Steel is mainly of two types: carbon steel and alloy steel. Carbon steel is mainly composed of two elements: iron and carbon. The proportion of carbon used varies between 0.02% and 1.7%. Alloy steel contains other elements apart from iron and carbon, such as manganese, silicon, boron, chromium, vanadium, and nickel. Depending on the degree of heat used while processing steel, it can be of two types: hot-rolled (1,000 °F) and cold-rolled (normal temperature). Different-shaped sections such as I-sections, angle sections, leg sections, etc. can be produced very precisely by hot-rolling. The cold-rolling method uses a temperature close to room temperature and can produce some basic shapes, such as pipe types, hollow structural sections, round, square, etc. ASTM grades a steel section based on its yield strength and ultimate strength, for different types of hot and cold-rolled sections, plates and bars, reinforcing bars, and fasteners.

The tension test is the most common test conducted on steel specimens. It determines some major properties of steel, such as the modulus of elasticity, yield stress, ultimate strength, and Poisson's ratio. Some other properties such as failure stress, elongation capacity, reduction in area, etc. are also very often determined. Using the modulus of elasticity and Poisson's ratio, the shear modulus can be determined using an equation. The torsion test is the direct method of determining the shear modulus. The impact test measures the toughness (energy absorbed up to failure) and the brittle-to-ductile transition temperature of steel materials. The hardness test measures the resistance to surface deformation where a spherical ball is indented by applying a force. The intended diameter is measured and used to calculate a hardness number.

Weldability is another important parameter of steel for construction purposes. Having too large an amount of alloy makes the steel unable to be welded. The weldability of steel can be calculated using the proportions of other metals such as carbon, copper, nickel, chromium, etc.

ORGANIZATIONS DEALING WITH STEELS

The Association for Iron & Steel Technology (AIST). The AIST is a non-profit organization with 17,500 members from more than 70 countries. With 30 Technology Committees and 22 Local Members Chapters, AIST represents an incomparable network of steel industry knowledge and expertise. Its mission is to advance the technical development, production, processing and application of iron and steel.

Location: Warrendale, PA.
Website: https://www.aist.org

The World Steel Association (Worldsteel). Worldsteel is a non-profit organization with headquarters in Brussels, Belgium. Worldsteel represents over 160 steel producers (including 9 of the world's 10 largest steel companies), national and regional steel industry associations, and steel research institutes. Worldsteel members cover around 85% of the world's steel production.

Location: Brussels, Belgium
Website: www.worldsteel.org

The American Institute of Steel Construction (AISC). The AISC is a non-partisan, not-for-profit technical institute and trade association established in 1921 to serve the structural steel design community and construction industry in the United States.

Location: Chicago, IL.
Website: https://www.aisc.org

Concrete Reinforcing Steel Institute (CRSI). The CRSI is an organization that develops standards and stands as the authoritative resource for information related to steel-reinforced concrete construction. CRSI offers many industry-trusted technical publications, standards documents, design aids, reference materials, and educational opportunities.

Location: Schaumburg, IL
Website: www.crsi.org

ASTM International is also a resource for steel materials.

REFERENCES

Aghayere, A. O., and Vigil, J. 2014. *Structural Steel Design: A Practice-Oriented Approach*, 2nd edition. Pearson, Upper Saddle River, NJ.
CRSI. 2019. Concrete Reinforcing Steel Institute (CRSI) website, www.crsi.org, Last Accessed 2/22/2019.
Domone, J. M. and Illston, P. 2010. *Construction Materials, Their Nature and Behavior*, 4th edition. Spon Press, New York, NY.
Mamlouk, M. and Zaniewski, J. 2014. *Materials for Civil and Construction Engineers*, 4th edition, Pearson, Upper Saddle River, NJ.
Mckee, B. and Hursely, T. 2007. "Structural steel: How it's done". *Modern Steel Construction*, August 2007: 22–29.
Tamboli, R. Akbar. 1997. *Steel Design Handbook: LRFD Method*. McGraw-Hill, New York, NY.

FUNDAMENTALS OF ENGINEERING (FE) EXAM STYLE QUESTIONS

FE PROBLEM 9.1

Which of the following statements is true when a circular shaft is subjected to torsion only?

A. No shear stress is present throughout the shaft
B. Constant shear stress occurs throughout the shaft
C. Maximum shear stress occurs at the center of the shaft
D. Maximum shear stress occurs at the outermost fiber

Solution: D

Consider the shear stress formula, $\tau = Tr/J$. As the distance from the center, r, increases, the shear stress, τ, increases.

FE Problem 9.2

The stress–strain (F–ε) diagram of a tension test is shown in Figure 9.18. The yield point is not distinct; therefore, a parallel line has been drawn at 0.2% strain value. A point (labeled 'A') is marked on the original F–ε curve. Which of the following statements is valid about the marked point, A?

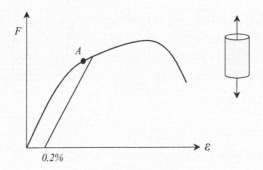

FIGURE 9.18   Stress–strain diagram for FE Problem 9.2.

A. The point, A, is located within the proportional limit
B. The point, A, is located within the non-linear elastic zone
C. The point, A, is located within the strain hardening zone
D. The point, A, is located within the strain softening zone

Solution: B

The point, A, is not located within the proportional (linear) region. However, it is before the yield point (elastic limit). Therefore, A is located within the non-linear elastic region. The strain hardening zone is beyond the yield point and before the ultimate stress. The strain softening region stays beyond ultimate stress and before the failure point.

FE Problem 9.3

If a tensile force is applied on a 2-m long rod having a 40-mm diameter. After applying the load, the diameter of the rod decreases to 39.8 mm and the length increases to 2.01 m. The lateral (transverse) strain developed is closest to:

A. 0.5%
B. 0.05%
C. 0.01%
D. 0.005%

Solution: A

Lateral (transverse) strain, $\varepsilon_t = \dfrac{\Delta d}{d_o} = \dfrac{d - d_o}{d_o} = \dfrac{39.8 - 40}{40} = -0.005 = -0.5\%$

FE PROBLEM 9.4

Which of the following statements is true for Figure 9.19?

A. P=Proportional limit, S=Failure stress
B. P=Yield point, S=Ultimate stress
C. Q=Proportional limit, O=No loading point
D. P=Linear elastic limit, R=Failure stress

Solution: A

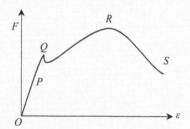

FIGURE 9.19 Stress–strain diagram for FE Problem 9.4.

PRACTICE PROBLEMS

PROBLEM 9.1

A tensile force of 100 kN is applied on a 2-m long rod having a 0.02-m diameter. After applying the load, the diameter of the rod decreases to 0.01998 m and the length increases to 2.01 m. Calculate the lateral (transverse) strain.

PROBLEM 9.2

A tensile force of 100 kN is applied on a 2-m long steel rod having a 0.02-m diameter. The steel has a modulus of elasticity of 210 GPa and a Poisson's ratio of 0.3. Calculate the final diameter and the final length of the specimen.

PROBLEM 9.3

A tensile force of 100 kN is applied on a 2-m long rod having a 0.02-m diameter. After applying the load, the diameter of the rod decreases to 0.01998 m and the length increases to 2.01 m. Calculate the true axial strain.

PROBLEM 9.4

A 2 ft circular shaft having an inner radius of 6 in. and outer radius of 7 in. is subjected to a torque of 10 kip.ft. The shear modulus of the shaft material is 10,000 ksi. Calculate the angle of twist of the shaft.

PROBLEM 9.5

A steel floor girder in an existing building needs to be strengthened by welding a structural member to its bottom flange. The steel grade is unknown, but materials testing has revealed the following percentages by weight of the following alloy chemical elements in the girder:

C = 0.16%
Cr = 0.10%
Cu = 0.20%
Mn = 0.8%
Mo = 0.15%
Ni = 0.25%
V = 0.06%
Si = 0.20%

Calculate the carbon equivalent (CE) and determine the weldability of the steel.

PROBLEM 9.6

A steel floor girder in an existing building needs to be strengthened by welding a structural member to its bottom flange. The steel grade is unknown, but materials testing has revealed the following percentages by weight of the following alloy chemical elements in the girder:

C = 0.12%
Cr = 0.11%
Cu = 0.21%
Mn = 0.7%
Mo = 0.12%

Ni = 0.20%
V = 0.07%
Si = 0.30%

Calculate the carbon equivalent (CE) and determine the weldability of the steel.

PROBLEM 9.7

Three 2-in. long steel rods with a diameter of 0.5 in. and carbon content of 0.15, 0.45, and 0.75%, respectively, are subjected to tensile force until failure. The test data are listed in Table 9.7.

TABLE 9.7

Load-Deformation Data for Problem 9.7

Carbon Content	0.15%	0.45%	0.75%
Deformation (in.)	Measured Load (kip)		
0	0	0	0
0.002	5.7	5.7	5.7
0.003	9.7	9.7	9.7
0.004	13.7	13.7	13.7
0.005	17.7	17.7	17.7
0.009	25	32	35
0.012	20	30	33
0.025	25	43	55
0.05	35	55	65
0.075	45	55	45
0.1	35	51	
0.2	15		

Determine the following:

a) Plot the stress–strain curve up to failure for each specimen
b) The proportional limit of each specimen
c) The yield strength by traditional approach
d) The modulus of elasticity of each specimen
e) The ultimate strength of each specimen
f) The failure stress of each specimen
g) Discuss the effect of carbon content based on the graphs

PROBLEM 9.8

Three 2-in.-long steel rods with a diameter of 0.5 in. are subjected to tensile force until failure. The test data are listed in Table 9.8.

TABLE 9.8

Load-Deformation Data for Problem 9.8

Deformation (in.)	Measured Load (kip)		
	Specimen #1	Specimen #2	Specimen #3
0	0	0	0
0	0	0	0
0.001	2.85	2.85	2.85
0.002	5.7	5.7	5.7
0.003	9.7	9.7	9.7
0.004	9	8.5	9.2
0.005	11	11	15
0.006	13	14	16
0.012	15	16	15
0.025	13	11	14
0.05	7	–	–

Determine the following:

a) Plot the stress–strain curve up to failure for each specimen
b) The proportional limit of each specimen
c) The yield strength of each specimen
d) The modulus of elasticity of each specimen
e) The ultimate strength of each specimen
f) The failure stress of each specimen

PROBLEM 9.9

A hardness test is conducted on a brass specimen using a tungsten sphere-ball of 10-mm diameter. If the HBW of the specimen is 65, calculate the indented diameter of the brass specimen if a force of 1,000 kg is applied.

PROBLEM 9.10

In a Rockwell hardness test, upon applying a force of 100 N, the ball indenter indents 0.0005 mm. When the force is increased to 25 kN, the indentation is 0.070 mm. Upon decrease in the force to 100 N, the indentation is measured to be 0.0005 mm again. Calculate the Rockwell hardness of the material.

Problem 9.11

A steel pipe has a length of 1 m, outer diameter of 0.22 m, and wall thickness of 12 mm. The modulus of elasticity of steel is 210 GPa, and Poisson's ratio is 0.30. If the pipe is subjected to an axial compression of 210 kN, determine the following:

a) The axial stress on the pipe
b) The axial strain of the pipe
c) The shortening of the pipe
d) The increase in the outside diameter
e) The increase in the wall thickness

Problem 9.12

A steel pipe has a length of 4 ft, outer diameter of 5 in., and wall thickness of 0.375 in. The modulus of elasticity of steel is 29,000 ksi and Poisson's ratio is 0.30. If the pipe is subjected to an axial compression of 80 kip, determine the following:

a) The axial stress on the pipe
b) The axial strain of the pipe
c) The shortening of the pipe
d) The increase in the outside diameter
e) The increase in the wall thickness

Problem 9.13

A drill rod with a diameter of 10 mm has a shear modulus of 80 GPa. The rod is subjected to a torque of T. If the allowable shear stress of rod is 100 MPa, calculate the minimum length of the rod so that one end of the rod rotates 10° with respect to the other end. Assume linear elastic behavior.

Problem 9.14

If a steel rod, 600 mm long with a cross-sectional area of 50 mm², is pulled with an axial stress of 500 MPa, calculate the final length and the final cross-sectional area of the rod. The modulus of elasticity and Poisson's ratio of steel are 210 GPa and 0.3, respectively. Assume linear elastic behavior.

PROBLEM 9.15

A 24-in. long steel rod with a cross-sectional area of 0.75 in.^{2} is being pulled. The modulus of elasticity of steel is 29,000 ksi and Poisson's ratio is 0.25. Calculate the amount of force required to produce a diameter change of 10^{-4} in.

10 Woods

This chapter discusses the species, production, seasoning, defects, grading, physical properties, mechanical properties, etc. of wood sections. Different types of engineered wood products, such as sawn lumber, glulam, prefabricated I-joist, veneer, plywood, etc., are also discussed. The standard laboratory characterization methods, such as tension, bending, shear, etc., are included. Finally, the design philosophy of wood structures for bending, shear, and tension is introduced.

10.1 BACKGROUND

Wood is a widely used civil engineering material and is the most abundant renewable building material provided by nature. The term *wood*, in this context, pertains to the hard and fibrous material which forms the trunk, large branches, and roots of a tree. In the United States, 90% of all residential buildings are made out of wood materials (Stone and Tyree 2015). Wood buildings are mostly single-family residences, retail buildings, offices, hotels, schools and colleges, healthcare and recreation facilities, senior living and retirement homes, and religious buildings. However, the presence of larger apartment complexes such as official, commercial, and industrial buildings is increasing day by day, due to their exceptional strength-to-weight ratio. Residential structures are usually up to three stories in height, while multifamily and hotel structures can be up to five or six stories in height. Sometimes, wood sections are used in conjunction with steel. Some other structures, such as recreational areas, pergolas, statues, etc., are being constructed with wood due to the low-cost and aesthetic options available. Figure 10.1 shows a four-storied hotel and a park structure in Colorado made with wood. Mid-rise buildings with four to six stories, similar to one shown in Figure 10.1, are common. Figure 10.1 also shows a very detailed park structure made with wood. This type of structure is cheaper if wood is used in place of steel or other materials. With the invention of artificial engineered wood sections, such as cross-laminated timber (CLT), glulam, oriented-strand board (OSB), etc., the popularity of wood buildings and structures is increasing very rapidly. Therefore, the wood industry has great potential for use in civil and construction engineering, especially in North America. Numerically, the wood industry values $67 billion, only in the United States. About 100,000 employees work in wood product manufacturing companies in Canada, which exports about $8 billion of wood products, mainly to the United States (NCASI 2018).

Wood is an environmentally friendly civil and construction engineering material. The production of wood, its usage, and demolition do not adversely affect the environment as greatly as many other materials. Figure 10.2 shows the amount of carbon (C) emitted to produce various different types of civil and construction engineering related materials (EPA 2006). In the production of one ton of framing lumber, wood lumber produces about 33 kg (73 lb) of carbon. Two major materials, concrete and steel, produce about 291 kg (642 lb) and 694 kg (1,530 lb) of carbon, respectively, during the production of one ton of materials; this is about nine and twenty-one times the amount of carbon compared to wood lumber, respectively (EAP 2006). This calculation is based on materials' full life-cycles, including gathering, processing, and transportation.

A Four-Story Hotel   A Park Structure

FIGURE 10.1   Two wood structures. *Photos taken in Colorado.*

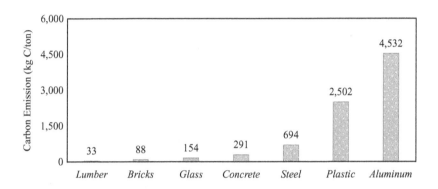

FIGURE 10.2   Carbon emissions in producing various materials. 'Adapted from *EPA (2006).* *Solid waste management and greenhouse gases—a life cycle assessment of emissions and sinks, 3rd Edition, US Environmental Protection Agency (EPA), Washington, DC.*

Several advantages of wood are as follows:

- It has a very high strength-to-weight ratio. Compared to its weight, the strength is great, which makes wood so popular in civil and construction engineering.
- It has elasticity and tenacity more than adequate to bear certain grades of bowing and shock waves (Zhang 2011).
- It has very minimal thermal conductivity when it is dry. Thus, it is a very good thermal insulating material.
- It can be very durable if it can be kept dry.
- It can be made into products of various shapes and finishes. Wood has varying grains and textures, a wide selection of tones and shades, and can be styled in many ways to fit many different demands and designs.

Despite its numerous advantages, wood has the following disadvantages:

- It is a very non-homogeneous material. It has different properties in different directions/orientations. In addition, natural defects require careful engineering judgment before using it. Engineered wood products can prevent these issues.
- It expands and shrinks due to wetness and dryness, respectively. Thus, it is liable to crack or warp when treated incorrectly. Proper seasoning prevents these problems.

- If not properly conserved, it may be corrupted, rot, or even eaten by parasites. Proper seasoning and treatment can be used to avoid these problems.
- It is very combustible and is liable to burn.

10.2 CLASSIFICATION OF WOODS

There are different bases for the classification of lumbers. The following are the important bases to be used in civil and construction engineering (Bhavikatti 2010):

(i) Mode of growth
(ii) Modulus of elasticity
(iii) Durability

10.2.1 CLASSIFICATION BASED ON MODE OF GROWTH

Based on the mode of growth, trees are classified as:

(a) Exogenous
(b) Endogenous

10.2.1.1 Exogenous Trees

Exogenous trees grow outward by adding a distinct, consecutive ring every year their life. These rings are known as *annual rings*. Hence, it is possible to find the age of lumber by counting these annual rings. These trees may be further divided into two types:

- Coniferous or gymnosperm trees
- Deciduous or angiosperm trees

Coniferous or *gymnosperm* trees are those which have cone-shaped leaves and fruits. Trees that bear cones are coniferous. This is the definition of coniferous: bearing cones. The leaves of these trees do not fall until new ones have grown (remains evergreen), and grow continuously throughout the year. They yield soft wood. Structural soft wood is readily available, light-weight, economical, and easy to use. Structural soft wood is widely used in both residential and commercial building frames as studs, joists, bearers, lintels, roof beams, and trusses. Examples of conifers include the Douglas fir, pine, redwood, spruce, larch, and cedar. These woods have desirable amounts of strength, are cost-effective, and are very popular for building construction. Soft woods make up 90% of wood volume (Sivakugan et al. 2018).

Deciduous or angiosperm trees are those which have broad leaves and produce seeds inside ripening fruits. These leaves fall in autumn and new ones appear in spring. They yield hard wood. Hard woods tend to be slower growing and are therefore usually denser. Examples of deciduous trees are oak, mahogany, maple, teak, aspen, ash, balsa, and hickory. Hard woods are strong, durable, tidy, and beautiful. Hard woods are more popular for high-quality furniture, flooring, boats, tools, musical instruments, barrels, etc. Hard woods are also used to make large-span floor joists, high-strength structural beams, bridge girders, and wharves. In summary, the differences between soft wood (coniferous) and hard wood (deciduous) are given below:

- In soft wood, annual rings are seen distinctly, whereas in hard wood, they are not.
- The color of soft wood is light, whereas the color of hard wood is dark.

- Soft woods have less strength in compression and shear compared to hard woods.
- Soft woods are lighter than hard woods.
- Fire resistance of soft wood is poor compared to that of hard wood.
- The structure of soft wood is resinous while the structure of hard wood is close-grained.

10.2.1.2 Endogenous Trees

Endogenous trees grow inwards without producing annual rings. The innermost portions of the trees have fresh fibers. They are not useful for structural works, and therefore are not discussed here in detail. Examples of endogenous trees are bamboo and cane.

10.2.2 CLASSIFICATION BASED ON MODULUS OF ELASTICITY

Young's modulus (E) is determined by conducting the bending test. On this basis, lumber is classified as:

Group A: $E = 12.5$ GPa (1,800 ksi) or higher
Group B: $E = 9.8$–12.5 GPa (1,400–1,800 ksi)
Group C: $E = 5.6$–9.8 GPa (800–1,400 ksi)

10.2.3 CLASSIFICATION BASED ON DURABILITY

Durability tests are conducted by the forest research establishment. They bury test specimens of size $24 \times 2 \times 2$ in. ($600 \times 50 \times 50$ mm) in the ground to half their length, and observe their conditions regularly over several years. Then, lumbers are classified as:

High durability: Average life is more than 10 years.
Moderate durability: Average life between 5 and 10 years.
Low durability: Average life less than 5 years.

10.3 CROSS-SECTION OF WOOD SECTION

The cross-section of an exogenous tree is as shown in Figure 10.3. The following components are visible to the naked eye:

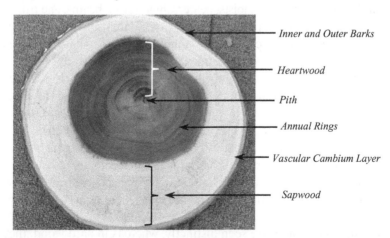

FIGURE 10.3 Cross-section of an exogenous tree. *Photo taken in Pueblo, Colorado.*

10.3.1 Pith

The *pith* is at the center of the tree and is a soft, pulpy zone which is usually about one centimeter in diameter. The pith is sometimes called the medulla, given its medullary rays. The pith is the oldest part of exogenous trees. When the plant becomes old, the pith dies and becomes fibrous and dark. It varies in size and shape.

10.3.2 Heartwood

Heartwood is the dark and dead part of wood surrounding the pith with many annular rings. This portion of wood provides the strength to support the tree and is useful for various engineering purposes.

10.3.3 Annual Rings

Annual rings are the circular rings of wood, grown each year of its age. The annual ring of a tree is created each year when a new layer of wood is added to the trunk. New wood grows from the cambium layer between the old wood and the bark. The annual growth of a tree can be measured by the distance between the growth rings. As the growth rate slows down in the winter months, the new layers of wood cells (late wood) are smaller and packed more closely together, forming a ring which is darker in color than the wood grain which grows at a faster rate earlier in the growing season (early wood).

10.3.4 Sapwood

Sapwood, also known as *alburnum*, is the layer immediately outer of the heartwood. It is the recently grown wood with some sap (wood moisture) and a non-distinct, lighter annual ring. Being immature, sap wood is weaker and less durable than heartwood. In fact, sap wood helps trees grow by allowing sap and mineral salts to move from the root of the tree to the leaves.

10.3.5 Vascular Cambium Layer

The vascular cambium layer is a thin layer of fresh sap bordering the edge of the sapwood. It contains sap which has not yet been converted into sapwood. If the bark is removed and the cambium layer is exposed to atmosphere, cells cease to be active and the tree dies. The *cambium layer* is the layer of thin cells, invisible to the naked eye, positioned inside the live bark. This layer of cells facilitates all growth inside of the tree trunk. The cambium grows wood cells on the inside and live barks cells on the outside.

10.3.6 Inner Bark

Inner bark, also called *live bark*, is the inner skin of the tree protecting the vascular cambium layer. Live bark is a layer of living tissue under the dead bark. Food materials produced by the leaves are conducted through the live bark to the branches, trunk, and roots of the tree.

10.3.7 Outer Bark

Outer bark, also called *dead bark*, is the outer skin of the tree, consisting of wood fibers and dead tissue, which acts as protection for the trunk and branches. Sometimes, it contains fissures and cracks. It also reduces water loss from the living cells of the tree.

10.4   SPECIES AND SPECIES GROUPS OF WOOD

There are about 60,000 tree species in this world, as reported by the BBC News (2017). It is not practical to list all of them for engineering purposes. The species which are particularly useful for civil and construction engineering are commonly discussed. There are some species which are very similar to each other, and so they are grouped together. Thus, a species group is a combination of two or more species. For example, Douglas Fir-Larch (DF-L) is a species group that is obtained from a combination of Douglas Fir (DF) and Western Larch species. Hem-Fir is a species group that can be obtained from a combination of Western Hemlock and White Fir. Spruce-Pine-Fir (SPF) is a species group obtained from a combination of spruces, pines, and firs commonly grown in Canada.

Structural wood members are processed from different stocks of trees. The selection of wood species for use in design and construction is typically a matter of cost-effectiveness, performance, and local availability. For a given location, only a few species groups might be readily available. The species groups that have the highest available strengths are DF-L and Southern Pine, also called Southern Yellow Pine (SYP). Examples of widely used species groups (i.e., combinations of different wood species) of structural lumber in wood buildings include DF-L, Hem-Fir, SPF, and SYP. Each species group has a different set of tabulated design stresses listed in the Nation Design Specification Supplement (NDS-S) by the American Wood Council. Wood species within a particular species group possess similar properties and have the same grading rules, which are also provided in the NDS-S. Some common species and their combinations are shown in Figure 10.4, and are listed in Table 10.1. A larger list of the species, their grading, strength information, modulus of elasticity, etc. can be obtained from the NDS-S (2018) and Wood Handbook (2010).

10.5   PRODUCTION OF WOOD

The production of wood is a very simple but labor-intensive process. Trees are cut down first when they have grown and matured. Then, is the trees are transported to the factory, sawn into the desired sizes and shapes, and seasoned in kilns to reduce moisture, in order to achieve dimensional stability. Three types of sawing can be conducted to produce square or rectangular sections: flat, quarter, and rift. In *flat*, or *plain sawing*, the log is cut with each slice parallel through the log, as shown in Figure 10.5. The lumber produced may twist, bow, warp, or shrink unevenly. In *quarter sawing*, the log is cut into four quadrants, which are at right angles to each other. This process is less vulnerable to distortion, but more expensive. In *rift sawing*, logs are milled radially. This procedure produces stable timbers, but is the most expensive technique. The sawing process is also called the *conversion*.

Seasoning of timber is the process by which the moisture content in the timber is reduced to a suitable level. Then, different wood products are prepared in the factory as desired. Plywood and composite panels are manufactured by peeling, slicing, or chipping wood; pressing and gluing these raw materials into panels, engineered wood shapes, and other wood products; and setting the glue with the application of heat. Particleboard and fiberboard panels can also be made from sawdust, planer shavings, and board trim from lumber operations.

10.6   SEASONING OF WOODS

Seasoning is a process by which moisture content in a freshly cut tree is reduced to a suitable level. By reducing moisture content, the strength, elasticity, and durability properties are

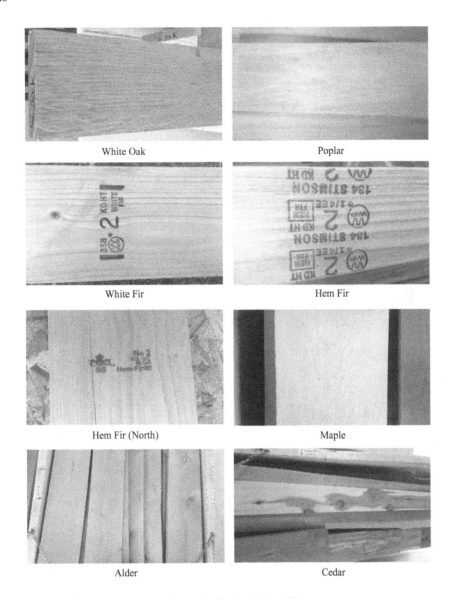

White Oak

Poplar

White Fir

Hem Fir

Hem Fir (North)

Maple

Alder

Cedar

FIGURE 10.4 Some commonly used species in the United States.

developed and/or improved. Seasoning prevents shrinkage, i.e., improves dimensional stability by minimizing the splitting and warping of a wood section. Seasoned woods are also less prone to decay and fungal attack. It is also easier to paint, glue, and apply preservatives on seasoned timber. The various methods of seasoning used may be classified into two categories:

(a) Natural seasoning (water and air)
(b) Artificial seasoning (boiling, chemical seasoning, kiln seasoning, electrical seasoning)

10.6.1 NATURAL SEASONING

Natural seasoning may be air seasoning or water seasoning.

Air seasoning is performed by arranging raw wood or sawn lumber in layers using stickers or strips to ensure vertical and horizontal separation between each member, so that

TABLE 10.1

Some Common Species and Their Combinations

Species or Species Combinations	Species Included in the Combination
Douglas Fir-Larch	• Douglas Fir • Western Larch
Southern Pine	• Loblolly Pine • Longleaf Pine • Shortleaf Pine • Slash Pine
Spruce-Pine Fir	• Alpine Fir • Balsam Fir • Black Spruce • Engelmann Spruce • Jack Pine • Lodgepole Pine • Red Spruce • White Spruce
White Oak	• Bur Oak • Chestnut Oak • Live Oak • Overlap Oak • Post Oak • Swamp Chestnut Oak • Swamp White Oak • White Oak

Reference: Wood Handbook (2010). Wood Handbook: Wood as an Engineering Material, Technical Report FPL-GTR-190, 2010, Forest Service, United States Department of Agriculture, Madison, WI.

Flat Sawing   Quarter Sawing   Rift Sawing

FIGURE 10.5   Three types of sawing procedures used for wood production.

ambient air can circulate around each piece. Some space is left between wood sections so that proper air circulation is achieved. The high moisture content in wood continues to decrease as the wood section loses moisture in the air. With time, the moisture content in the wood reaches the equilibrium moisture with air (no in or out of moisture between air and wood sections). This seasoning is very slow, but cost-effective.

Water seasoning is conducted close to water ponds or rivers. Woods can be simply submerged into water, or with the thicker end of the lumber kept pointing upstream. The sap in the wood is washed out within weeks. However, a short period of air seasoning is required soon after the water seasoning is completed, so as to dry the remaining water out.

10.6.2 ARTIFICIAL SEASONING

The natural seasoning procedures discussed previously take a long time to finish (weeks to months), and a large space is required during that time. To expedite the process, *artificial seasoning* is performed. In this method, lumber is seasoned in a chamber with regulated heat, controlled humidity, and proper air circulation, and allows for seasoning of bulk amounts of wood. This way of seasoning can be completed in 4–5 days. The different methods of artificial seasoning include:

- Boiling
- Kiln seasoning
- Chemical seasoning
- Electrical seasoning

Boiling. Lumber is boiled for 3–4 hours after being submerged into water. After boiling, a short period of air seasoning is required. Hot steam may be circulated on lumber instead of boiling water. The process of seasoning is very fast. The lumber may be ready for usage in 1–2 days.

Kiln Seasoning. A kiln is an airtight chamber where heat, humidity, and the amount and duration of air can be controlled for optimum drying. Then, fully saturated steam and hot air with a temperature of about 95°F–100°F (35°C–38°C) are forced into the kiln where lumber is placed. Once the heat reaches inside the lumber, the humidity is gradually reduced, and the temperature is increased and maintained until the desired moisture content is achieved. The kiln used may be stationary or progressive. In progressive kilns, the carriages carrying lumber travel from one end of the kiln to another end gradually. The hot air is supplied from the discharging end so that the temperature increase is gradual from the charging end to discharging end. This method is used for seasoning on a larger scale.

Chemical Seasoning. Lumber is immersed in a hygroscopic solution of suitable salt (sodium chloride) solution for a few hours; because of this, it is also called salt seasoning. Then the lumber is dried in a kiln or in air. The preliminary treatment by the chemical seasoning ensures uniform seasoning of the outer and the inner parts of lumber. Other dehydrating agents commonly used in the chemical seasoning process are polyethylene glycol, urea, and sodium nitrate (Sivakugan et al. 2018). After the lumber is removed from the solutions, it is seasoned in a natural process like the air seasoning.

Electrical Seasoning. A high-frequency alternate electric current is passed through lumber. Resistance to the electric current is low when the moisture content in lumber is high. As the moisture content reduces, so does the resistance. The measure of resistance can be used to stop seasoning at the appropriate level and time.

10.7 PRESSURE TREATMENT OR INCISING OF WOOD

In addition to seasoning, sometimes pressure treatment or incising of chemicals is required when woods are to be used for outdoor applications, such as roadside barriers, retaining walls, or woods having close-proximity to soil, concrete, or water, to prevent the decay of wood or to protect it from insects or fungus attack. This process includes permeating the wood with chemical preservatives to prevent decay-causing fungus to enter the wood. Wood is placed inside a closed cylinder, and then vacuum and pressure are applied to force the preservatives into the wood. In most cases, pressure treatment is adequate to prevent decay or fungus attack. However, certain species of wood, such as Coastal Douglas-Fir, Hemlock, Hem-fir, Spruce, Lodgepole Pine, Jack Pine, Redwood, Spruce-Pine-Fir, etc., must also be *incised*, to ensure chemicals reach inside the wood.

In incising, sharp steel teeth (similar to stapling) are pressed into all sides of lumber and timbers to increase the penetration of the chemical into the wood during the pressure-treatment process. The purpose of incising is to force preservative chemicals deeper into the wood. Pressure-treated woods look blueish color, and incised woods have a blueish color with many stapling-type penetrations, as shown in Figure 10.6.

Pressure-Treated Lumbers   Incised Lumbers

FIGURE 10.6   Pressure-treated and incised woods sections. *Photos taken in Colorado Spring.*

Three broad classes of preservatives are commonly used in pressure treating and/or incising wood. These are mentioned below:

- Waterborne
- Creosote
- Oil-borne (penta)

Wood treated with waterborne preservatives is typically used in residential, commercial, and industrial building structures. Creosote is primarily used for treating railroad ties, guardrail posts, and timbers used in marine structures. Oil-borne (penta) is most often used for treating utility poles and cross arms.

10.8   DEFECTS IN WOOD

Various defects which are likely to occur in lumber may be grouped into the following three sources:

- Due to natural forces

- Due to defective seasoning and conversions
- Due to fungi and insects attack

10.8.1 Defects due to Natural Forces

The following primary defects are caused by natural forces:

- Knots
- Shakes
- Wind cracks
- Upsets

Knots. When a tree grows, many of its branches fall and the stump of these branches in the trunk is covered. In the sawn pieces of lumber, the stumps of fallen branches appear as *knots*, as shown in Figure 10.7. Knots are dark and hard pieces. Grains are distorted in this portion of the wood. If the knot is intact with the surrounding wood, it is called a live knot. If it is not held firmly, it is a dead knot.

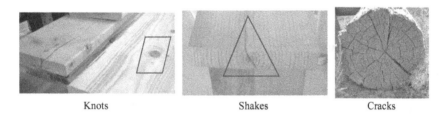

Knots   Shakes   Cracks

FIGURE 10.7   Some natural wood defects. *Photos taken in Colorado Spring.*

Shakes. The *shakes* are cracks in the lumber which appear due to excessive heat, frost, or twisting due to wind during the growth of a tree. Depending on the shape and position, shakes can be classified as star shakes, cup shakes, ring shakes, and heart shakes.

Wind Cracks. These are the cracks which occur on the outside of a log due to the shrinkage of the exterior surface.

Upsets. This type of defect is due to excessive compression in the tree when it was young. Upset is an injury caused by crushing, and is also known as *rupture*.

10.8.2 Defects due to Defective Seasoning and Conversion

If seasoning is not uniform, the converted lumber may warp and twist in various directions. Sometimes, honey combining and even cracks appear. These types of defects are more susceptible in the case of kiln seasoning. In the process of converting lumber to commercial sizes and shapes, the following types of defects (Figure 10.8) are likely to arise: bow, spring, cup, and twist.

- *Bow* – A curvature in the longitudinal direction of the plank.
- *Spring* – A curvature of a plank in the plane of its wide face.
- *Cup* – A concave curvature of wood across its width in the transverse direction.
- *Twist* – A spiral distortion of a wood member along its length.

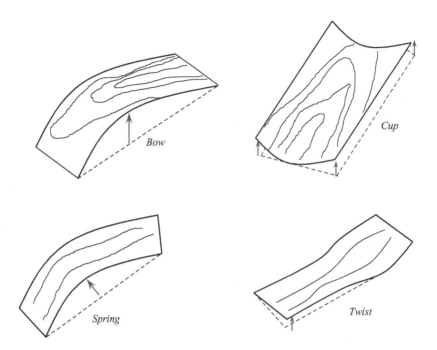

FIGURE 10.8   Defects during seasoning and conversion.

10.8.3   Defects due to Fungi and Insects Attack

Fungi are minute, microscopic plant organisms. They grow in wood if the moisture content is greater than about 20% and exposed to air. When wood experiences fungal attack, it will begin to rot. Wood becomes weak and stains will appear on it. Beetles, marine borers, and termites (white ants) are insects which eat wood and weaken the lumber. Some woods, like teak, have chemicals in their compositions naturally, and resist such attacks. Other woods must be protected by chemical treatment.

10.9   SIZES AND DRESSING

10.9.1   Size Provision

Lumber is commonly available on the market with its sawn dimension, known as *nominal size*, *stated size*, or *full-sawn size*. The lumber is then further dressed or surfaced to have a smooth surface and uniform size. Depending on the smoothness level, it may be *rough-sawn* (rough-smooth) or *dressed* (full-smooth). This dressed dimension is less than the nominal dimension.

Typically, lumber is S4S (*surfaced four sides*) before it is used in a structure, especially if it is to be exposed after construction. Some other finishes are also possible. For example, S2S1E, meaning surfaced two sides and one edge, may be used.

Full-sawn lumber is first rough sawn and finally dressed to have a smooth surface. The cross-sectional dimensions of rough-sawn lumber and standard dressed lumber are also dependent on the green or dry conditions. NDS-S (2018) Tables 1A and 1B list the dressed dimensions of different sections. Consider an 8×12 member, i.e., nominal size $= 8 \times 12$ in. Then, its:

Dressed lumber = 7-½ in. × 11-½ in.
Rough-sawn lumber = 7–5/8 in. × 11–5/8 in.
Full-sawn lumber = 8 in. × 12 in.

10.9.2 SIZE CLASSIFICATION

Depending on the size, sawn lumbers have different names, as listed in Figure 10.9 and Table 10.2.

Dimension lumber, which usually comes in lengths of 8–20 ft, is typically used for floor joists or roof rafters/trusses, and 2 × 8, 2 × 10, and 2 × 12 are the most frequently used floor joist sizes. For light-frame residential construction, dimension lumber is generally used.

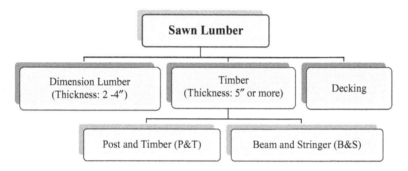

FIGURE 10.9   Types of wood sections based on size.

TABLE 10.2
Size Classifications for Sawn Lumber

Lumber classification	Size
Dimension lumber	Nominal thickness: 2–4 in.
	Nominal width: ≥2 in. but ≤16 in.
	Examples: 2×6, 3×8, 2×12, 4×10
Post and timber (P&T)	Approximately square cross-section
	Nominal thickness: ≥5 in.
	Nominal width: ≤2 in. + nominal thickness
	Examples: 5×5, 6×6, 10×12
Beam and stringer (B&S)	Rectangular cross-section
	Nominal thickness: >5 in.
	Nominal width: >2 in. + nominal thickness
	Examples: 6×10, 6×18, 8×20
Decking	Wide face applied directly in contact with framing
	Sometimes, tongue and grooved
	Used as roof or floor sheathing
	Example: 2×12 lumber used in a flatwise direction, 5/4×6, and 3×6

Note: Mentioning no unit in structural wood sections means inches. For example, a 2×6 section means a 2 in. × 6 in. nominal-sized section.

Beam and stringer (B&S) lumber is used as floor beams or girders, as well as door or window headers. *Post and timber (P&T) lumber* are used for columns or posts. Dimension lumber and many other lumber products are often supplied in length increments of 2 ft (e.g., 10 ft, 12 ft, etc.).

10.10   GRADING OF LUMBER

Grading is an essential step in lumber production. It provides and assures the engineer, architect, and contractor of the quality of the lumber in terms of strength, visual quality, moisture content, etc. Examples of grading agencies in the United States include:

- Western Wood Products (WWP) Association
- West Coast Lumber Inspection Bureau
- Northern Softwood Lumber Bureau
- Northeastern Lumber Manufacturers Association
- Southern Pine Inspection Bureau

Two types of grading systems are used for structural lumber:

a) Visual grading
b) Mechanical grading

After grading, a grade stamp is placed on the lumber section. A grade stamp includes the mill number where the wood was produced, and the responsible grading agency, moisture content, seasoning method, stress grading, etc.

10.10.1   Visual Grading

Visual grading is the oldest and most common grading system for grading lumber by species. It involves the visual inspection of lumbers by an experienced and certified grader following the grading rules of the corresponding grading agency. Then, a grade stamp is placed on the lumber, as shown in Figure 10.10. Basically, the lumber stress quality is judged by the presence of defects, severity, and locations. For example, a knot near the neutral axis

Showing Western Wood Products(WWP) as the agency, the grade as No. 2, Kiln-Dried (KD), Heat Treated (HT), and producing mill lumber 358. The wood species is White Fir.

Showing Western Wood Products(WWP) as the agency, the grade as No. 2, Kiln-Dried (KD), Heat Treated (HT), and producing mill lumber 134. The wood species is Hem Fir. 1/4 EE means ¼-in. eased edge (rounded) to help in handling.

FIGURE 10.10   Visual grading stamps. *Photos taken in Pueblo, Colorado.*

TABLE 10.3

Reference Design Values for Visually Graded Dimension Lumber of Douglas Fir–Larch Dimension Lumber

Grading	Bending Strength, psi (MPa)	Tensile Strength Parallel to Grain, psi (MPa)	Shear Strength Parallel to Grain, psi (MPa)	Compression Strength Perpendicular to Grain, psi (MPa)	Compression Strength Parallel to Grain, psi (MPa)	Modulus of Elasticity, ksi (MPa)
Select Structural	1,500 (10.3)	1,000 (6.9)	180 (1.2)	625 (4.3)	1,700 (11.7)	1,900 (13,100)
No. 1 and Better	1,200 (8.3)	800 (5.5)	180 (1.2)	625 (4.3)	1,550 (10.7)	1,800 (12,400)
No. 1	1,000 (6.9)	675 (4.7)	180 (1.2)	625 (4.3)	1,500 (10.3)	1,700 (11,700)
No. 2	900 (6.2)	575 (4.0)	180 (1.2)	625 (4.3)	1,350 (9.3)	1,600 (11,000)
No. 3	525 (3.6)	325 (2.2)	180 (1.2)	625 (4.3)	775 (5.3)	1,400 (9,700)
Stud	700 (4.8)	450 (3.1)	180 (1.2)	625 (4.3)	850 (5.9)	1,400 (9,700)
Construction	1,000 (6.9)	650 (4.5)	180 (1.2)	625 (4.3)	1,650 (11.4)	1,500 (10,300)
Standard	575 (4.0)	375 (2.6)	180 (1.2)	625 (4.3)	1,400 (9.7)	1,400 (9,700)
Utility	275 (1.9)	175 (1.2)	180 (1.2)	625 (4.3)	900 (6.2)	1,300 (9,000)

Adapted from NDS-S 2018, Table 4A, National Specifications for Wood Construction Supplement (NDS-S): Design Values for Wood Construction, American Wood Council, Leesburg, Virginia. Used with permission.

of a wood member that may be subjected to bending is not as critical as a knot near the tension face of the member. The stress grade of lumber also decreases as the number and size of defects increases for a certain species. Therefore, the effectiveness of the grading system is fully dependent on the experience of the professional grader. Grading agencies usually have certification exams that lumber graders have to take and pass annually to maintain their certification and to ensure accurate and consistent grading of sawn lumber.

The various lumber stress grades, along with their few reference design values, are listed in Table 10.3 for a wood species, Douglas-Fir-Larch (DF-L), Dimension Lumber (2–4 in. thick). These values are adopted from NDS-S (2018). All these values are for permanent load duration and dry service conditions. The details of grain condition (parallel to grain and perpendicular to grain) are discussed in detail later in this chapter. The first column of Table 10.3 shows the stress grading such as Select structural, No. 1 and better, No. 1, etc. The bending strength of DF-L Dimension Lumber varies from 275 to 1,500 psi (1.90 to 10.34 MPa), depending on the size, number, and location of defects. Similarly, the modulus of elasticity varies from 1,300 to 1,900 ksi (8,963 to 13,100 MPa), from one grade to another. This comparison shows the importance of grading, and how the strength and properties of a lumber piece of same species vary based on defect size, number, and location.

10.10.2 Mechanical Grading

Mechanical grading is a nondestructive grading system that is based on the relationship between the stiffness and strength of wood members, regardless of wood species. The lumber section is mechanically tested in the laboratory to determine its strength and stiffness, in addition to visual inspections. There are two types of machine-graded lumber used in North America:

a) Machine stress-rated (MSR) lumber
b) Machine-evaluated lumber (MEL)

The mechanical grading process for machine stress-rated (MSR) lumber involves the use of mechanical stress-rating equipment to obtain measurements of deflections of the sawn lumber, from which the stiffness is calculated. From the strength information obtained, a strength grade is assigned (Canadian Wood Council 2011). In addition to the nondestructive test, each piece of wood is also subjected to a visual check. The grading is expressed as reference flexural/bending strength, F_b, in psi, and the modulus of elasticity, E, in mega-pound per square inch. (Msi). For example, the designation, '750f-1.4E' means its reference flexural/bending strength value is 750 psi and its modulus of elasticity is 1.4 Mpsi, or 1,400,000 psi. NDS-S (2018) lists reference design values for MSR, from 750f-1.4E to 3000f-2.4E.

The grade stamp on mechanically graded lumber includes the value of the reference bending stress, F_b, in psi for single member usage and the modulus of elasticity, E, in mega-pound per square in. (Msi), the designation 'Machine Rated' or 'MSR,' species, specific gravity (SG), producing mill number, etc. For example, a stamp shows Machine Rated by the Western Wood Products (WWP) (the grading agency), the grade is 2100f-1.8E (meaning flexural/bending strength of 2,100 psi and elasticity of 1,800,000 psi), specific gravity of 0.47, service dry, and producing mill number 12. Additionally, the wood species is Southern Pine-Fir Spruce (SPFS). Some MSR grading, along with their flexural/bending strength, tensile strength, compression strength, and modulus of elasticity from the Wood Handbook (2010) are shown in Table 10.4. The full list is available at the NDS-S (2018).

In machine-evaluated lumber (MEL), a nondestructive X-ray inspection technique that measures density is used, in addition to a visual check. The grade stamp on MEL

TABLE 10.4
Examples of Machine Stress Rated (MSR) Grading System with Some References

Machine Stress-Rated (MSR) Commercial Grading	Bending Strength, psi (MPa)	Modulus of Elasticity, ksi (GPa)	Tensile Strength Parallel to Grain, psi (MPa)	Compression Strength Parallel to Grain, psi (MPa)
1350f-1.3E	1,350 (9.3)	1,300 (9.0)	750 (5.2)	1,600 (11.0)
1450f-1.3E	1,450 (10.3)	1,300 (9.0)	800 (5.5)	1,625 (11.2)
1650f-1.5E	1,650 (11.4)	1,500 (10.3)	1,020 (7.0)	1,700 (11.7)
1800f-1.6E	1,800 (12.4)	1,600 (11.0)	1,175 (8.1)	1,750 (12.1)
1950f-1.7E	1,950 (13.4)	1,700 (11.7)	1,375 (9.5)	1,800 (12.4)
2100f-1.8E	2,100 (14.5)	1,800 (12.4)	1,575 (10.9)	1,875 (12.9)
2250f-1.9E	2,250 (15.5)	1,900 (13.1)	1,750 (12.1)	1,925 (13.3)
2400f-2.0E	2,400 (16.5)	2,000 (13.8)	1,925 (13.3)	1,975 (13.6)

Reference: Wood Handbook (2010). Wood Handbook: Wood as an Engineering Material, Technical Report FPL-GTR-190, 2010, Forest Service, United States Department of Agriculture, Madison, WI.

includes a strength grade designation 'M' followed by a number (e.g., M-10 or M-24), and includes the reference bending stress, the reference tension stress, and the modulus of elasticity. NDS-S lists reference design values for MEL from M-5 to M-40 in Table 4C. Some MEL grading, along with their flexural/bending strength, tensile strength, compression strength, and modulus of elasticity from the Wood Handbook (2010) are shown in Table 10.5. The full list is available at the NDS-S (2018).

TABLE 10.5

Examples of Machine Evaluated Lumber (MEL) Grading System with Some References

Machine-Evaluated Lumber (MEL) Commercial Grading	Bending Strength, psi (MPa)	Modulus of Elasticity, ksi (GPa)	Tensile Strength Parallel to Grain, psi (MPa)	Compression Strength Parallel to Grain, psi (MPa)
M-10	1,400 (9.7)	1,200 (8.3)	800 (5.5)	1,600 (11.0)
M-11	1,550 (10.7)	1,500 (10.3)	850 (5.9)	1,675 (11.5)
M-14	1,800 (12.4)	1,700 (11.7)	1,000 (6.9)	1,750 (12.1)
M-19	2,000 (13.8)	1,600 (11.0)	1,300 (9.0)	1,825 (12.6)
M-21	2,300 (15.9)	1,900 (13.1)	1,400 (9.7)	1,950 (13.4)
M-23	2,400 (16.5)	1,800 (12.4)	1,900 (13.1)	1,975 (13.6)
M-24	2,700 (18.6)	1,900 (13.1)	1,800 (12.4)	2,100 (14.5)

Reference: Wood Handbook (2010). Wood Handbook: Wood as an Engineering Material, Technical Report FPL-GTR-190, 2010, Forest Service, United States Department of Agriculture, Madison, WI.

10.11 UNITS OF MEASUREMENT

The standard unit of measurement for lumber in the United States is the board foot (bf), which is equal to 144 in.^{3} of lumber using the nominal dimensions. A thousand board feet can be abbreviated as Mbf. Similarly, a million board feet can be abbreviated as MMbf.

144 in.^{3} of lumber $= 1$ bf
1,000 bf $= 1$ Mbf
1,000,000 bf $= 1$ MMbf

This is the common unit in North America. Different local units may be used in different regions.

Example 10.1 Board Foot

What is the board foot (bf) of a 4×8 1umber that is 20 ft long?

Note: No unit in structural wood section means inch.

Solution

$$\frac{4\times8\times\left(20\times12\ \frac{in.}{ft}\right)}{144\ in.^3} = 53.333\ bf$$

Answer: 53 bf

10.12 PHYSICAL PROPERTIES

Some of the major physical properties of wood are described below:

10.12.1 ANISOTROPY

Wood is an *orthotropic* and *anisotropic* material. Orthotropic materials have material properties different along the three mutually perpendicular directions. Anisotropic means a physical property has a different value when measured in different directions.

A tree increases in diameter by adding an annual ring every year as it grows. The annual rings are *not* tightly bonded, in most species. The wood fibers are also oriented in the longitudinal direction (vertically upward). Due to these two conditions, properties vary along the three mutually perpendicular axes: longitudinal, radial, and tangential (Figure 10.11). The longitudinal axis is parallel to the fiber (grain) direction, the radial axis is perpendicular to the grain direction and normal to the growth rings, and the tangential axis is perpendicular to the grain direction and tangent to the growth rings. Most wood properties differ in each of these three directions. However, the differences between the radial and tangential axes are minor compared to the differences between any other two axes. Consider the examples of strength and properties listed in Tables 10.3 to 10.5.

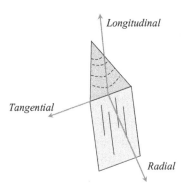

FIGURE 10.11   Anisotropy of a wood slice.

10.12.2 MOISTURE CONTENT AND SHRINKAGE

Moisture is the most detrimental factor reducing the strength, performance, and service life of wood and wood products. More specifically,

- Moisture variation causes expansion and contraction of wood and wood products.
- The integrity of bonded wood products can be destroyed by swelling-induced stresses when wet.
- Continuing deflection over time of wood members increases with repetitive moisture variations.
- Mechanical connections, such as nailing and bolting, can be hampered with repetitive moisture variations.
- Excess moisture attracts biological attack.

Therefore, excess moisture or repeated moisture variation is an important topic to study. Wood products dried to an appropriate level and maintained within a reasonable range of moisture content provide desirable service for a prolonged amount of time. On the other hand, inadequate drying or high service moisture may cause a very premature failure.

The solid portion of wood is made up of a complex cellulose-lignin compound. The cellulose makes the framework of the cell walls and the lignin binds the cells together by cementing. In addition to the cellulose-lignin, the wood of living trees can contain up to 200% moisture (Breyer et al. 2014). Moisture content (MC) can be defined as the ratio of water to the solid part of wood, and is expressed as shown in Eq. 10.1:

$$MC = \frac{\text{Moist Weight} - \text{Oven Dry Weight}}{\text{Oven Dry Weight}} \times 100 \text{ percent} \qquad (10.1)$$

The average moisture contents of some species are listed in Table 10.6. A more detailed list is available in the Wood Handbook (2010).

TABLE 10.6
Average Moisture Content of Green Wood

Species	Moisture Content (%)		Species	Moisture Content (%)	
	Heartwood	Sapwood		Heartwood	Sapwood
Aspen	95	113	Douglas Fir	37	115
Maple	58	97	White Fir	98	160
Oak (White)	64	78	Western Larch	54	119
Cottonwood	162	146	Pine (Red)	32	134
Cedar (Western Red)	58	249	Balsam Fir	88	173
Black Spruce	52	113	Western Hemlock	85	170

Reference: Wood Handbook (2010). Wood Handbook: Wood as an Engineering Material, Technical Report FPL-GTR-190, 2010, Forest Service, United States Department of Agriculture, Madison, WI.

Although living trees have very high MC, structure lumber while in service has a lot less moisture. The average moisture that a structural lumber contains while in service is called the *equilibrium moisture content* (EMC). The EMC varies between and 2 and 24%, depending on ambient temperature, relative humidity, and species (Wood Handbook, 2010). At the time of construction, the MC is about twice of the EMC, though it decreases with service age and eventually reaches the EMC (Breyer et al. 2014). EMC also depends on the ambient humidity of areas to be used. For example, the EMC of a lumber in New Mexico (dry area) is not the same as that lumber piece of the same species in Florida (humid area). This is why lumbers processed in wet areas should be used in construction in wet areas, and vice versa. In fact, it is a common practice in the construction industry that wood sections are to be stored close to the construction site for months while preparing the site.

Wood holds moisture in two ways: *free water* and *bound water.* Water contained in the cell cavity is known as free water, and water contained within the cell walls is known as bound water. As wood dries, the first water to be driven off is the free water. When all the free water is dried off (i.e., only bound water remaining), the condition is known as the *fiber saturation point* (FSP). No loss of bound water occurs as lumber dries off above the FSP, as shown in Figure 10.12. In addition, neither a volume change nor a change in structural properties occurs when lumber dries off above the FSP. However, wood shrinks or swells if moisture dries off or increases, respectively, below the FSP. A decrease in moisture below the FSP accompanies a gain in strength, and the reaches the peak strength at some MC, roughly 10–15%. For MC below that point, strength remains constant. The FSP changes with the species and, on average, is 30% (Wood Handbook 2010).

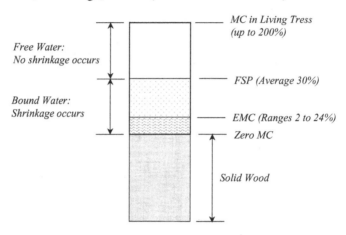

FIGURE 10.12 Phase diagram of a wood piece.

Two important engineering properties can be summarized as follows:

a) Wood shrinks if MC decreases below the FSP. This is a volume loss.
b) Strength of wood increases, reaches a peak, and then remains constant if MC decreases below the FSP.

Above the fiber saturation point, wood does not shrink or swell from changes in moisture content. This is because free water is found only in the cell cavity and is not associated within the cell walls. However, wood changes in dimension, as moisture content varies below the fiber saturation point. Wood shrinks as it loses moisture below the fiber saturation point and swells as it gains moisture up to the fiber saturation point. These dimensional changes may result in splitting, checking, and warping. The phenomena of dimensional stability and EMC must be understood; dimensional stability of wood is one of the few properties that significantly differs in each of the three axis directions. Dimensional changes in the longitudinal direction between the fiber saturation point and oven-dry state are between 0.1% and 0.2%, and are of no practical significance; however, in reality, these percentages may be significantly higher. The combined effects of shrinkage in the tangential and radial axes can distort the shape of wood pieces because of the difference in shrinkage and the curvature of the annual rings. Generally, tangential shrinkage (varying from 4.4% to 7.8%, depending on species) is twice that of radial shrinkage (from 2.2% to 5.6%). Longitudinal shrinkage of wood (shrinkage parallel to the grain) is generally quite small and very often neglected.

There are several different methods of calculating shrinkage. These are described as follows:

Wood Handbook (2010) Method: The Wood Handbook (2010) assumes that no shrinkage occurs at the FSP of 30% MC, and full shrinkage occurs at zero MC. A linear interpolation is used for shrinkage at intermediate MC values. Values of tangential, radial, and volumetric shrinkage from clear woods are listed for many species. The maximum shrinkage can be estimated using the tangential shrinkage, and the minimum can be evaluated with the radial value. Thus, the Wood Handbook (2010) method can be used to determine the range of shrinkage values for numerous species. Table 10.7 lists shrinkage values of a few common species.

TABLE 10.7
Shrinkage Values of a Few Common Species

Species	Shrinkage (%) from Green to Oven-Dry Moisture Content		
	Radial	Tangential	Volumetric
Aspen (Bigtooth)	3.3	7.9	11.8
Beech (American)	5.5	11.9	17.2
White Oak	5.6	10.5	16.3
Cedar (Yellow)	2.8	6.0	9.2
Douglas Fir (Coast)	4.8	7.6	12.4
White Fir	3.3	7.0	9.8
Red Pine	3.8	7.2	11.3
Black Spruce	4.1	6.8	11.3
Slash Pine	5.4	7.6	12.1
Western Larch	4.5	9.1	14.0

Reference: Wood Handbook (2010). Wood Handbook: Wood as an Engineering Material, Technical Report FPL-GTR-190, 2010, Forest Service, United States Department of Agriculture, Madison, WI.

Rummelhart and Fantozzi (1992) Simplified Method: A simplified method is proposed by Rummelhart and Fantozzi (1992). In this method, a constant shrinkage value of 6% is used for both the width and the thickness of a member. The shrinkage is taken as zero, as FSP of 30% and the full 6% shrinkage is assumed to occur at zero MC. That means a uniform, constant shrinkage value of $6/30 = 0.002$ (0.2%) per 1% change in MC is assumed in this method. Then, a linear relationship is used for shrinkage at any MC value between 0 and 30%. More specifically, the amount of shrinkage (S) can be calculated using Eq. 10.2:

$$S = SV \times d \times \Delta_{MC} \tag{10.2}$$

where,
S = Shrinkage (same unit of the dimension)
SV = Shrinkage value, say 0.002 in./in. per 1% decrease in MC
d = Dimension of the section
Δ_{MC} = Change in MC (%)

Green (1989) Method: Another method is proposed by Green (1989). It provides formulas for calculating the percentage of shrinkage for the width and thickness of a piece of lumber, and it is included in the appendix of the ASTM D 1990 standard.

Despite these many methods of shrinkage calculation, the simplified one proposed by Rummelhart and Fantozzi (1992) is very often preferred because of the following:

- Shrinkage varies for a given member in radial and tangential directions
- As the orientation of the annual rings in a real piece of lumber is unknown, an exact calculation of shrinkage is nearly impossible
- The actual species of a member may not be known, although the designer may know the species groups

Example 10.2 Moisture Content of Wood

A 520-g wet wood is oven-dried to 500 g. What is the moisture content of the wood piece?

Solution

Moist weight = 520 g
Dry weight = 500 g

From Eq. 10.1: Moisture content, $MC = \dfrac{\text{Moist Weight} - \text{Oven Dry Weight}}{\text{Oven Dry Weight}} \times 100$ percent

Therefore, $MC = \left(\dfrac{520\ g - 500\ g}{500\ g}\right) \times 100 = 4\%$

Answer: 4.0%

Example 10.3 Shrinkage of Wood

The profile of a three-storied wood frame (neglecting the floor sheathing) is shown in Figure 10.13. The shrinkage value of the wood used is 0.002 in./in per 1% change in MC. If MC decreases by 14%, determine the settlement of the building. Ignore the longitudinal shrinkage and the floor sheathing.

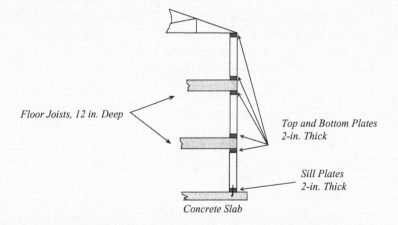

FIGURE 10.13   A short profile of a wood building for Example 10.3.

NDS-S (2018) lists that the dressed (smoothed) sizes of nominal sizes of 2 in. and 12 in. are 1.5 × 11.25 in., respectively.

Solution

Shrinkage of wall-studs ≈ 0, as longitudinal shrinkage is neglected.

Dressed size of 12-in. joist = 11.25 in.

Dressed size of 2-in. joist = 1.5 in.

From Eq. 10.2:

Shrinkage of a joist = $SV \times d \times \Delta_{MC}$ = 0.002 in./in per 1% change in MC × 11.25 in. × 14% = 0.315 in.

Shrinkage of 2 floor joists = 2 × 0.315 = 0.63 in.

Shrinkage of a plate = $SV \times d \times \Delta_{MC}$ = 0.002 in./in per 1% change in MC × 1.50 in. × 14% = 0.042 in.

Shrinkage of 6 plates = 6 × 0.042 in. = 0.252 in.

Total settlement = 0.63 in. + 0.252 in. = 0.882 in.

Answer: 0.882 in.

10.12.3 RESIDUAL/GROWTH STRESS

Longitudinal growth stresses are present in all standing trees which maintain a vertical position. They may also be present in cut large logs but may be released when trees are processed into smaller lumbers. Sometimes, growth stress causes splitting, cracking, and warping, thus resulting in loss of potential engineering value. The residual stresses can be removed by wetting the lumber surface with steam or steaming by water spraying.

10.12.4 DENSITY AND SPECIFIC GRAVITY

The density of a material is the amount of mass per unit volume of the material. For a hydroscopic (water-loving) material like wood, density depends on two factors: the weight of the wood structure and moisture retained in the wood. The density of wood at various moisture contents varies greatly, and thus should be reported along with the corresponding moisture content as well. The higher the density is, the higher the strength of the wood member. Sometimes, unit weight is used in place of density. Unit weight is the weight (mass × gravitational acceleration) of material per unit volume of the material. Specific gravity is a dimensionless ratio of the mass of an oven-dry volume of wood to the mass of an equal volume of water. Specific gravity is thus related to density, as specific gravity is the ratio of density of the material to the density of water.

The density of wood is a function of the MC of the wood, and the weight of the wood substance or cellulose present in a unit volume of wood. The density of wood can vary widely between species, from as low as 20 lb/ft³ (320 kg/m³) to as high as 65 lb/ft³ (1,040 kg/m³), for the most commonly used species (Faherty and Williamson 1995). The density of wood *up to* 30% MC can be obtained from Eq. 10.3 (AITC 2012):

$$\gamma_{wood} = \frac{SG_o}{1 + 0.265\, SG_o \left(\dfrac{MC}{30}\right)} \left(1 + \frac{MC}{100}\right)\left(62.4\,\frac{\text{lb}}{\text{ft}^3}\right) \tag{10.3}$$

where,

SG_o = Specific gravity of oven-dried wood (i.e., at MC = 0%)

γ_{wood} = Specific gravity of wood at the specified MC (1b/ft³)

MC = Specified moisture content between 0% and 30%

For MC *above* 30%, the density of wood is calculated using Eq. 10.4:

$$\gamma_{wood} = \frac{1.3\,SG_o}{1+0.265\,SG_o}\left(1+\frac{MC-30}{30}\right)\left(62.4\frac{\text{lb}}{\text{ft}^3}\right) \qquad (10.4)$$

The density of wood at a MC of 12% is typically used in design to calculate the self-weight of the wood member and dead loads (AITC 2012). A wood density of 32 pcf (510 kg/m³) is commonly used for seasoned wood unless otherwise noted. The dry density of wood ranges from 10 pcf (160 kg/m³) for balsa, to 83 pcf (1,330 kg/m³) for black ironwood. The majority of wood types have densities in the range of 20–45 pcf (320–720 kg/m³) (Mamlouk and Zaniewski 2017).

Example 10.4 Specific Gravity of Wood

The specific gravity of a wood piece at zero moisture content is 0.50. What is the unit weight of that wood piece at 10% moisture content and at 40% moisture content?

Solution

At 10% moisture content (from Eq. 10.3):

$$\gamma_{wood} = \frac{SG_o}{1+0.265\,SG_o\left(\dfrac{MC}{30}\right)}\left(1+\frac{MC}{100}\right)\left(62.4\frac{\text{lb}}{\text{ft}^3}\right)$$

$$= \frac{0.50}{1+0.265(0.50)\left(\dfrac{10}{30}\right)}\left(1+\frac{10}{100}\right)\left(62.4\frac{\text{lb}}{\text{ft}^3}\right)$$

$$= 32.9\frac{\text{lb}}{\text{ft}^3}$$

At 40% moisture content (from Eq. 10.4):

$$\gamma_{wood} = \frac{1.3\,SG_o}{1+0.265\,SG_o}\left(1+\frac{MC-30}{30}\right)\left(62.4\frac{\text{lb}}{\text{ft}^3}\right)$$

$$= \frac{1.3(0.50)}{1+0.265(0.50)}\left(1+\frac{40-30}{30}\right)\left(62.4\frac{\text{lb}}{\text{ft}^3}\right)$$

$$= 47.8\frac{\text{lb}}{\text{ft}^3}$$

Answers: 32.9 pcf and 47.8 pcf

10.12.5 Electrical Property

The electrical conductivity of wood is determined by its resistance to the flow of electrical current. It depends on wood species, temperature, humidity, and the grain of the wood. Electrical conductivity of dry wood is negligible, which allows it to be used as an insulating material. However, as the moisture content increases, the electrical conductivity increases dramatically. As the moisture content of wood increases from near zero to the fiber saturation, resistivity can decrease by a factor of over 10^{10}. Resistivity is about 10^{15}–10^{16} Ωm for oven-dry wood, and 10^{3}–10^{4} Ωm for wood at the fiber saturation point (Wood Handbook, 2010). As the moisture content increases from fiber saturation to full saturation, the decrease in resistivity is smaller, less than a hundredfold.

10.12.6 Thermal Property

Wood has a low thermal conductivity, since it has a lot of voids filled with air. The thermal conductivity of wood depends on its moisture, grain orientation, density, extractive content, and structural integrity. The less the amount of moisture present, the lower the thermal conductivity is. As the moisture content increases from 0% to 40%, the thermal conductivity increases by about 30%. Heat flow in the longitudinal direction is roughly 2.0–2.8 times faster than that in the radial direction. Heat flow across the radial and tangential directions is almost equal. The thermal conductivity ranges from 0.34 Btu/(h-ft-°F) (0.06 W/(m°K) for balsa, to 1.16 Btu/(h-ft-°F) (0.17 W/(m°K) for rock elm (Mamlouk and Zaniewski 2017). The coefficient of thermal expansion measures the fractional change in size per degree change in temperature. For both hard and soft woods, the longitudinal thermal expansion coefficient value ranges from 0.0000017 to 0.0000025 per °F. The thermal expansion of oven-dry wood parallel to the grain is not dependent on the specific gravity and species.

10.13 MECHANICAL PROPERTIES

Mechanical properties are the behaviors related to an applied load and the resulting deformation. In civil and construction engineering, mechanical properties are the primary consideration when selecting a material for a job. Some commonly used mechanical properties of woods are discussed in this section. Remember that the definitions of these mechanical properties are discussed throughout this book.

10.13.1 Modulus of Elasticity and Modulus of Rigidity

As previously discussed in this chapter, wood is an orthotropic material, meaning it has different properties along the three mutually perpendicular directions (longitudinal, radial and tangential). The modulus of elasticity (E) and modulus of rigidity (G) also change with the direction.

Assume the following:

- L is the longitudinal axis parallel to the fiber (grain)
- R is the radial axis normal to the growth rings (perpendicular to the grain in the radial direction)
- T is the tangential axis (perpendicular to the grain but tangent to the growth rings)

The three moduli of elasticity denoted by E_L, E_R, and E_T are the elastic moduli along the longitudinal, radial, and tangential axes of wood, respectively. Average values of E_R and

E_T for samples from a few species are presented in Table 10.8 as ratios with respect to E_L. NDS-S (2018) lists the modulus of elasticity values of different species according to their visual or mechanical grading.

TABLE 10.8
Elasticity Ratios of Woods at 12% Moisture Content

Species	E_T/E_L	E_R/E_L	G_{LR}/E_L	G_{LT}/E_L	G_{RT}/E_L
Ash (white)	0.080	0.125	0.109	0.077	–
Balsa	0.015	0.046	0.054	0.037	0.005
Yellow Birch	0.050	0.078	0.074	0.068	0.017
Red Oak	0.082	0.154	0.089	0.081	–
Red Maple	0.067	0.140	0.133	0.074	–
Douglas Fir	0.050	0.068	0.064	0.078	0.007
Western Red Cedar	0.055	0.081	0.087	0.086	0.005
Western Hemlock	0.031	0.058	0.038	0.032	0.003
Red Pine	0.044	0.088	0.096	0.081	0.011
Redwood	0.089	0.087	0.066	0.077	0.011

Reference: Wood Handbook (2010). Wood Handbook: Wood as an Engineering Material, Technical Report FPL-GTR-190, 2010, Forest Service, United States Department of Agriculture, Madison, WI.

The modulus of rigidity, also called shear modulus, indicates the resistance to deflection of a member caused by shear stresses. The three moduli of rigidity denoted by *GLR*, *GLT*, and *GRT* are the elastic constants in the *LR*, *LT*, and *RT* planes, respectively. For example, *GLR* is the modulus of rigidity based on shear strain in the *LR* plane and shear stresses in the *LT* and *RT* planes. Average values of shear moduli of a few species expressed as ratios with respect to E_L are listed in Table 10.8. Wood Handbook (2010) has a bigger list with numerous species.

Example 10.5 Modulus of Wood

The elastic modulus of Balsa in the longitudinal direction is 1,500,000 psi. What are its elastic moduli values along the tangential and radial directions, and shear modulus of rigidity based on shear strain in the LR plane?

Solution
From Table 10.8, $E_T/E_L = 0.015$
Therefore, $E_T = 0.015(E_L) = 0.015(1,500\ \text{ksi}) = 22.5\ \text{ksi}$
From Table 10.8, $\dfrac{E_R}{E_L} = 0.046$

Therefore, $E_R = 0.046(E_L) = 0.046(1,500\ \text{ksi}) = 69\ \text{ksi}$
From Table 10.8, $\dfrac{G_{LR}}{E_L} = 0.054$

Therefore, $G_{LR} = 0.054(E_L) = 0.054(1,500\ \text{ksi}) = 81\ \text{ksi}$

Answers: 22.5 ksi, 69.0 ksi, and 81.0 ksi

10.13.2 POISSON'S RATIO

Average values of experimentally determined Poisson's ratios for samples of a few species are given in Table 10.9. Wood Handbook (2010) has the full list. The Poisson's ratios are denoted by μ_{LR}, μ_{RL}, μ_{LT}, μ_{TL}, μ_{RT}, and μ_{TR}. The first letter of the subscript refers to the direction of applied stress and the second letter to the direction of lateral deformation. For example, μ_{LR} is the Poisson's ratio for deformation along the radial axis caused by stress along the longitudinal axis.

TABLE 10.9
Poisson's Ratios of Woods at 12% Moisture

Species	μ_{LR}	μ_{LT}	μ_{RT}	μ_{TR}	μ_{RL}	μ_{TL}
Ash (White)	0.371	0.440	–	0.360	0.059	0.051
Balsa	0.229	0.488	–	0.231	0.018	0.009
Yellow Birch	0.426	0.451	–	0.426	0.043	0.024
Red Oak	0.350	0.448	–	0.292	0.064	0.033
Red Maple	0.434	–	–	0.354	0.063	0.044
Douglas Fir	0.292	0.449	0.390	0.374	0.036	0.029
Western Red Cedar	0.378	0.296	0.484	0.403	–	–
Western Hemlock	0.485	0.423	0.442	0.382	–	–
Red Pine	0.347	0.315	0.408	0.308	–	–
Redwood	0.360	0.346	0.373	0.400	–	–
Sitka Spruce	0.372	0.467	0.435	0.245	0.040	0.025

Reference: Wood Handbook (2010). Wood Handbook: Wood as an Engineering Material, Technical Report FPL-GTR-190, 2010, Forest Service, United States Department of Agriculture, Madison, WI.

10.13.3 STRENGTH PROPERTIES

Strength properties of wood depend on species, grading, grain direction, moisture condition, etc. NDS-S (2018) lists the strength values of different species according to their visual or mechanical grading. While designing wood structures, it is convenient to use the reference design values listed in NDS-S (2018). Tables 10.3 to 10.5 list a few examples of strength properties of woods. Readers are referred to NDS-S (2018) for a larger list of strength properties. Some other minor properties such as torsional strength, toughness, creep properties, fatigue, shear strength, nanoindentation hardness, fracture toughness, etc. can also be found in the Wood Handbook (2010).

10.14 CHARACTERIZATIONS OF WOOD

10.14.1 FLEXURE TEST ON WOOD

The ASTM D 198 test method covers the determination of the flexural strength of wood beams made of solid or laminated wood, or of composite constructions. ASTM D 198 can also be used for compression (for a short column), compression (for a long member), tension, torsion, and shear modulus. This test method is intended primarily for beams of rectangular cross-section, but is also applicable to beams of other cross-sections. The size of the beams can be as large as 12 in. (300 mm) deep by 6 in. (150 mm) wide. The beam is subjected to a bending moment by supporting it near its ends and applying loads, as shown in Figure 10.14. The beam is deflected at a prescribed rate, and continuous observations of loads and deflections are made until rupture occurs.

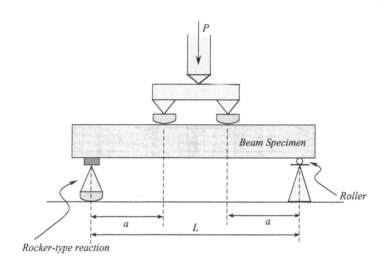

FIGURE 10.14   Schematics for bending test on wooden beam.

The flexural strength can be calculated using the mechanics-based computation method as follows (Eq. 10.5):

$$S_R = \frac{Mc}{I} = \frac{3Pa}{bd^2} \tag{10.5}$$

The modulus of elasticity can be calculated using Eq. 10.6:

$$E = \frac{P_p a}{48 \Delta I} \left(3L^2 - 4a^2 \right) \tag{10.6}$$

where,
S_R = flexural strength (psi or MPa)
E = modulus of elasticity (psi or MPa)
d = Depth of the section (in. or mm)
b = Width of the section (in. or mm)
L = Beam span (in. or mm)
c = Distance from neutral axis to the extreme point = $h/2$ (in. or mm)
I = Moment of inertia = $bh^3/12$ (in.^{4} or mm^4)
P = Maximum load at failure (lb or N)
a = Distance from reaction to nearest load point (in. or mm)
P_p = Load at proportional limit (lb or N)
M = Maximum moment = Pa (lb.in. or N.mm)

10.14.2   Flexural Test of Small-Clear Specimen

The ASTM D 143 test method covers the determination of various strength and related properties of wood by testing small, clear specimens. The static bending tests shall be made on 2 by 2 by 30 in. (50 by 50 by 760 mm) primary method specimens, or on 1.0 by 1.0 by 16 in. (25 by 25 by 410 mm) secondary method specimens. A *center point* load is applied, and the resulting deformation is recorded. The load is continued up to a 6.0 in. (150 mm) deflection, or until the specimen fails to support a load of 200 lb (890 N) for primary method specimens, and to a 3 in. (76 mm) deflection, or until the specimen fails to support a load of 50 lb (220 N) for secondary method specimens. The flexural strength, or modulus of rupture, (R) is calculated using Eq. 10.7:

$$R = \frac{Mc}{I} = \frac{\left(\dfrac{PL}{4}\right)\left(\dfrac{d}{2}\right)}{\dfrac{bd^3}{12}} = \frac{3}{2}\frac{PL}{bd^2} \qquad (10.7)$$

where,

R = Modulus of rupture (psi or MPa)

M = Developed maximum moment, (lb.in. or N.mm)

c = The distance from the neutral axis to the extreme tensile fiber, half of the depth for rectangular section (in. or mm)

I = Moment of inertia of the beam section = $bd^3/12$ (in.^{4} or mm^4)

b = Width of the beam at the fracture (in. or mm)

d = Height of the beam at the fracture (in. or mm)

ASTM D 198 can also be used for:

- Compression parallel to the grain
- Impact bending,
- Toughness
- Compression perpendicular to grain
- Hardness
- Shear parallel to grain
- Cleavage
- Tension parallel to grain
- Tension perpendicular to grain
- Nail withdrawal
- Specific gravity and shrinkage in volume, radial and tangential shrinkage, and
- Moisture determination.

10.14.3   Compression Test

The ASTM D 3501 test methods cover the determination of the compression properties of wood-based structural panels. Wood-based structural panels in use include plywood, waferboard, oriented strand board, and composites of veneer and wood-based layers.

Method A: Compression Test for Small Specimens. This method is applicable to small specimens (nominal width of 1 in. [25 mm] and length of 4 in. [100 mm]) that are uniform with respect to elastic and strength properties. Two types of compression tests are employed: one to evaluate both elastic and compressive strength properties, and one to evaluate maximum compressive strength only.

An axial load is applied through a spherical loading block with a continuous rate of cross-head movement so as to produce failure within 3–10 minutes after initiation of loading. A loading rate of 0.003 in./in. (mm/mm) of length per minute is suggested. The loading rate shall be modified if times fall outside the 3–10-minute range. The maximum crushing force over the initial cross-sectional area is the maximum crushing stress. If both compressive strength and elastic properties are to be sought out, then lateral support is required to prevent buckling during the test. In addition, larger load-deflection data are to be recorded to accurately determine the modulus of elasticity and proportional limit.

Method B: Compression Test for Large Specimens. This method employs large specimens and responds well to manufacturing variables and growth characteristics that influence compression properties of structural panels. The dimension of the specimen shall be 7.5 in. (190 mm) wide by 15 in. (381 mm) long. The specimen is held loosely by the side restraining rail to avoid lateral buckling.

The load is applied continuously throughout the test at a rate of movable head motion which will produce failure within 3–10 minutes after initiation of loading. A head speed rate of 0.035 in./min (0.9 mm/min) is suggested. A large amount of load-deformation data is recorded until the proportional limit if the modulus of elasticity and proportional limit are to be sought out. The compressive strength is the peak load over the initial cross-sectional area.

10.14.4 TENSION TEST

The ASTM D 4761 test method provides procedures for the determination of the tensile strength of lumber and other wood products. The specimen is clamped at the ends and is subjected to an axial tensile load. The specimen is loaded at a prescribed rate and observation of the load is made until failure occurs. The failure load should not be reached in less than 10 seconds, nor more than 10 minutes. The tensile strength is the peak load over the initial cross-sectional area. The ASTM D 4761 test method can also be used for bending edge-wise, bending flat-wise using center point loading or third point loading, and axial strength in compression.

10.14.5 SHEAR PARALLEL TO GRAIN

The shear parallel-to-grain tests shall be made on a specimen notched in accordance with the ASTM D 143, as shown in Figure 10.15, to produce a shear failure on a 2 by 2 in. (50 by 50 mm) surface.

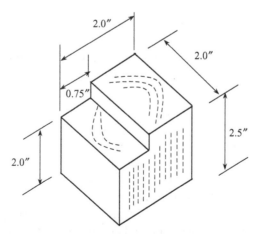

FIGURE 10.15 Shear parallel-to-grain test specimen.

The load shall be applied continuously throughout the test at a rate of motion of the movable crosshead of 0.024 in. (0.60 mm) per minute. Apply the load to, and support, the specimen on end-grain surfaces. Observe the maximum load only.

The reference design values for the visually graded dimension lumber of 2–4 in. thick for No. 1 Grading for some species are listed in Table 10.10.

TABLE 10.10

Reference Design Values for Visually Graded Dimension Lumber of 2–4 in. Thick for No. 1 Grading

Species (Grading No. 1)	Bending Strength, psi (MPa)	Tensile Strength Parallel to Grain, psi (MPa)	Shear Strength Parallel to Grain, psi (MPa)	Compression Strength Perpendicular to Grain, psi (MPa)	Compression Strength Parallel to Grain, psi (MPa)	Modulus of Elasticity, ksi (MPa)
Alaska Cedar	975 (6.7)	525 (3.6)	165 (1.1)	525 (3.6)	900 (6.2)	1,300 (9,000)
Alaska Hemlock	900 (6.2)	550 (3.8)	185 (1.3)	440 (3.0)	1,100 (7.6)	1,600 (11,000)
Alaska Spruce	950 (6.6)	600 (4.1)	160 (1.1)	330 (2.3)	1,100 (7.6)	1,500 (10,300)
Aspen	625 (4.3)	375 (2.6)	120 (0.8)	265 (1.8)	600 (4.1)	1,100 (7,600)
Cottonwood	625 (4.3)	375 (2.6)	125 (0.9)	320 (2.2)	625 (4.3)	1,200 (8,300)
Douglas Fir-Larch	1,000 (6.9)	675 (4.7)	180 (1.2)	625 (4.3)	1,500 (10.3)	1,700 (11,700)
Douglas Fir-Larch (North)	850 (5.9)	500 (3.5)	180 (1.2)	625 (4.3)	1,400 (9.7)	1,600 (11,000)
Douglas Fir-Larch (South)	925 (6.4)	600 (4.1)	180 (1.2)	520 (3.6)	1,450 (10.0)	1,300 (9,000)
Hem-Fir	975 (6.7)	625 (4.3)	150 (1.0)	405 (2.8)	1,350 (9.3)	1,500 (10,300)
Hem-Fir (North)	1,000 (6.9)	575 (4.0)	145 (1.0)	405 (2.8)	1,450 (10.0)	1,600 (11,000)
Mixed Maple	725 (5.0)	425 (2.9)	195 (1.3)	620 (4.3)	700 (4.8)	1,200 (8,300)
Red Maple	925 (6.4)	550 (3.8)	210 (1.4)	615 (4.2)	900 (6.2)	1,600 (11,000)
Red Oak	825 (5.7)	500 (3.5)	170 (1.2)	820 (5.7)	825 (5.7)	1,300 (9,000)
White Oak	875 (6.0)	500 (3.5)	220 (1.5)	800 (5.5)	900 (6.2)	1,000 (6,900)

Adapted from NDS-S 2018, Table 4A, National Specifications for Wood Construction Supplement (NDS-S): Design Values for Wood Construction, American Wood Council, Leesburg, Virginia. Used with permission.

10.15   MAJOR WOOD SECTIONS

10.15.1   SAWN LUMBER

Common sawn timber products include solid timber beams and more rectangular timber sections, as shown in Figure 10.16. Sawn timber is timber that is cut from logs into different shapes and sizes. Sawn timber is generally cut into varying rectangular widths and lengths, but may also be wedge shaped. Sawn lumber consists of dimensions lumber, beams and stringers, posts and timbers, and decking. NDS-S (2018) lists the standard specification for the dimensions of lumber sections.

FIGURE 10.16   Some sawn lumbers at a construction site (2×6 sections). *Photo taken in Pueblo, Colorado.*

10.15.2   GLUED LAMINATED TIMBER

Glued laminated timber (glulam) is composed of several layers of dimensional lumbers glued together with moisture-resistant adhesives, as shown in Figure 10.17. The combined section is a large, strong, structural member that can be used as columns, beams, or in curved shapes with extensive design flexibility. The grain of all laminations is approximately parallel longitudinally. The separate lamination should not exceed 2 in. in net

FIGURE 10.17   Glued laminated timber (glulam). *Photo taken in Colorado Spring.*

thickness. NDS-S (2018) lists the standard specifications for the dimension of glulam sections, the reference design strengths, and other information.

10.15.3 ROUND TIMBER AND POLES

Wooden sections can also be round, similar to naturally grown trees. Round timbers and poles are very often used as columns, foundation piles, land and fresh water piles in ranch areas, or for decorative purposes. Two wooden structures using the round poles are shown in Figure 10.18.

FIGURE 10.18 Two structures made with wooden poles. *Photo by Kyle Witzman, taken in Colorado Spring.*

10.15.4 PREFABRICATED WOOD I-JOISTS

Prefabricated wood I-joists are currently very popular for use in floors, ceilings, and roofs. I-joists are very effective in beam action, as they have larger moment of inertia values. They also help the member to resist warping, twisting, and shrinking which can lead to squeaky floors. The flanges of I-joists are made with laminated veneer lumber (LVL) and the web is made up of oriented strand board (OSB), as shown in Figure 10.19. LVL and OSB are discussed later in this chapter.

FIGURE 10.19 Prefabricated wood I-joists. *Photos taken in Pueblo, Colorado.*

10.15.5 STRUCTURAL COMPOSITE LUMBERS (SCLs)

Three types of structural composite lumber are included in the NDS (2018). These are discussed here.

Laminated Veneer Lumber (LVL). LVL refers to a composite of wood veneer sheet elements with wood fiber primarily oriented along the length of the member, as shown in Figure 10.20. Veneer thickness should not exceed 0.25 in. (6 mm). It is produced by bonding thin wood veneers together in a large billet. The grain of all veneers in the LVL billet is parallel to the long direction. The resulting section has improved mechanical properties and structural stability, offering a broader range in product width, depth, and length than the conventional lumbers. LVL is commonly used in the same structural applications as conventional sawn lumber and timber, including rafters, headers, beams, joists, rim boards, studs, and columns (APA 2016).

FIGURE 10.20   Laminated veneer lumber (LVL). *Photos taken in Pueblo, Colorado.*

Parallel Strand Lumber (PSL). PSL is prepared from veneers clipped into long strands laid in parallel formation and bonded together with an adhesive. The length-to-thickness ratio of the strands in PSL is about 300. PSL is used mainly for beam and header beam applications (APA 2016).

Laminated Strand Lumber (LSL). LSL is prepared from flaked wood strands that have a high length-to-thickness ratio. The minimum dimension of the strands should not exceed 0.1 in., and the average length should be at least 150 times the minimum dimension. Combined with an adhesive, the strands are oriented and formed into a large mat or billet and are pressed, as shown in Figure 10.21. LSL offers good fastener-holding strength and mechanical connector performance, and is commonly used in a variety of applications, such as beams, headers, studs, rim boards, and millwork components.

FIGURE 10.21   A laminated strand lumber (LSL). *Photo taken at the Colorado State University–Pueblo.*

10.15.6 WOOD STRUCTURAL PANELS

A wood panel refers to a wood-based panel product bonded with a waterproof adhesive. This includes the following:

- Plywood
- Oriented Strand Board (OSB)
- Fiberboard
- Particleboard, etc.

Plywood. Plywood refers to wood structural panels comprised of wood veneer arranged in cross-aligned layers. The plies are bonded with an adhesive that cures upon the application of heat and pressure. By alternating the grain direction of the veneers from layer to layer, or cross-orienting, the panel strength and stiffness in both directions are maximized.

Oriented Strand Board (OSB). OSB refers to a mat-formed wood structural panel comprising thin, rectangular wood strands arranged in cross-aligned layers, with surface layers normally arranged in the long panel direction and bonded with waterproof adhesive, as shown in Figure 10.22. The individual layers can be cross-oriented to provide strength and stiffness to the panel. The wood strands in the outmost layer on each side of the board are normally aligned into the strongest direction of the board. Produced in enormous, continuous mats, OSB is a solid panel product of consistent quality with no laps, gaps, or voids (APA 2017).

FIGURE 10.22   Oriented strand board (OSB). *Photos taken in Pueblo, Colorado.*

Fiberboard. *Fiberboard* is made by breaking down hardwood or softwood residuals into wood fibers, then combining them with wax and a resin binder, and forming panels by applying high temperature and pressure (Binggeli 2013).

Particle Board. Particle board is prepared from wood chips, sawmill shavings, or even sawdust, and a synthetic resin or other suitable binder, which is pressed and extruded. Particle board is cheaper, denser, and more uniform than conventional wood and plywood, and is substituted for them when cost is more important than strength and appearance. However, it is very prone to expansion and discoloration due to moisture, particularly when it is not covered with paint or sealant.

10.15.7 CROSS LAMINATED TIMBERS (CLTs)

Cross Laminated Timber (CLT) is a versatile, multilayered panel made of lumber placed crosswise to adjacent layers for increased rigidity and strength, as shown in Figure 10.23.

FIGURE 10.23   Cross-laminated timbers. *Photo by Nick Wibb.*

CLT can be used for long spans and all assemblies, such as floors, walls, or roofs. CLT is produced in a factory and supplied, ready to fit and screw together, as a flat pack assembly project (APA 2016).

10.15.8   DENSIFIED WOODS

Densified wood is produced by using a mechanical hot press to compress wood fibers and increase the density by a factor of three (Ashby and Medalist 1983). This increase in density improves the strength and stiffness of the wood by a proportional amount.

10.15.9   CHEMICALLY DENSIFIED WOODS

In this method, both chemical process and mechanical hot press are used together to densify wood. Chemical processes break down lignin and hemicellulose, which are found naturally in wood. Following dissolution, the cellulose strands are mechanically hot compressed. Compared to the three-fold increase in strength observed from hot pressing alone, chemically processed wood has been shown to yield an 11-fold improvement. This extra strength comes from hydrogen bonds formed between the aligned cellulose nanofibers. The densified wood is used in situations where the regular strength of wood is not sufficient.

10.15.10   TRANSPARENT WOOD COMPOSITES

Transparent wood composites are new composites made at the laboratory scale which combine transparency and stiffness. They are not yet available on the market.

10.16   FACTORS AFFECTING PERFORMANCE OF WOOD

There are many factors which affect the strength of lumber, such as:

- Species and species group
- Grading
- Orientation of the wood grain
- Duration of loading
- Moisture content
- Ambient temperature
- Size and shape of the wood member

- Chemical (pressure or incising) treatment
- Post-grading defects

However, these factors are not discussed here in detail, as they are commonly covered in a senior-level course, such as structural wood design.

10.17 WOOD DESIGN PHILOSOPHY

The design methods are not the scope of this text book. However, wood design is not covered in many civil and construction engineering programs. Thus, the design philosophy is very concisely described here. The design philosophy for wood structures is similar for both the Allowable-Stress Design (ASD) and the Load and Resistance Factor Design (LRFD).

The basic civil engineering design philosophy is that the capacity stress of a section must be equal to or greater than the probable stress (that may occur in the member while in service) in the section. For example, a beam in a building might have the maximum bending stress of 750 psi and the maximum shear stress of 800 psi (5.52 MPa). A section is to be selected which can resist the bending stress of at least 750 psi (5.17 MPa) and shear stress of 800 psi (5.52 MPa).

Other classes such as Strength of Materials, Structural Analysis, etc. discuss how to calculate the probable maximum stress that may occur in a member while in service. Here, it is discussed how to calculate the maximum capacity of a section as follows:

Step 1. Obtain the reference design values for the desired species, grading, and loading mode. Figure 10.24 gives an example for visually graded dimension lumber.

Step 2. Multiply by the appropriate factors as suggested by the 8NDS (2018)

Table 4A   Reference Design Values for Visually Graded Dimension Lumber (2" - 4" thick)[1,2,3]

(All species except Southern Pine—see Table 4B) (Tabulated design values are fo duration and dry service conditions. See NDS 4.3 for a comprehensive description of des adjustment factors.)

USE WITH TABLE 4A ADJUSTMENT FACTORS

Species and commercial grade	Size classification	Design values in pounds per square inch (psi)						
		Bending	Tension parallel to grain	Shear parallel to grain	Compression perpendicular to grain	Compression parallel to grain	Modulus of Elasticity	
		F_b	F_t	F_v	$F_{c\perp}$	F_c	E	E_{min}
ALASKA CEDAR								
Select Structural	2" & wider	1,150	625	165	525	1,000	1,400,000	510,000
No. 1		975	525	165	525	900	1,300,000	470,000
No. 2		800	425	165	525	750	1,200,000	440,000
No. 3		450	250	165	525	425	1,100,000	400,000
Stud	2" & wider	625	350	165	525	475	1,100,000	400,000
Construction		900	500	165	525	950	1,200,000	440,000
Standard	2" - 4" wide	500	275	165	525	775	1,100,000	400,000
Utility		250	125	165	525	500	1,000,000	370,000
ALASKA HEMLOCK								
Select Structural	2" & wider	1,300	825	185	440	1,200	1,700,000	620,000
No. 1		900	550	185	440	1,100	1,600,000	580,000
No. 2		825	475	185	440	1,050	1,500,000	550,000
No. 3		475	275	185	440	600	1,400,000	510,000

FIGURE 10.24 A screenshot from NDS-S (2018), table 4A. *Adapted from NDS-S 2018, table 4A, National Specifications for Wood Construction Supplement (NDS-S): Design Values for Wood Construction, American Wood Council, Leesburg, Virginia. Used with permission.*

For example, an Alaska Cedar beam is being designed for bending. The maximum bending stress is 500 psi (calculated based on the knowledge from other classes, such as Strength of Materials, Structural Analysis, etc.). The grading of the Alaska Cedar section to be used is No. 2, and the section thickness is 2.0 in. Then, from Figure 10.24, the reference design bending stress value is 800 psi. This reference value is to be multiplied by the factors listed in NDS (2018) appropriate for bending. The applicable factors include load factor (C_D), wet service factor (C_M), temperature factor (C_t), beam stability factor (C_L), size factor (C_F), etc. These factors are to be selected from tables or determined using very simple equations listed in the manual. Once obtained, the product of these factors is multiplied with the reference design bending stress value of 800 psi. Then, the capacity of this section for bending will be established. This capacity is to be compared with the probable maximum bending stress of 500 psi (as given). If the section is adequate to support this bending stress, then this section can be selected. If the capacity of this section is less than the probable maximum bending stress of 500 psi, then a larger section is to be tried.

If the LRFD design method is used, then the probable maximum stress is determined based on the factored load, i.e., the loads are to be increased by multiplying with factors recommended by the ASCE 7–16 Minimum Design Loads and Associated Criteria for Buildings and Other Structures, or any other appropriate design manual.

Reference Design Bending Stress can be obtained from different design manuals, such as National Specifications for Wood Construction (NDS) Supplement: Design Values for Wood Construction, published by American Wood Council, Leesburg, Virginia. The NDS supplement contains the list of grading agencies, species combination, section properties, reference design values, and some adjustment factors for sawn lumbers and structural glued laminated timbers of different sizes.

10.18   CHAPTER SUMMARY

Almost all of the residential buildings in the United States are made with wood. Nowadays, the construction of commercial buildings having up to four-five stories is increasing rapidly. The details regarding the material properties of wood can be obtained from the Wood Handbook (2010) or the latest publication.

Wood is an environmentally friendly, renewable, and versatile material. Two major materials, concrete and steel, produce about nine and twenty-one times the amount of carbon compared to that of wood lumber, respectively. Several other advantages of wood are its high strength-to-weight ratio, great elasticity, small thermal conductivity, durability, and ability to be made into products of various shapes and aesthetic finishes. Wood has some disadvantages, however, including non-homogeneity, shrinking, expanding, rotting or even being eaten by insects, and combustibility, making it liable to burn.

Engineering woods come from exogenous trees. The cross-section of an exogenous tree has different parts, such as the pith, heartwood, annual rings, sapwood, vascular cambium, inner bark, and outer bark. Heartwood is the most desirable part of wood for engineering uses.

Several hundreds of wood species are used in engineering. Each species has its own characteristics and performance. Some similar species are grouped together for easy handling.

Trees are cut down first when they are mature enough. Then, they are transported to a factory, sawn into desired sizes and shapes, and are seasoned in kilns to reduce moisture to achieve dimensional stability. The various methods of seasoning used may be classified into natural seasoning (water and air) and artificial seasoning (boiling, chemical seasoning, kiln seasoning, electrical seasoning). In addition to seasoning, sometimes pressure treatment or incising of chemicals is required when woods are to be used for outdoor applications. In incising, sharp steel teeth (similar to stapling) are pressed into all sides of lumber and timbers. This is done to increase the penetration of the chemical into the wood during the pressure-treatment process, in order to force preservative chemicals deeper into the structure of the wood.

Lumber is commonly available with its sawn dimension, known as nominal size, stated size, or full-sawn size. The lumber is then further dressed or surfaced to have a smooth surface and uniform size. Dimension lumber is the sawn section with thickness in between 2–4 in. Timber has a thickness of 5 in. or more. Timber can be further divided into beam and stringer (B&S), and post and timber (P&T).

Visual grading includes the visual inspection of the wood section by a certified person and stamps on the section. The stamp includes the agency, the grade, drying type, amount of treatment, producing mill lumber, wood species, and presence of a rounded edge. Mechanical grading conducts nondestructive testing on wood sections and grades accordingly. The visual grading method is still more popular than mechanical grading.

Strength and many other properties of wood vary along three mutually perpendicular axes: longitudinal, radial, and tangential. Living trees have very high moisture contents (up to 200%). With the decrease in moisture content until the fiber saturation point (free water), wood does not shrink. However, in a decrease in moisture below the fiber saturation point, wood shrinks and is accompanied by a gain in strength, and then reaches the peak strength at some moisture, roughly 10–15%. Wood shrinkage is different along the three different mutually perpendicular directions, although the shrinkage along the longitudinal direction is negligible. In the simplified method of the shrinkage calculation, a constant shrinkage value of 6% is used for both the width and the thickness (not the length) of a member. The shrinkage is taken as zero at the fiber saturation point of 30%, and the full 6% shrinkage is assumed to occur at zero moisture content. The specific gravity of wood depends on the moisture content of the wood and the weight of the wood substance present in the wood. Electrical conductivity of dry wood is negligible. However, with an increase in moisture content, electrical conductivity of wood increases dramatically. The thermal conductivity of wood depends on its moisture, grain orientation, density, extractive content, and structural integrity. The less the moisture content of wood, the lower the thermal conductivity.

Wood has different mechanical properties, such as modulus of elasticity, modulus of rigidity, Poisson's Ratio, tensile strength, etc., along the three mutually perpendicular directions. However, it has the highest strength along the longitudinal direction and the lowest strength along the tangential direction. Different standard laboratory tests, such as the flexural test, compression test, shear test, etc., are performed to fully characterize a wood section.

Common sawn timber products include solid timber beams and more rectangular timber sections. Glued laminated timber (glulam) is composed of several layers of dimensional timber, glued together with moisture-resistant adhesives. Round timbers and poles are very often used as columns, foundation piles, land and fresh water piles in ranch areas, or for decorative purposes. Prefabricated wood I-joists are very popular in floors, ceilings, and roofs nowadays. Laminated veneer lumber (LVL) refers to a composite of wood veneer sheet elements, with wood fiber primarily oriented along the length of the member. Parallel strand lumber (PSL) refers to a composite of wood strand elements with the wood fiber primarily oriented along the length of the member. Laminated strand lumber (LSL) is manufactured from flaked wood strands that have a high length-to-thickness ratio.

Several types of wood panels are commonly available, such as plywood, oriented strand board (OSB), and composite panels. OSB is very popular for floor and wall sheathing. Plywood refers to wood structural panels comprising wood veneer arranged in cross-aligned layers. Fiberboard is made by breaking down hardwood or softwood residuals into wood fibers, and combining them with wax and a resin binder. Particle board is manufactured from wood chips, sawmill shavings, or even sawdust, and a synthetic resin or other suitable binder, which is pressed and extruded. For cross-laminated timber (CLT), each layer of boards is placed cross-wise to adjacent layers for increased rigidity and strength. Densified wood is made by using a mechanical hot press to compress wood fibers and increase the density, to enhance the strength and stiffness of the wood by a proportional amount. Currently, chemically densified wood is also available, and transparent wood composites will be available on the market, in the future.

ORGANIZATIONS DEALING WITH WOOD

American Wood Council (AWC). AWC supports the utilization of wood products by developing and disseminating consensus standards, comprehensive technical guidelines, and tools for wood design and construction. It contributes to the development of sound public policies, codes, and regulations for manufacture and use of wood products.

Location: Leesburg, VA
Website: https://awc.org

APA – The Engineered Wood Association. APA has focused on helping the industry create structural wood products of exceptional strength, versatility, and reliability. Combining the research efforts of scientists with the knowledge gained from decades of field work, and cooperation with its member manufacturers.

Location: Tacoma, Washington.
Website: https://www.apawood.org

United States Forest Service (USFS). The USFS is an agency of the US Department of Agriculture that administers the nation's 154 national forests and 20 national grasslands, which encompass 193 million acres. Managing approximately 25% of federal lands, it is the only major national land agency.

Location: Washington, DC
Website: https://www.fs.fed.us/
ASTM International is also a resource for woods.

REFERENCES

AITC. 2012. *Timber Construction Manual*, 6th edition. Wiley, Hoboken, NJ.
APA. 2016. *Structural Composite Lumber Selection and Specification, Form E30W*. APA, Tacoma, Washington.
APA. 2017. *Oriented Strand Product Guide, Form No. W410E*. APA, Tacoma, Washington.
Ashby M. F., Medalist R. F. M. The mechanical properties of cellular solids. *Metall. Trans. A*, 14 (9) (1983), pp.1755–1769.
BBC News, World is home to 60,000 tree species, By Mark Kinver, Environment reporter, 5 April 2017, https://www.bbc.com/news/science-environment-39492977. Last accessed: 12/29/18.

Bhavikatti, S. S. 2010. *Basic Civil Engineering*, New Age International Publishers, New Delhi.

Binggeli, C. 2013. *Materials for Interior Environments*, 2nd edition. Wiley.

Breyer, D. E., Cobeen, K., Fridley, K. J. and Pollock, D. 2014. *Design of Wood Structures-ASD/LRFD*, 7th edition, McGraw-Hill Education.

Canadian Wood Council. 2011. *Introduction to Wood Design*. Ottawa, Canada.

EPA. 2006. *Solid Waste Management and Greenhouse Gases—A Life Cycle Assessment of Emissions and Sinks*, 3rd Edition. US Environmental Protection Agency, Washington, DC.

Faherty, Keith F. and Williamson, Thomas G. 1995. *Wood Engineering and Construction*. McGraw-Hill, New York, NY.

Green, D. W. 1989. Moisture content and the shrinkage of lumber, FPL RP 489, Forest Service, United States Department of Agriculture, Madison, WI.

Mamlouk, M. and Zaniewski, J. 2014. *Materials for Civil and Construction Engineers*, 4th edition, Pearson, Upper Saddle River, NJ.

NCASI. 2018. National Council for Air and Stream Improvement, www.ncasi.org. Last accessed August 31, 2018.

NDS. 2018. National Specifications for Wood Construction (NDS), American Wood Council, Leesburg, Virginia. Latest publication June 2018.

NDS-S. 2018. National Specifications for Wood Construction Supplement (NDS-S): Design Values for Wood Construction, American Wood Council, Leesburg, Virginia. Latest publication June 2018.

Rummelhart, R. and Fantozzi, J. A. 1992. Multistory Wood-Frame Structures: Shrinkage Considerations and Calculations, Proc. of the ASCE Structures Congress, Reston, VA.

Sivakugan, N., Gnanendran, C. T., Tuladhar, R. and Kannan, M. B. 2018. *Civil Engineering Materials*, 1st edition, Cengage Learning, Boston, MA.

Stone, Jeffrey B. and Tyree, David 2015. Fire Protection in Wood Buildings—Expanding the Possibilities of Wood Design, American Wood Council. https://www.awc.org/pdf/education/bcd/ReThinkMag-BCD200A1-DesigningForFireProtection-150801.pdf Last accessed August 31, 2018.

Wood Handbook. 2010. *Forest Products Laboratory (FPL), Wood Handbook: Wood as an Engineering Material*, Technical Report FPL-GTR-190, Forest Service, United States Department of Agriculture, Madison, WI.

Zhang, Haimei. 2011. *Building Materials in Civil Engineering*. Woodhead Publishing Limited, Sawston, Cambridge.

FUNDAMENTALS OF ENGINEERING (FE) EXAM STYLE QUESTIONS

FE PROBLEM 10.1

In which direction is wood the strongest?

A. Transverse direction
B. Radial direction
C. Equal in any direction
D. Longitudinal direction

Solution: D

Wood is the strongest along the longitudinal direction.

FE Problem 10.2

Seasoning of wood is performed:

A. To increase strength
B. To increase durability
C. To remove natural defects
D. To reduce moisture to a suitable level

Solution: D

Seasoning is a process by which moisture content in a freshly cut tree is reduced to a suitable level. By doing so, the durability and strength of lumber are increased. Seasoning cannot remove natural defects but can prevent further decaying. Considering all, the best answer is option D.

FE Problem 10.3

Which of the following parts of a wood section is desirable for engineering use?

A. Pith
B. Heartwood
C. Sapwood
D. Vascular cambium layer

Solution: B

When the plant becomes old, the pith dies and becomes fibrous, which is not useful in engineering. Heartwood is the matured and solid portion of a tree, and is useful for various engineering purposes. Sapwood is the recently grown wood with some sap (wood moisture), and is weaker and less durable than heartwood. The vascular cambium layer is a thin layer of fresh sap just outside of the sapwood, which helps in wood's growth.

FE Problem 10.4

The mechanical grading, '750f-1.4E' means

A. Its compressive strength is 750 psi
B. Its modulus of elasticity is 1.4 ksi
C. Its modulus of elasticity is 1,400,000 psi
D. It is just a symbolic number

Solution: C

The mechanical grading, '750f-1.4E' means its reference flexural/bending strength value is 750 psi and its modulus of elasticity is 1.4 Mpsi, or 1,400,000 psi.

FE PROBLEM 10.5

A grading stamp on a wood piece is shown in Figure 10.25. Which of the following statements is true for this stamp?

A. It is a machine stress-rated (MSR) grading
B. It is a machine-evaluated lumber (MEL) grading
C. It is the best quality visually graded section
D. The edges of this section are sharp

Solution: C

MSR grade shows a number at the beginning, followed by the letter 'f', then a dash followed by another number, and finally the letter 'E'. For example, 750f-1.4E. MEL grade includes the letter 'M' followed by a number. For example, M-5, M-40, etc. 1/4 EE means a ¼-in. eased edge (rounded) to help in handling. Therefore, Options A, B, and D are not correct statements.

Visual grading grades lumber such as 'Select Structural,' 'No. 1 and Better,' 'No. 1,' etc., where 'Select Structural' is the best quality section. Thus, Option C is the correct answer.

FIGURE 10.25 Grading stamp on a lumber for FE Problem 10.5. *Photo taken in Pueblo, Colorado.*

FE PROBLEM 10.6

What is the difference between timber and lumber?

A. Timber is the processed wood to be used in engineering; lumber is the raw wood
B. Lumber is the processed wood to be used in engineering; timber is the raw wood
C. Timber is standing or felled trees, before they are milled into boards
D. Timber is a lumber section with dimension 5×5 in. or larger

Solution: D

Sawn lumber has different size categories, dimension lumber has a thickness up to 4 in., and timber has dimensions of 5×5 in. or larger. Timber has two subsections (Table 10.2): post and timber (P&T), and beams and stringers (P&S).

FE PROBLEM 10.7

Which of the following statements is true?

A. Specific gravity of wood is independent of moisture content
B. Shrinkage keeps increasing as moisture content decreases to zero
C. Shrinkage is negligible in a longitudinal direction
D. The performance of a lumber section increases as the thickness of the section increases

Solution: C

Option C is correct. Shrinkage is negligible in a longitudinal direction.

Specific gravity of wood is dependent on moisture content. This is why there are two different equations to calculate specific gravity based on moisture content, one above 30%, and the other below 30%. Shrinkage keeps on increasing as moisture content decreases, reaches the peak, and again decreases as moisture approaches to zero. The performance of a lumber section decreases as the thickness of the section increases, and the size factor decreases as thickness increases.

PRACTICE PROBLEMS

PROBLEM 10.1

What is the difference between lumber and timber?

PROBLEM 10.2

Based on the natural properties of wood, do you think in any direction, such as the longitudinal direction, wood material is homogeneous? Discuss your opinion based on the structural capacity of wood. More clearly, wood has grains parallel to the longitudinal direction. In that case, does wood behave the same in the longitudinal direction and in the transverse direction?

PROBLEM 10.3

A 2,520-g wet wood is oven-dried to 2,300 g. Calculate the moisture content of the wood piece.

PROBLEM 10.4

The elastic modulus of red oak in the longitudinal direction is 2,700,000 psi. Determine its moduli values along the tangential and radial directions, and the shear modulus of rigidity based on shear strain in the LR plane.

PROBLEM 10.5

A 12-in. wood plank has a cross-section of 4×1 in. If this plank is pulled and it fails at the load of 1,850 lb, determine the tensile strength of the wood piece.

PROBLEM 10.6

A center-point loading flexural test was performed on a wood beam of 28-in. span. The beam's cross-section is 12 in. in width and 8 in. in height. The beam failed at the load of 900 lb. Determine the flexural strength of the concrete beam.

PROBLEM 10.7

A White Fir pole has a diameter of 12 in. and length of 6 ft when green. Calculate its final volume if the green pole is oven dried.

11 Masonry

Two common types of masonry: concrete masonry and clay bricks (including mortar, grout, and plaster) are discussed in this chapter. Their properties, production methods, laboratory testing, etc. have been presented with numerical, worked-out examples. The ASTM requirements of different types of masonry units are also discussed, as well as some standard laboratory characterization methods.

11.1 BACKGROUND

Considering the primary civil engineering materials, *masonry* is the oldest one but is still in use with full swing. In fact, in developing countries, it is still the most widely used civil engineering material, followed by concrete and steel. Masonry is a broad term, and it refers to the combined individual pieces of stone, brick, or block, using a binder material such as cement, mortar, or lime mortar. Low-rise residential buildings in wet areas where woods are vulnerable, and partition walls in concrete structures such as arches, bridges, etc. can be made with masonry. Some famous examples of the world's masonry constructions include the pyramids of Egypt, the Great Wall of China, the Taj Mahal of India, the cathedrals of Europe, etc., which are still considered some of the most attractive structures in the world. The United States has been widely using masonry structures for such things as sewers, tunnel lining, bridge piers, walkways, and retaining walls for the last 200 years.

Masonry walls can be located inside a building as well as outside of a building. Although masonry is very popular for partition walls, it can be used in a load-bearing structure; sometimes, the wall is built as a hollow wall, reinforced wall, or a frame-type wall. Masonry walls are very popular for fire resistance, reduced vibration due to wind, and good resistance to heat and cold, depending on the circumstances. Partition walls are commonly a single wythe in thickness (the thickness of one masonry unit). Double wythes or one-and-a-half wythes are used for load-bearing walls. Insulation materials are very often used inside the wall to provide better resistance to heat and cold.

11.2 MASONRY UNITS

Masonry units are hollow or solid, as shown in Figure 11.1. Concrete masonry can be of both hollow and solid types. A clay brick is a solid rectangular piece of burnt clay. Clay bricks are solid, although structural clay tiles are commonly hollow. Glass blocks and stone blocks are always solid. A *solid masonry unit* is defined as having the net cross-sectional area of less than 75% of the area parallel to the bearing surface in every plane. Solid block has the net cross-sectional of equal to or greater than 75% of area parallel to the bearing surface in every plane. Concrete masonry (either solid or hollow), and clay bricks are commonly used in the United States. Thus, these two materials are discussed here in detail.

Gross area and net area of hollow units are very often used when calculating compression strengths of different units. Gross area does not consider the hollow space. The net area

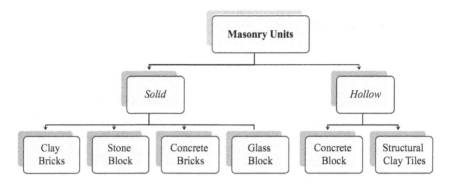

FIGURE 11.1   Types of masonry units.

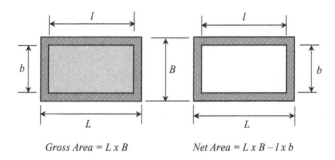

Gross Area = L x B   Net Area = L x B – l x b

FIGURE 11.2   Gross and net areas of a hollow masonry unit.

is calculated as the gross area minus the hollow area. Consider, for a masonry unit, that the outer length is L, the inner length is l, the outer width is B, and the inner width is b. Then, the gross area and the net area can be calculated as shown in Figure 11.2.

Some relationships between the gross area and the net area can be presented as follows (Eqs. 11.1–11.3):

$$\text{Gross Area} = \text{Actual Width} \times \text{Actual Length} \tag{11.1}$$

$$\text{Net Area} = \text{Gross Area} \times \text{Percent of Solid} / 100 \tag{11.2}$$

$$\text{Percent of Solid} = \text{Net Volume} / \text{Gross Volume} \times 100 \tag{11.3}$$

If the net area is more than 75% of the gross area, then the gross area is considered the net area, i.e., a hollow area of equal to or less than 25% is ignored for calculations. More specifically,

$$\text{Gross Area} \approx \text{Net Area} \left[\text{if the hollow area is equal to or less than 25\%}\right]$$

11.2.1 CONCRETE MASONRY

Concrete masonry units have different names, such as cinderblocks, hollow blocks, or concrete blocks. Concrete masonry units are stronger than clay bricks and are mostly used for load-bearing structures. They are made with aggregate, cement, and water. Sometimes, other materials are added to make it lightweight. Any type of cement can be used, although

Type I (general purpose) cement is commonly used. Depending on the demand of such factors as workability, time constraint, and compaction procedure, other types of cements are acceptable.

11.2.1.1 Types of Concrete Masonry Units
Depending on the usages of units, they can be of two types (shown in Figure 11.3):

- Load-bearing concrete masonry units
- Concrete building units

A load-bearing concrete masonry unit is either solid or hollow. A hollow unit has the net cross-sectional area of 75% less in every plane parallel to the bearing surface. A solid unit (although it may not be completely solid) has the net cross-sectional of equal to or more than 75% in every plane parallel to the bearing surface. Concrete building units are not load-bearing materials and are intended primarily for partition walls or similar. They are made with cement, water, and preferably lightweight aggregate, although normal-weight aggregate is common. Thus, both the load-bearing and concrete building units are of three types based on weight. These are bulleted below:

- Lightweight units
- Medium-weight units
- Normal-weight units

Lightweight units use lighter aggregates, such as expanded clay, shale, scoria, and cinders, which make the dry unit weights between 85–105 pcf (1,360–1,680 kg/m³). Medium-weight

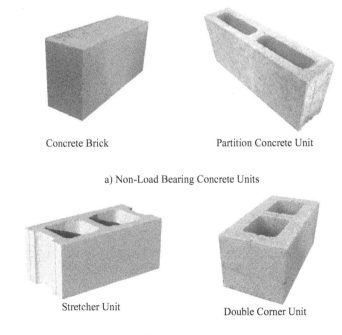

Concrete Brick   Partition Concrete Unit

a) Non-Load Bearing Concrete Units

Stretcher Unit   Double Corner Unit

b) Load Bearing Concrete Units

FIGURE 11.3 Concrete masonry units.

aggregates have dry unit weights between 105–125 pcf (1,680–2,000 kg/m³). Small proportions of lighter aggregates are used to decrease the unit weight. Normal-weight aggregates use conventional crushed stones and gravel, and their dry unit weights can be greater than 125 pcf (2,000 kg/m³). ASTM C 90 specifies this weight classification for concrete masonry units.

Load-bearing concrete masonry units are produced in two ways:

- Type I: Moisture-controlled units
- Type II: Non-moisture-controlled units

Moisture-controlled units are prepared to control the shrinkage limiting the moisture content, as recommended by ASTM C 90. Type II does not consider the moisture controlling. The ASTM C 90 requirements for strength, absorption, and density classification are presented in Table 11.1.

TABLE 11.1

Strength, Absorption, and Density Classification Requirements

Density Classification	Oven-Dry Density of Concrete, pcf (kg/m³) Average of 3 Units	Maximum Water Absorption, pcf (kg/m³)		Minimum Net Area Compressive Strength, psf (MPa)	
		Average of 3 Units	Individual Unit	Average of 3 Units	Individual Unit
Lightweight	<105 (1,680)	18 (288)	20 (320)	1,900 (13.1)	1,700 (11.7)
Medium-weight	105 to less than 125 (1,680 to less than 2,000)	15 (240)	17 (272)	1,900 (13.1)	1,700 (11.7)
Normal-weight	≥125 (≥2,000)	13 (208)	15 (240)	1,900 (13.1)	1,700 (11.7)

Courtesy of ASTM International, 100 Barr Harbor Drive, PO Box C700, West Conshohocken, PA. ASTM C 90, Used with permission.

Linear shrinkage of a unit is dependent on the moisture content. Once moisture goes off, the unit shrinks. If moisture content is controlled, shrinkage can be controlled. In addition to the moisture content, ASTM C 90 specifies that the minimum compressive strength of Type I and II concrete units, based on net area (NOT the gross area), must have 1,900 psi (13.1 MPa) for three units, with an individual value of at least 1,700 psi (11.7 MPa). The maximum water absorptions with respect to the oven-dry weight are 20 pcf (320 kg/m³), 17 pcf (272 kg/m³), and 15 pcf (240 kg/m³) for lightweight, medium-weight, and normal-weight concrete units, respectively. Lightweight units are allowed to have higher absorption, as they contain porous aggregates. Water absorption can be calculated using Eq. 11.4:

$$\text{Water Absorption} = \text{Saturated Weight} - \text{Oven dry Weight} \qquad (11.4)$$

The full classification of concrete masonry units is shown in Figure 11.4.

11.2.1.2 Grades of Concrete Masonry Units

Two types of grading are proposed by the ASTM C 55. These are mentioned below and are shown in Figure 11.4:

- Grade N (Types I and II)
- Grade S (Types I and II)

ASTM C 55 specifies the minimum compression strength based on the average gross area tested flatwise and maximum water absorption by weight classification, with respect to oven-dry weight for these two grades (N and S), listed in Table 11.2. Grade S is intended for general and has less strength and higher absorption compared to Grade N. Grade N is intended for architectural purposes in an exterior environment. Both of these grades can be Type I: moisture-controlled units and Type II: non-moisture-controlled units.

The 24-hour absorption is the absorbed water ratio with respect to the dry weight of the masonry unit, if it is submerged under normal water for 24 hours (ASTM C 140). It can be calculated using the following equations (Eqs. 11.5–11.7):

$$\text{Absorption (\%)} = \frac{W_S - W_d}{W_d} \times 100 \tag{11.5}$$

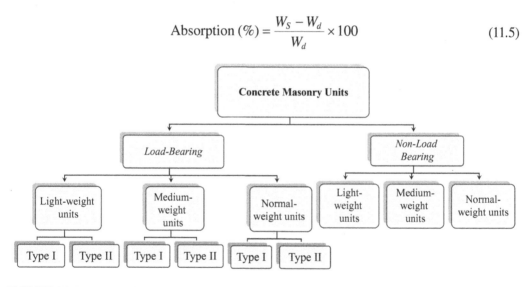

FIGURE 11.4 Types of concrete masonry units.

TABLE 11.2

Strength and Absorption Requirements for Concrete Masonry Units

	Minimum Compressive Strength, psi (MPa) Based on Average Gross Area Tested Flatwise		Maximum Water Absorption, pcf (kg/m³)		
Grade	Average of 3 Bricks	Individual Brick	Lightweight (<105 pcf)	Medium-Weight (105–125 pcf)	Normal-Weight (>125 pcf)
N (I and II)	3,500 (24.1)	3,000 (20.7)	15 (243)	13 (210)	10 (162)
S (I and II)	2,500 (17.3)	2,000 (13.8)	18 (292)	18 (292)	13(210)

Courtesy of ASTM International, 100 Barr Harbor Drive, PO Box C700, West Conshohocken, PA. ASTM C 55, Used with permission.

$$\text{Absorption (lb/ft}^3 \text{ or pcf)} = \frac{W_S - W_d}{W_d} \times 62.4 \qquad (11.6)$$

$$\text{Absorption (kg/m}^3) = \frac{W_S - W_d}{W_d} \times 1{,}000 \qquad (11.7)$$

where,

W_d = Dry weight of the specimen
W_s = Saturated surface-dry weight
$W_s - W_d$ = Amount of water absorbed

Moisture content of a sample is defined as the moisture present in the sample over the dry weight of the sample. Some equations related to moisture content are presented in Eqs. 11.8–11.10.

Thus, moisture content of the received sample can be determined as

$$\frac{W_r - W_d}{W_d} \times 100 \qquad (11.8)$$

$$\text{Moisture content as a percentage of total absorption} = \frac{W_r - W_d}{W_S - W_d} \times 100 \qquad (11.9)$$

$$\text{Dry density} = \frac{\text{Dry Weight}}{\text{Volume Occupied}} = \frac{\text{Dry Weight}}{\text{Volume Displaced in Water}}$$

$$= \frac{\text{Dry Weight}}{\dfrac{\text{Weight Displaced in Water}}{\text{Unit Weight}}}$$

$$= \frac{W_d}{\left(\dfrac{W_d - W_{\text{sub}}}{\gamma_w}\right)} = \left(\frac{W_d}{W_d - W_{\text{sub}}}\right)\gamma_w \qquad (11.10)$$

where,

W_{sub} = Submerged weight
W_r = As received weight
$W_r - W_d$ = Amount of water present in the sample as received condition

Example 11.1 Characterization of Concrete Masonry

A concrete masonry unit weighing 7.265 lb was received from a client. The sample was then tested according to the ASTM C 140. After oven drying, the weight was found to be 7.155 lb. The sample was then submerged under normal water, weighing in at 3.775 lb. The sample was taken out and the surfaces were dried using a damp towel. It was then weighed again, and found to be 7.995 lb.

Determine the following properties of the concrete masonry unit using the above test data:

a) The moisture content of the received sample
b) Absorption of the sample in percent of the dry weight
c) Absorption of the sample in lb/ft^3
d) Moisture content as a percent of total absorption
e) Weight classification
f) If the unit is a Grade N, Type I sample, does it satisfy the ASTM C 55 requirement for the absorption?

Solution

Measured Test Data:

W_d = Dry weight of the specimen = 7.155 lb
W_s = Saturated surface-dry weight = 7.995 lb
W_{sub} = Submerged weight = 3.775 lb
W_r = As received weight = 7.265 lb

a) The moisture content of the received sample (from Eq. 11.8):

$$\frac{W_r - W_d}{W_d} \times 100 = \frac{7.265\,\text{lb} - 7.155\,\text{lb}}{7.155\,\text{lb}} \times 100 = 1.54\% \text{ (Answer)}$$

b) Absorption of the sample in percent of the dry weight (from Eq. 11.5):

$$\text{Absorption (\%)} = \frac{W_S - W_d}{W_d} \times 100 = \frac{7.995\,\text{lb} - 7.155\,\text{lb}}{7.155\,\text{lb}} \times 100$$

$$= 11.74\% \text{ (Answer)}$$

c) Absorption of the sample in lb/ft^3 (from Eq. 11.6):

$$\text{Absorption (pcf)} = \frac{W_S - W_d}{W_d} \times 62.4$$

$$= \frac{7.995\,\text{lb} - 7.155\,\text{lb}}{7.155\,\text{lb}} \times 62.4\,\text{pcf}$$

$$= 7.33 \text{ (Answer)}$$

d) Moisture content as a percent of total absorption (from Eq. 11.9):

$$\frac{W_r - W_d}{W_S - W_d} \times 100 = \frac{7.265\,\text{lb} - 7.155\,\text{lb}}{7.995\,\text{lb} - 7.155\,\text{lb}} \times 100 = 13.1\% \text{ (Answer)}$$

e) Weight classification
From Eq. 11.10:

$$\text{Dry density} = \left(\frac{W_d}{W_d - W_{sub}} \right) \gamma_w$$

$$= \frac{7.155\,\text{lb}}{7.155\,\text{lb} - 3.775\,\text{lb}} \times 62.4\,\text{pcf} = 132\,\text{pcf (Answer)}$$

The dry weight is more than 125 pcf; therefore, it is a normal-weight concrete masonry unit.

f) From Table 11.2, for Grade N, Type I sample, ASTM C 55 requires the maximum absorption of 10 pcf. The current sample has absorption of 7.33 pcf. Therefore, the sample satisfies the ASTM C 55 requirement.

Example 11.2 Strength of Concrete Masonry

A hollow concrete masonry unit has the dimensions $12 \times 8 \times 4$ in. This unit has been tested for the compressive strength by applying load on the largest face (flatwise). The unit failed at the load of 210 kip. If the net volume of the unit is 230 in.^{3}, determine its compressive strength based on the gross area and the net area.

Solution

From Eq. 11.1: Gross Area = Actual Width × Actual Length = $8 \times 12 = 96$ in.^{2}

Net Area = Gross Area – Hollow Area.

However, hollow area is not known. Another known equation is as follows:

From Eq. 11.2: Net Area = Gross Area × Percent of Solid/100

The percent of solid can be calculated as follows:

From Eq. 11.3: Percent of Solid = Net Volume/Gross Volume × 100 = 230 in.^{3}/($12 \times 8 \times 4$ in.^{3}) × 100 = 59.9%

Therefore, Net Area = Gross Area × Percent of Solid/100 = (12×8 in.^{2}) × 59.9/100 = 57.5 in.^{2}

Compressive Strength (Gross Area) = Load/Gross Area = 210 kip/96 in.^{2} = 2.188 ksi = 2,188 psi

Compressive Strength (Net Area) = Load/Net Area = 210 kip/57.5 in.^{2} = 3.652 ksi = 3,652 psi

Answers: 2,188 psi and 3,652 psi

11.2.2 CLAY BRICKS

Two types of clay masonry are available: clay bricks and structural clay tile, as shown in Figure 11.5. Common clay bricks are solid and modular clay bricks have a hollow area equal to or less than 25%. Structural clay tiles are hollow and used for structural load-bearing structures.

11.2.2.1 Production of Clay Bricks

Figure 11.6 shows the production of clay bricks. First, clays (surface clay, shales, or fire clay) are collected. Sand or silt of about 30% of the total mass is preferred. Too much sand/silt provides less strength, and too little sand/silt causes excessive shrinkage during burning. After collection, the required amount of water is mixed with this clay. About 7–10% of water is mixed for low plasticity clay. This amount of water is very low, and this process is called the *dry press process*. The clay is shaped into a brick by pressing lightly. Due to its dry state, it takes time and physical labor to shape it. However, the shape remains intact while drying in the sun and being placed in the furnace. Another type of process is called *stiff-mud*, where about 12–15% water is added. Then, the mix is pressed in a die, after which the top surface is leveled using a cutting bar. Another process is used for clay with a high amount of water in its natural state. The inside of the mold is sanded first to prevent sticking. Then the wet clay is pressed inside, and the suitable shape is formed, as shown in Figure 11.6.

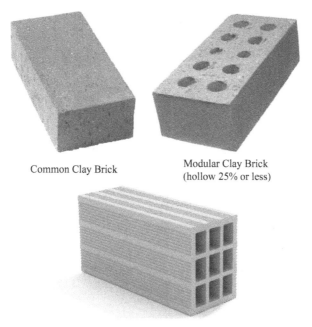

Common Clay Brick

Modular Clay Brick
(hollow 25% or less)

Structural Clay Tile

FIGURE 11.5 Clay bricks and structural clay tile.

Pressing Clay in the Mold   Drying Raw Bricks in the Sun   Burning Bricks in Fire

FIGURE 11.6 Manufacturing clay bricks.

11.2.2.2 Size of Clay Bricks

There are different sizes of bricks available. The height or depth of clay bricks vary from 2 to 8 in., width or thickness vary from 3 to 12 in., and the length can be up to 16 in. The bricks are specified by their nominal width by the nominal height by the nominal length. For example, a $4 \times 2\text{-}2/3 \times 8$ brick has a nominal width of 4 in., nominal height of 2-2/3 in., and nominal length of 8 in.

11.2.2.3 Grading and Types of Clay Bricks

In engineering, bricks are graded following the ASTM C 216 specification. It grades bricks based on the minimum compressive strength, maximum water absorption, and saturation coefficient. Three types of grading are available:

- Grade SW (severe weathering). This type of brick has the highest compressive strength, and absorption of water is very limited. It is suitable for areas subjected to severe weathering, which is determined by the product of the number of freeze–thaw days in a year and the winter rainfall.

- Grade MW (moderate weathering). This type of brick has a moderate compressive strength and moderate absorption limit. It is suitable for areas subjected to moderate weathering.
- Grade NW (negligible weathering). This type of brick has the lowest compressive strength, and absorption of water is not limited. It is suitable for areas subjected to minimal weathering.

The determination of minimum compressive strength, maximum water absorption, and saturation coefficient of bricks is discussed in the next subsection. Table 11.3 lists the ASTM C 216 requirements of these properties for different grading of bricks.

TABLE 11.3
Physical Requirements for Building Bricks

Grade	Minimum Compressive Strength, psi (MPa)		Maximum Water Absorption by 5-h Boiling (%)		Maximum Saturation Coefficient	
	Average of 5 Bricks	Individual	Average of 5 Bricks	Individual	Average of 5 Bricks	Individual
SW (Severe Weathering)	3,000 (20.7)	2,500 (17.2)	17	20	0.78	0.80
MW (Moderate Weathering)	2,500 (17.2)	2,200 (15.2)	22	25	0.88	0.90
NW (Negligible Weathering)	1,500 (10.3)	1,250 (8.6)	No Limit	No Limit	No Limit	No Limit

Courtesy of ASTM International, 100 Barr Harbor Drive, PO Box C700, West Conshohocken, PA. ASTM C 216, Used with permission.

Based on the usages of clay bricks, there are four types of clay bricks:

a) Building bricks
b) Facing bricks
c) Flooring bricks
d) Paving bricks

Bricks for *building* are prepared considering mainly the strength. *Facing bricks* must have good sizes, texture, and uniformity; appearance is the main concern. *Floor bricks* must be highly resistant to abrasion and/or have a smooth texture. *Paving bricks* may not be smooth and uniform. However, with the improvement in technology and machine production, high-quality bricks are being produced day to day with different colors, sizes, etc.

Facing bricks are of three types considering the factors affecting their appearance:

a) Type FBS
b) Type FBX
c) Type FBA

The first one, Type FBS (meaning Face Brick Standard), is the most common facing brick, and it is for general use in exposed buildings. The second one, FBX (meaning Face Brick Extra), is used for interior construction. FBX is superior to FBS in terms of precision and quality. Type FBA, meaning Face Brick Architecture, has some architectural characteristics in size and appearance.

11.2.2.4 Strength of Clay Bricks

Two types of strength are commonly used for clay bricks: compressive and flexural. The strength of clay bricks depends on the following factors:

- Composition of clay
- Proportions of sand/silt present in the clay
- Type of clay processing (dry or wet)
- Degree of burning, etc.

Compressive strength is determined by loading a brick until failure occurs. The load is applied flatwise, i.e., on the wider surface, and the compressive strength is determined following the ASTM C 67 as the peak load over the net area. If the net area is more than 75% of the gross area, then the gross area is used as the net area.

The flexural strength or the modulus of rupture (MOR) is determined by placing the brick flatwise, similar to a flat beam, and applying load at the mid span. Then, the flexural strength or the MOR is calculated using Eq. 11.11:

$$MOR = \frac{3}{2}\frac{WL}{bd^2} \tag{11.11}$$

where,

W = Maximum load, lb
L = Span of specimen, in.
b = Average width, in.
d = Average depth, in.

The span of the brick is taken as 1 in. less than the length of it. The flexural strength of clay brick varies between 500 and 3,800 psi (3.5 MPa and 26.2 MPa). The tensile strength of bricks is about 30–40% of the compressive strength.

Example 11.3 Modulus of Rupture of Masonry

A clay brick unit has the dimensions of 7.625 × 3.625 × 2.25 in. This unit has been tested for the flexural strength by applying load at the center on the largest face (flatwise). The unit failed at a load of 4.5 kip. Determine its modulus of rupture (MOR).

Solution

Given data:

Maximum load, W = 4.5 kip = 4,500 lb
Span, L = 1 in. less than the length = 7.625 – 1.0 = 6.625 in.

Average width, $b = 3.625$ in.
Average depth, $d = 2.25$ in.

From Eq. 11.11: Modulus of rupture,

$$MOR = \frac{3}{2}\frac{WL}{bd^2} = \frac{3}{2}\frac{(4,500 \text{ lb})(6.625 \text{ in.})}{(3.625 \text{ in.})(2.25 \text{ in.})^2} = 2,437 \text{ psi}$$

Answer: 2,437 psi

11.2.2.5 Absorption of Clay Bricks

Absorption is the physical process of sucking water to saturate pore spaces. If the absorption is high, then efflorescence can happen. ASTM C 62 limits the absorption of water for different grading of clay bricks. Commonly, most clay bricks have absorption of about 4–10% (Somayaji 2001). Two types of absorptions are recommended by the ASTM C 67:

a) 24-hour absorption
b) 5-hour boiling absorption

The 24-hour absorption is the absorbed water ratio with respect to the dry weight of the clay brick if it is submerged under normal water for 24 hours. It can be calculated using Eq. 11.12:

$$\text{Absorption } (\%) = \frac{W_S - W_d}{W_d} \times 100 \tag{11.12}$$

where,
  W_d = Dry weight of the specimen
  W_s = Saturated surface-dry weight

The 5-hour absorption is the absorbed water ratio with respect to the dry weight of the clay brick if it is submerged under boiling water for 5 hours. It can be calculated using Eq. 11.13.

$$\text{Absorption } (\%) = \frac{W_b - W_d}{W_d} \times 100 \tag{11.13}$$

where,
  W_d = Dry weight of the specimen
  W_b = Saturated surface-dry weight after boiling for 5 hours

The saturation coefficient can be calculated using Eq. 11.14.

$$\text{Saturation coefficient} = \frac{W_S - W_d}{W_b - W_d} \times 100 \tag{11.14}$$

Example 11.4 Absorption of Clay Brick

A piece of clay brick is measured in oven-dry condition and found to weigh 4.15 lb. Then, this brick is submerged into water for 24 hours, after which it weighed 4.70 lb. The brick is then dried again using an oven. After that, the brick is boiled underwater for 5 hours, taken out, its surface dried, and then weighed one last time. This final weight was found to be 4.75 lb.

Calculate the following:

a) The 5-hour absorption by boiling
b) The saturation coefficient
c) The grade possible (SW, MW, or NW)

Solution

Test data:

Dry weight of the specimen, $W_d = 4.15$ lb
Saturated surface-dry after boiling for 5 hours, $W_b = 4.75$ lb
Saturated surface-dry after 24 hours of submerging, $W_s = 4.70$ lb

a) From Eq. 11.13:

$$\text{The 5-hour absorption (\%)} = \frac{W_b - W_d}{W_d} \times 100$$

$$= \frac{4.75 \text{ lb} - 4.15 \text{ lb}}{4.15 \text{ lb}} \times 100 = 14.5\% \text{ (Answer)}$$

b) From Eq. 11.14:

$$\text{Saturation coefficient} = \frac{W_s - W_d}{W_b - W_d} \times 100$$

$$= \frac{4.70 \text{ lb} - 4.15 \text{ lb}}{4.75 \text{ lb} - 4.15 \text{ lb}} \times 100 = 91.7\% \text{ (Answer)}$$

c) From Table 11.3, considering the 5-hour absorption of 14.5%, the brick can be any grade. The saturation coefficient of 0.917 proves it can be NW, as the maximum limit of individual brick for MW and SW are 0.90 and 0.80, respectively.

11.3 MORTAR

Masonry units are individual pieces. They must be bonded to each other to make a large, cohesive structure. *Mortar* is the material used to bind two masonry units' surfaces. Mortar is a plastic mixture of cementitious materials, sand, and water. In addition to binding two masonry units, it also provides a uniform bed to place a masonry unit. Cement, lime, or a combination of both cement and lime can be used as cementitious materials. Specifically, mortar performs the following tasks:

- Bonds two masonry surfaces together, and may also bind reinforcement
- Provides a smooth bed for new masonry units

- Fills irregularities and voids in masonry
- Helps in constructing a single, large mass

There are four standard types of mortar:

- Type M
- Type S
- Type N
- Type O

Type M has the least volume of lime, while Type O has the greatest. ASTM C 270 provides the specifications of different types of cement–lime mortars and mortars of just cement.

The compressive strength of mortar is evaluated in the laboratory using the ASTM C 190 test standard with a 2-in cube. The cube is compressed until failure occurs, and the maximum load over the cross-sectional area is considered the compression strength. The minimum compression strengths of Type M, S, N, and O at 28 days of conditioning are 2,500, 1,800, 750, and 350 psi (17.2, 12.4, 5.2, 2.4 MPa), respectively (ASTM C 270).

11.4   GROUT

Grout is a high-slump (essentially fluid) concrete mixture that is used to fill the voids or cores of masonry, as well as bind reinforcing bars. Grout is a mixture of cement and water with or without fine aggregates proportioned so as to produce a pourable consistency without segregation of the components. The presence of gravel makes it capable of anchoring the reinforcing bars. In addition, it provides increased density, integrity of the section, and strength.

ASTM C 404 provides the specifications conforming to masonry grouting. The maximum aggregate of 3/8 in. is used in grout. The ratio of cement and lime used is 1:0.10, with some sand, fine gravel, and water included, as required by the mason's judgment. Sand is commonly used as 2.25 times to 3 times that of the total cementitious materials.

11.5   PLASTER AND STUCCO

Plaster, which is more fluidic compared to mortar, is a mixture of cement, sand, and water. Sometimes, lime is added to it.

Stucco is a plaster used for exterior walls. Stucco is applied in three coats. The first coat is the scratch coat, then the brown coat, and finally the finish coat. The first two coats are coarser (have larger fines) than the finish coat. Color may be added if desired. One part of cementitious material to 3–4 parts of loose sand is added together with water. The average compression strength of plaster and stucco is about 2,000 psi (13.8 MPa).

11.6   CHAPTER SUMMARY

Masonry is a very old material used in civil and construction engineering. This chapter discusses the basics of commonly used masonry units, such as concrete units and clay bricks. Masonry can be solid or hollow. Solid masonry includes clay bricks, stone blocks, concrete bricks, and glass blocks. Hollow masonry can also include the concrete blocks, as well as structural clay tiles.

Concrete masonry units can be load-bearing concrete masonry units and concrete building units. Both the load-bearing and concrete building units are of three types based on weight: lightweight units, medium-weight units, and normal-weight units. Load-bearing concrete masonry units are produced in two ways: Type I: moisture-controlled units and Type II: non-moisture-controlled units. Type I units are prepared to control the shrinkage limiting the moisture content. Type II does not consider moisture controlling. Two types of proposed grading are available for concrete masonry units: Grade N and Grade S. Grade S is intended for general use and has less strength and higher absorption compared to the Grade N. Grade N is intended for architectural purposes in exterior environments. Both of these grades can be of either Type I or Type II.

Two types of clay masonry are available: clay bricks and structural clay tiles. Common clay bricks are solid, and modular clay bricks have hollow spaces equal to or less than 25%. Structural clay tiles are hollow and used for structural load-bearing structures. Three types of grading of clay bricks are available: Grade SW (severe weathering), Grade MW (moderate weathering) and Grade NW (negligible weathering). Different graded bricks have different compressive strengths, water absorptions, and saturation coefficients. Based on the usages of clay bricks, there are four types: building bricks, facing bricks, flooring bricks, and paving bricks.

Compressive strength is determined by loading a brick in compression until failure. The load is applied flatwise, i.e., on the wider surface, and the compressive strength is determined as the peak load over the net area. The flexural strength or the modulus of rupture is determined by placing the brick flatwise, similar to a flat beam, and applying load at the mid span. Then, the flexural strength is calculated using the strength of materials rule. Absorption of clay brick is the sucking capacity of water to saturate its pore spaces. Two types of absorptions are used to characterize clay bricks: 24-hour absorption and 5-hour boiling absorption.

Mortar is used to bind these masonry units. The integrity of a masonry structure is mainly dependent on the strength of mortar. Plaster and stucco are used to cover the masonry units for beautification and protection from the environment. Masonry units with good architectural value are also available nowadays.

ORGANIZATIONS DEALING WITH MASONRY

The Masonry Society (TMS) is an international gathering of people interested in the art and science of masonry with advancing the knowledge of masonry. TMS is an educational, scientific, and technical society dedicated to the advancement of scientific, engineering, architectural, and construction knowledge of masonry. It is a not-for-profit, membership-driven organization that gathers, correlates, and disseminates information for the improvement of the design, construction, manufacture, use, and maintenance of masonry products and structures.

Location: Longmont, Colorado, USA.
Website: https://masonrysociety.org
ASTM International is also a resource for masonry characterizations.

REFERENCE

Somayaji, S. 2001. *Civil Engineering Materials*, 2nd Edition. Pearson, Upper Saddle River, NJ.

FUNDAMENTALS OF ENGINEERING (FE) EXAM STYLE QUESTIONS

FE Problem 11.1

If the net area of a masonry unit is more than 75% of the gross area, then it is considered

- A. A hollow masonry unit
- B. A solid masonry unit
- C. An unusual masonry unit
- D. Structural clay tiles

Solution: B

If the net area is more than 75% of the gross area, then it is considered a solid masonry unit. Hollow spaces of equal to or less than 25% are ignored for calculation.

FE Problem 11.2

A clay brick unit has the in-situ moisture content of 4.5% and the absorption capacity of 10.5%. If the wet weight of the brick is 12.5 lb, then the amount of additional water the unit absorbs upon soaking in water (in lb) is most nearly:

- A. 0.75
- B. 0.72
- C. 0.68
- D. 1.31

Solution: B

Additional amount of water it can absorb = Dry weight (Absorption – Moisture Content)
Dry brick + moisture = 100 + 4.5% = 104.5% = 1.045
Dry weight = 12.5 lb/1.045 = 11.96 lb
Additional amount of water it can absorb = 11.96 lb (0.105–0.045) = 0.72 lb

FE Problem 11.3

Which of the following properties of clay bricks is responsible for the efflorescence?

- A. Hollow space
- B. Less compression strength
- C. Less burning
- D. High absorption

Solution: D

High absorptive clay bricks suck in too much water. The excess water disintegrates the clay particles and causes the efflorescence.

PRACTICE PROBLEMS

PROBLEM 11.1

A hollow concrete masonry unit (Figure 11.7) has the following dimensions with a thickness of 6.625 in. This unit has been tested for the compressive strength by applying load on the largest face (flatwise). The unit failed at the load of 165 kip. Determine its compressive strength based on the gross area and the net area.

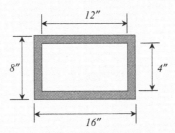

FIGURE 11.7   Hollow concrete masonry unit for Problem 11.1.

PROBLEM 11.2

A hollow concrete masonry unit (Figure 11.8) has the following dimensions with a thickness of 7.625 in. This unit has been tested for the compressive strength by applying load on the largest face (flatwise). The unit failed at the load of 110 kip. Determine its compressive strength based on the gross area and the net area.

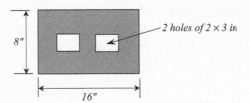

FIGURE 11.8   Hollow concrete masonry unit for Problem 11.2.

PROBLEM 11.3

A concrete masonry unit weighing 6.95 kg was received from a client. The sample was tested according to ASTM C 140. After oven drying, its weight was found to be 6.25 kg. The sample was then submerged under normal water, after which it weighed 2.75 lb. The sample was then taken out and the surfaces were dried using a damp towel. Afterward, it was weighed once more, and found to be 7.15 kg. Determine the following properties of the concrete masonry unit using the above test data:

a) The moisture content of the received sample
b) Absorption of the sample in percent of the dry weight
c) Absorption of the sample in kg/m³
d) Moisture content as a percent of total absorption
e) Weight classification
f) If the unit is a Grade S, Type II sample, does it satisfy the ASTM C 55 requirement for the absorption?

PROBLEM 11.4

A hollow concrete masonry unit was tested for compression and the following results were obtained:

> Failure load = 1,450 kN
> Gross area = 0.16 m²
> Gross volume = 0.032 m³
> Net volume = 0.018 m³

Determine the compressive strength of the unit based on the net area.

PROBLEM 11.5

A hollow concrete masonry unit has the dimensions of $7.625 \times 7.625 \times 15.625$ in. This unit has been tested for the compressive strength by applying load on the largest face (flatwise). The unit failed at the load of 190 kip. If the net volume of the unit is 750 in.³, then determine its compressive strength based on the gross area and the net area.

PROBLEM 11.6

A half-block concrete masonry unit is subjected to compression until failure following the appropriate ASTM standard. The outside dimensions of the unit are $166 \times 190 \times 320$ mm. There is a hollow rectangle which makes the wall 63 mm thick all around. The load was applied on the larger surface, and the block failed at the load of 236 kN. Determine its compressive strength based on the gross area and the net area.

PROBLEM 11.7

A clay brick unit has the dimensions $8 \times 4 \times 2.67$ in. This unit has been tested for flexural strength by applying a load at the center on the largest face (flatwise). The unit failed at a load of 2.22 kip. Determine its modulus of rupture (MOR).

PROBLEM 11.8

A piece of clay brick is measured in oven-dry condition and found to weigh 6.75 lb. Then, this brick is submerged into water for 24 hours, after which it weighed 7.75 lb. The brick is then dried again using an oven. After that, the brick is boiled underwater for 5 hours, taken out, its surface dried, and then weighed one last time. This final weight was found to be 7.90 lb. Calculate the following:

a) The 5-hour absorption by boiling
b) The saturation coefficient
c) The grade possible (SW, MW, or NW)

PROBLEM 11.9

A piece of clay brick weighing 5.95 kg was received from a client, After oven drying, its weight was found to be 5.48 kg. The absorption capacity of the brick is 17.5%. How much water will it absorb if it is submerged underwater for a prolonged period?

PROBLEM 11.10

Define solid and hollow masonry units based on the ASTM recommendation.

PROBLEM 11.11

Why are absorption capacities of concrete and clay masonry controlled?

PROBLEM 11.12

Define mortar, plaster, and grout. What are the main differences among them?

12 Sustainability in Materials

This chapter discusses the statistics of some common civil engineering materials such as aggregates, asphalt, and concrete and different strategies to adopt sustainability while using these materials.

12.1 CONCEPT OF SUSTAINABILITY

The term *sustainability* means satisfying the current needs and providing scopes for future users to satisfy their needs, considering economic, environmental, and social impacts. More specifically, sustainability is the concept of saving resources for future use without compromising the needs and resources of the present. Sustainability has been a very concerning topic in recent years because of the continued growth of the population, increased use of natural resources affecting the environment (e.g., global warming, more greenhouse gases, etc.), social impacts, and economic considerations (net benefits, life-cycle costs).

12.2 SUSTAINABILITY IN AGGREGATES

Although aggregate is cheaper than many other civil engineering materials, it has a huge cost impact as it is used in a bulk amount. Aggregates account for 80–85% by volume of typical asphalt concrete, and 62–68% by volume of hydraulic cement concrete. In 2012, approximately 1.3 billion tons of crushed stone, worth approximately $12 billion, was produced in the United States. Of the total crushed stone, 82% was used as a construction material, mainly for road construction and maintenance. For sand and gravel, approximately 927 million tons, worth about $6.4 billion, was produced, of which 26% was used for road base, coverings, and road stabilization, and 12% was used as asphalt concrete aggregates and in other asphalt–aggregate products (USGS 2013a).

Manufacturing of crushed stone requires several steps, such as drilling, blasting, crushing, transportation, and so on. About 42% of the aggregates consumed by weight in the United States have been processed through crushing (Moray et al. 2006). Energy is required in every phase of this production. Transportation involves the burning of engine fuels by transport vehicles, damage to pavements, noise production, etc. Other environmental issues arising from aggregate mining, processing, and transportation include dust pollution, groundwater use, noise, and issues concerning traffic safety on roads. Aggregates readily available in nature must also be transported and processed (Figure 12.1). Thus, there are scopes in adopting or improving sustainability in aggregate usages In all phases of aggregate work, some general approaches to improve sustainability are listed as follows (Van Dam et al. 2015):

- Reduce the amount of virgin aggregate used in different civil engineering structures
- Reduce the impact of virgin aggregate extracting and processing
- Minimize transportation impact by maximizing the use of marine/barge and rail transport, and minimizing truck transport
- Use locally available materials as much as possible

FIGURE 12.1   A batch of aggregate ready to be used in a roadway. *Photo taken in Pueblo, Colorado.*

- Encourage the use of recycled aggregates, co-products, and wastages aggregates, depending on the project type, i.e., wastage used for landfilling
- Improve aggregate durability to decrease the demand

12.3   SUSTAINABILITY IN ASPHALTIC MATERIALS

Asphalt material is typically produced by removing the lighter hydrocarbon molecules through a combination of vacuum and heat, or by mixing with a solvent, such as propane. The source of crude oil can have a significant effect on the energy and environmental impact of a specific asphalt material, as the processes needed to extract, process, transport, and refine it to produce asphalt material, and other products vary with the source. All these steps are costly and have social impacts. Although asphalt is used primarily in roadways and parking lots, the volume of other materials used for these applications in the United States is still great. For example, the United States used approximately 130 million barrels (23 million tons) of asphalt binder and road oil in 2011, costing $7.7 billion (EIA 2011). The value of asphalt paving mixtures produced in the United States was estimated at $11.5 billion in 2007 (United States Census Bureau 2007a).

The following considerations can be adopted for improving sustainability with regard to asphalt materials (Van Dam et al. 2015):

- Use greater quantities of recycled asphalt pavement (RAP) materials (Figure 12.2) RAP is being used all over the United States currently
- Use rubberized asphalt for asphalt concrete to reduce the usage of natural binder
- Use recycled asphalt shingles (RAS) as a partial replacement for asphalt binder if the same or better performance can be realized
- Reduce energy needed for and emissions from mixing asphalt concrete by using asphalt mixtures that require less heat, such as warm-mix, cold-mix, etc.
- Extend the life of asphalt concrete materials by conducting proper maintenance, using polymers or rubberized asphalt, and so on
- Reduce the need for virgin materials and transportation through in-place recycling
- Develop alternatives to petroleum-based binders
- Use more locally available materials

FIGURE 12.2 Collection of asphalt millings from a highway. *Photo taken in Pueblo, Colorado.*

12.4 SUSTAINABILITY IN CONCRETE MATERIALS

Concrete is produced from mixing Portland cement, aggregates, water, and optional admixtures. How aggregates can contribute to sustainability has been discussed in the previous section. Portland cement, although used in small proportions, contributes to unsustainable practices, as its production is energy consuming, impacts the environment by emitting greenhouse gases, and affects society with dust, noise, etc. The first step in manufacturing cement is to mine and process the necessary raw materials. Then, raw materials are ground and heated in a rotary cement kiln at about 2,640° F (1,450° C). The cements are finally transported to dealers and job sites. Each step of this process is energy intensive. Cement also incurs other socioeconomic impacts, such as transportation, concrete production, etc.

The United States used approximately 111 million tons of hydraulic cement in 2005, worth about $9.1 billion, according to the United States Geological Survey. Approximately 5% of cement used in the United States in 2011 was used for road-paving purposes (USGS 2013b). The United States had about 5,500 ready-mixed concrete plants in 2011 (United States Census Bureau 2013). The value of ready-mixed concrete produced in the United States was estimated at $34.7 billion in 2007, meaning the value of concrete used for road paving was about $1.7 billion, based on 5% of cement used (United States Census Bureau 2007b). Therefore, the value of concrete is high for this country. In addition, the production of Portland cement consumes huge amounts of energy and produces greenhouse gases. Reductions in these energy consumption and emission levels can be achieved by reducing the use of Portland cement in paving mixtures. Old concrete can be demolished, crushed, and processed to produce recycled concrete aggregate (RCA), as shown in Figure 12.3. Some advantages of using RCA are saving natural aggregates sources, landfill space, energy, and money (Hall et al. 2007). Concrete recycling offers 20–30% of the cost of pavement construction materials and supplies, and 10–15% of total construction costs (Halm 1980). One case study showed a saving of $5 million on a single project, all through cost savings from concrete pavement recycling (CMRA 2008). The basic steps of concrete pavement recycling are listed below.

- Evaluation of the source concrete
- Breaking and removing the concrete
- Removal of any steel mesh, rebar, or dowels

FIGURE 12.3   Processing of RAC from a demolished bridge. *Photo by Armando Perez, taken in Pueblo, Colorado.*

- Crushing and screening the RCA
- Beneficiation or quality control to remove any additional contaminants or improve properties

Some other approaches for improving sustainability with regard to concrete material usages are listed below (Van Dam et al. 2015):

- Improved cement plant efficiency through better energy harvesting and improved grinding
- Utilization of renewable energy, including wind and solar
- Utilization of more efficient fossil fuels
- Utilization of waste fuels
- Minimized clinker content in Portland cement through allowable limestone additions and inorganic processing additions
- Increased concrete mixing plant efficiency and reduced emissions
- Utilization of renewable energy
- Utilization of less cement in concrete mixtures without compromising performance
- Utilization of recycled washout water
- Utilization of recycled water used to process aggregates
- Improved durability of concrete

12.5   SUSTAINABILITY IN OTHER MATERIALS

Some other materials used in civil and construction engineering are steel, reinforcing fibers, and wood. Steel production is also very energy intensive, more so than the production of cement.

The production of steel from raw materials involves extracting the raw materials from ores, separating the impurities, melting, and shaping. The production of steel from recycled steel preserves raw materials. Wood is comparatively green compared to steel and concrete. It is also a renewable resource. Usages of steel or concrete can be avoided when wood structures are adequate in strength and durability, as substituting with wood can be a more environmentally sustainable course of action.

12.6   CHAPTER SUMMARY

Sustainability is an ongoing improvement process with no end. This chapter provides guidance to civil engineering students on how sustainability can be incorporated when selecting a material. The strategies listed in this chapter for various materials should be incorporated into the material analysis. Some common strategies to improve sustainability are using local and recycled materials, and extending the service life by conducting proper maintenance.

REFERENCES

Construction Materials Recycling Association (CMRA). 2008. *Case Histories.* Construction Materials Recycling Association, Aurora, IL.

Energy Information Agency (EIA). 2011. *Asphalt and Road Oil Consumption, Price, and Expenditure Estimates.* Energy Information Agency, Washington, DC.

Hall, K. T., Dawood, D. Vanikar, S. et al. 2007. Long-Life Concrete Pavements in Europe and Canada. FHWA-PL-07-027. Federal Highway Administration, Washington, DC.

Halm, H. J. 1980. "Concrete recycling." *Transportation Research News.* Volume 89. Transportation Research Board, Washington, DC.

Moray, S., Throop, N., Seryak, J., Schmidt, C., Fisher, and D'Antonio, M. 2006. Energy Efficiency Opportunities in the Stone and Asphalt Industry. *Proceedings of the Twenty-Eighth Industrial Energy Technology Conference*, New Orleans, LA.

United States Census Bureau. 2007a. *EC0731SX1: Manufacturing: Subject Series: Industry-Product Analysis: Industry Shipments by Products: 2007 Economic Census.* United States Census Bureau, Washington, DC.

United States Census Bureau. 2007b. *Manufacturing: Industry Series: Detailed Statistics by Industry for the United States: 2007 Economic Census. Ready Mixed Concrete Manufacturing.* United States Census Bureau, Washington, DC.

United States Census Bureau. 2013. *Number of Firms, Number of Establishments, Employment, and Annual Payroll by Enterprise Employment Size for the United States, All Industries: 2011.* United States Census Bureau, Washington, DC.

United States Geological Survey (USGS). 2013a. *Mineral Commodity Summaries 2013.* United States Department of the Interior, United States Geological Survey, Reston, VA.

United States Geological Survey (USGS). 2013b. *2011 Minerals Yearbook, Cement. Advance Release.* United States Department of the Interior, United States Geological Survey, Reston, VA.

Van Dam et al. 2015. Towards Sustainable Pavement Systems: A Reference Document. Report FHWA-HIF-15-002, Federal Highway Administration (FHWA), Washington, DC.

FUNDAMENTALS OF ENGINEERING (FE) EXAM STYLE QUESTIONS

FE PROBLEM 12.1

The term 'sustainability' means:

A. Satisfying the needs of the present demand
B. Ensure the ability of future users to satisfy their own needs

C. Both A and B
D. None of the above

Solution: C

FE PROBLEM 12.2

Some strategies for improving sustainability with regard to asphalt materials production are: (Select all that apply)

A. Reduce the amount of virgin asphalt and virgin aggregate by plant recycling
B. Reduce the energy needed for and emissions from mixing asphalt concrete
C. Extend the life of asphalt concrete materials
D. Reduce the need for virgin materials and transportation through in-place recycling
E. Use concrete pavement instead of asphalt pavement

Solution: A, B, C, and D

Usage of concrete pavement instead of asphalt pavement may not be a sustainable option, as cement production is very costly and risky to environments.

FE PROBLEM 12.3

Which of the following materials is the most environmentally friendly?

A. Concrete
B. Asphalt
C. Steel
D. Wood

Solution: D

The production of wood does not require a great amount of energy and labor. It is a renewable and natural material, with good strength and durability.

PRACTICE PROBLEMS

PROBLEM 12.1

List some strategies to adopt sustainability in pavements in terms of aggregate usages.

PROBLEM 12.2

List some strategies for improving sustainability with regard to asphalt materials.

PROBLEM 12.3

List some strategies for improving sustainability with regard to concrete materials.

Laboratory Testing

PREFACE TO LABORATORY TESTING

This concise laboratory testing guide is intended for classroom purposes, for undergraduate civil engineering, construction engineering, civil engineering technology, construction management engineering technology, and construction management programs. The worldwide, popular standards by ASTM International has been primarily used to demonstrate these laboratory testing procedures. The testing equipment may differ slightly from manufacturer to manufacturer, region to region, and so on. The instructor will need to be familiar with the equipment before demonstrating the tests to students. Without providing full details, an effort has been made to provide insight into the commonly performed testing methods in undergraduate studies. Please visit the ASTM website, www.astm.org for detailed information.

Students are also introduced to a certification program by the American Concrete Institute (ACI), specifically the Concrete Field Testing Technician—Grade I certification. Students can be certified ACI field technicians, just after completing this course. The tests which are included in the certification exams are especially discussed, and some sample study questions are provided to give students an idea of possible questions in the exam.

PREREQUISITE KNOWLEDGE TO LABORATORY TESTING

UNIT CONVERSION

In most (if not all) laboratories, it is required to convert some basic units, such as pounds to grams, centimeter to in., Fahrenheit to Celsius, etc. or vice-versa. Therefore, it is advised to memorize, or keep handy, the following (Table A.1) conversions:

TABLE A.1
Useful Conversion Factors

From	To	From	To
1 in.	25.4 mm	1 kg	2.2046 lb
1 ft	0.3048 m	1 lb	453.592 g
1 m	39.37 in.	1 ton	2,000 lb
1 m	3.2808 ft	1 psi	6,895 Pa
1 cm	0.3937 in.	25°C	77°F

SAFETY KNOWLEDGE

The following must be followed:

- All students must wear safety goggles while engaged in all laboratory work.
- All students are to wear thermal gloves and safety goggles when handling the oven or drying furnace

- Full-sleeved shirts, full-length pants, and at least closed-toe shoes are compulsory.
- Masks are required when dealing with dry soils. Allergies and/or sneezing due to dust is common.

Some other basic safety instructions:

- The EMERGENCY telephone number is 9-1-1 (or university emergency number)
- No eating or drinking in laboratories
- Clutter causes accidents: Neatness = safety
- If there is an active fire alarm, everyone MUST, by law, leave the building immediately
- If there is an alarm, please exit the building to let others know you are safe
- Avoid working alone in the laboratory
- While working, always keep the laboratory door unlocked (preferably, keep the door open)

DATA RECORDING

Immediately after measuring or reading some data, write it down clearly in the laboratory book or in any appropriate place(s). Consider significant figures after decimals to meet the standard. Also, make sure the recording includes the units. Most of the laboratories have equipment showing different units. Students may forget the unit once they leave the laboratory. Be aware of possible human errors, such as 789 versus 879, etc. Erasing the wrong entry and rewriting is not a good practice. Rather, cross through it clearly and rewrite nearby. This practice is more authentic. Never regress a number (predict the future value) unless it has actually been measured. However, after measuring a parameter, it is a good practice to judge the value, to see whether it is logical.

LABORATORY OBJECTIVES

Each laboratory session has been organized for an opportunity to:

- Review the procedures used in testing civil engineering materials.
- Observe, through actual testing, the consistencies and differences that various materials exhibit while being subjected to the same testing.
- Undergo the testing practices to observe if actual test results are consistent with predicted or historical data for the same type of material.
- Analyze why results found may not have been the same as those from historical data and review the potential environmental and personal reasons for possible inconsistencies and errors.
- Organize the above program into a comprehensive laboratory report, which will lead to a future reference file for the student.

LABORATORY REPORT WRITING

Dr. Fawad S. Niazi
Department of Civil and Mechanical Engineering
Purdue University Fort Wayne

At all stages of their careers, engineers are expected to have effective written technical communication skills that enable them to write project reports and proposals, and allow them to communicate with supervisors, clients, and co-workers. Because technical writing skills are critical for career advancement, effective report writing will be emphasized in this course. To aid in report preparation for this course, and in future technical writings, style and organization guidelines that convey the *formality*, *precision*, and *organization* important in technical writing are provided below.

1.0   GENERAL REQUIREMENTS

Raw data collected during the laboratory testing:

- Must be *checked with the teaching assistant or instructor* before leaving the laboratory
- Need not be typed
- Must be included in an appendix, except for lengthy, computer-generated data

Reports for the laboratory testing must:

- *Be entirely the student's own work* (Failure to comply with this requirement may be treated as suspected academic misconduct)
- Follow the formatting as advised by the instructor (commonly used format is to be typed with text and printed on 8½ × 11 in. size paper (one-face only), using 10–12 point font size in Times New Roman style, with titles and headings in 12–18 point bold font, and be single spaced)
- Include page numbers (except on the cover page, abstract, and table of contents)
- Follow the organizational guidelines provided in Section 2.0 of this document
- Include, in this order: a title page, abstract, table of contents, objective or introduction, materials, equipment, procedure, results, discussion, conclusions, references (if applicable), and appendices containing raw data and calculation sheets

2.0   REPORT ORGANIZATION

Laboratory reports are to be written in the style of a professional engineering report and should be neatly prepared and well organized. In addition to the style guidelines described in Section 3.0, specific guidelines for report preparation and organization are described below. Furthermore, the instructor may provide supplementary instructions pertaining to each laboratory exercise.

2.1   TITLE PAGE

The title page should include the following information:

- Laboratory # and title of laboratory test
- Course #, name, and section
- Student name

- Group members' names (if laboratory is conducted in groups)
- Date

2.2 ABSTRACT

An abstract is a brief summary of the main facts and conclusions in the report. It should be viewed as a stand-alone document—something that can be read and understood independently of the rest of the report. This is because it often *is* read independently by readers who want only a general idea of the report's content. Therefore, one must craft the abstract carefully, making sure to include only pertinent information in a logical order.

The abstract must generally include the following information, in the following order:

- The objective of the experiment
- A *brief* explanation of the methodology (typically 1–3 sentences will suffice)
- A summary of the results (give numerical values, if applicable)
- An explanation of the larger significance of the work done ('The results suggest…,' 'The results indicate…,' or 'The results demonstrate…')

Two optional elements of an abstract are 'Background' and 'Scope.' Background information often explains the motivation or engineering context of the experiment. Scope defines the limits or parameters of the experiment. If one should choose to include a background statement, place it before the objective statement. If one should choose to include a scope statement, place it after the objective statement.

Abstracts should:

- Be one paragraph in length
- Be written in the present tense (except methodology, which can be written in the past tense)
- Be concise, but still include articles like 'a' and 'the'

Abstracts should not:

- Include extensive background material, i.e., no developed discussion of the problem (or need) that led to this work; however, as stated above, one *might* include a *brief* reference to the problem in relation to the objective of the experiment
- Include overly broad conclusions which are not developed in the report
- Cite references
- Refer to any part of the report
- Include figures, tables, equations, or footnotes

2.3 TABLE OF CONTENTS

The table of contents should include titles of the sections in the laboratory report (i.e., Introduction, Materials, Equipment, etc.) and corresponding page numbers. The title page, abstract page, table of contents, and appendices are *not* numbered with ordinal numbers. Ordinal page numbering (i.e., 1, 2, 3…) begins with the 'Introduction' section and ends with the 'Conclusion' section. Thus, having a table of contents:

- Requires that the entire report be paginated, with the exceptions stated above
- Requires that the first ordinal-numbered item will be the 'Introduction,' and the last the 'Conclusion'

The 'Abstract,' 'Table of Contents,' and 'Appendix' may be page numbered separately, typically by roman numerals (i.e., I, II, III, IV…), but this is not required.

2.4   INTRODUCTION OR OBJECTIVES

For this course, a brief statement of the purpose or objectives of the experiment, and significance or usefulness of the test will suffice for the 'Introduction.' Additional background information, describing the motivation for and context of the experiment, may also be included.

Example: Young et al. (1998) have reported that green gumballs are preferred to red gumballs. Thus, it is of interest to know the relative percentages of each color gumball in a given sample. The objective of this experiment is to sort gumballs by color, to determine if a higher percentage of green gumballs are present in the sample as compared to red gumballs.

2.5   MATERIALS

Identify and specify the characteristics of materials used in the experiment, including type of material, sample dimensions, and unique features. A sketch or photo is often useful to describe the materials, and a list often provides organization when many materials must be described.

2.6   EQUIPMENT

Describe all the equipment and testing apparatus required for the experiment. Aim to provide detailed descriptions that would enable the reader to replicate the experiment described. Again, a sketch or photo of the laboratory set up is often useful.

2.7   PROCEDURE

This section provides a summary of what was done in the laboratory test(s).

- This may be written in paragraphs, or numbered steps.
- Remember to use *passive* voice—do not include 'I' or 'we' statements.
- In an effort to faithfully recount what transpired in the laboratory test, don't go overboard and try to account for every little action that was done. Report activities central to the successful completion of the laboratory test, not ones that are part of most experimental procedures.
- Do not refer the reader to a laboratory handout for 'further information'; the reader may not have it. Fully describe, or visually display, all pertinent information. If figures are utilized, make sure to number and caption them, and label all significant parts.

- While the experimental procedure should be a concise description of the steps performed in the laboratory test, it must not explain *why* certain steps were taken. These explanations should be included in the 'Discussion' section of the report.

2.8 RESULTS

The results section is a combination of visual and written elements. However, the written elements merely describe the visual elements; they do not interpret or analyze them.

- All of the data collected during the laboratory test must be reported here.
- It is often useful to create summary tables, plots, and figures to illustrate comparisons among data. *See section 3.0, regarding the use of tables and figures.*
- Place the bulk of the raw data in an appendix.
- *Do not* analyze, interpret, or explain any data. Analysis and interpretation are done in the 'Discussion' section.
- A set of sample calculations, for each type of calculation performed, should be included in this section.
- If qualitative observations were made during the laboratory test, then the description of these observations is also considered a result and should be included here.
- Do not give 8 and 10 decimal place readouts unless very small quantities are encountered, such as strain readings. Do not make unnecessary subtotal computations.

2.9 DISCUSSION

This is the section in which critical thinking skills are required. In this section, one may now interpret, explain, and analyze the data. If not done thoroughly, the reader might a) not see or understand what was found; or b) misconstrue what was found.

- The data should be discussed in relation to what is expected in theory or based on prior research.
- Lecture notes, the course laboratory manual, course textbook, and other textbooks are good references to determine if the experiment results agree or disagree with anticipated results.
- Remember, discrepancies in the data are not incorrect; they should be noted and discussed. For example, 'This feature does not correlate with what is expected, but could be due to...'

The instructor may provide specific questions to be addressed in the 'Discussion' section. However, in general, comparisons between expected and actual results, comparisons between behavior of different materials (e.g., tension behavior of steel vs. aluminum), description of possible sources of error, and discussion of the limitations of the experiment are appropriate.

2.10 CONCLUSIONS

- Provide two or three conclusions based on the data collected in the laboratory test.
- Make sure the conclusions are directly related to the data collected in the laboratory test. For example, if microstructural characterization was not performed, then

it cannot be concluded that the small grains measured are good for mechanical properties.

- Make sure that conclusions are not just reiterated results, but rather *insights* that can be drawn from the results.
- Avoid making overly broad statements and generalizations.

2.11 REFERENCES

The references cited in the laboratory test must be listed here (either alphabetically or by order of citation) in standard documentation form.

For books:

Author(s) Last name, First and Middle initials (Year of publication). *Title of book in italics*, Place of publication, Publisher.

Example: Islam, M.R. and Tarefder, R.A. (2020). *Pavement Design—Materials, Analysis and Highways*, McGraw Hill, New York.

For articles:

Author(s) (Year of publication) 'Title of the article with quotation marks, but in normal type', *Italicized title of magazine/journal*, V.# (insert number X, if known), pp. x–xx.

Example: Niazi, F.S. and Mayne, P.W. (2016). 'CPTu-based enhanced UniCone method for pile capacity', *Engrg. Geol.*, 212: 21–34.

Make sure that all of the references included at the end of the report have been cited in the body of the report. References may be cited any of the following ways:

- Numbered by order of reference: '...agrees with values for toughness given by (1).' or 'According to Islam and Tarefder (1) ...'
- By author's last name and date of publication '... agrees with values for toughness given by (Islam and Tarefder, 2020).'

Be sure to be consistent in the way citations are made in the text and the way the references are given at the end of the paper. That is, if citing by number, be sure to number the references. If citing by author name and date, be sure to list the references alphabetically.

2.12 APPENDICES

Appendices may include the original raw data collected during the experiment, the laboratory data sheet (signed by the teaching assistant or instructor), and your calculations. Unlike the rest of the report, the appendices do not need to be typed. Also, it is not necessary to include lengthy printed versions of digital data acquired during the experiment.

3.0 VISUAL ELEMENTS

All visuals, such as figures and tables, should be introduced in the text. And, they must be referred to in the text <u>before</u> they appear in the report. One cannot refer to figure 1, for example, <u>after</u> the figure appears in the text; this is too late and must be adjusted accordingly. The reader must be told, in the text, that this figure is following.

All figures must be numbered (e.g., Figure 1) and captioned *below* the figure. The caption must concisely describe the content of the figure, but also be detailed enough so that the reader can understand the figure without reading the text.

All tables must be numbered (e.g., Table 1) and titled *above* the table.

<u>Example</u>: Stress–strain plots resulting from the steel and aluminum tension tests are shown in Figure 2.

<u>Example</u>: Table 1 gives the mixture proportions for the trial batch and ACI mix designs.

4.0 SENTENCE STYLE

4.1 USE THIRD PERSON

In technical documents, such as laboratory reports, sentences need to be impersonal, empirical, and objective. The reader knows that the author(s) (the laboratory test group/design team) have performed the described work; therefore, the author(s) don't need to constantly appear as the subject in the sentences. What should appear as the subjects of sentences are the things being studied.

<u>Example</u>:

Not this: We found that the pressure varied with changes in temperature.
But this: Pressure varied with changes in temperature.

4.2 USE PASSIVE VOICE AND DESCRIPTIVE VERBS

Furthermore, experiments require observations, so one should expect to use some verbs associated with seeing. Phenomena are 'observed,' 'seen,' 'found,' and 'shown,' and values are 'calculated' and 'determined.' These are the things that people do, but because the author(s) must not appear as the subjects of sentences (see above), one must often *use the passive voice* in order to maintain an impersonal stance while still recounting work faithfully.

<u>Example</u>:

Not this: We used Equation 4 (or Eq. 4) to determine the compressive strength.
But this: Equation 4 (or Eq. 4) was used to determine the compressive strength.
Or this: The compressive strength was determined using Equation 4.

4.3 USE DESCRIPTIVE AND PRECISE MODIFIERS

Finally, technical documents must be written with an eye to specificity. Therefore, modifiers (adjectives, adverbs) which are too general such as 'large,' 'greater,' and 'quickly' are not only inappropriate but ultimately useless in scientific reporting. Be sure to include numbers in the writing.

<u>Example</u>:

Not this: A small amount of aggregate was weighed out and then heated in the oven overnight.

> *But this:* A 250-g sample of saturated surface dry fine aggregate was heated to 200°F for 14 hours.

Example:

> *Not this:* The steel was much stronger than the aluminum.
>
> *But this:* Uniaxial tension testing showed that the yield strength of the steel is 20% greater than the yield strength of 2024-T351 aluminum.

AMERICAN CONCRETE INSTITUTE (ACI) CERTIFICATION PROGRAM

CONCRETE FIELD TESTING TECHNICIAN—GRADE I

Soon after completing this course, it is recommended to take the exam on Concrete Field Testing Technician—Grade I. One can take this exam while in school. There is huge career potential in the concrete industry for those who have the certification. This can be a very good starting point for a career for those who are in school. A Concrete Field Testing Technician—Grade I is an individual who has demonstrated the knowledge and ability to properly perform and record the results of seven basic field tests on freshly mixed concrete. The program requires a working knowledge of the following ASTM test methods and practices:

- ASTM C 1064/C1064M–Temperature of Freshly Mixed Hydraulic-Cement Concrete
- ASTM C 172/C172M–Sampling Freshly Mixed Concrete
- ASTM C 143/C143M–Slump of Hydraulic-Cement Concrete
- ASTM C 138/C138M–Density (Unit Weight), Yield, and Air Content (Gravimetric) of Concrete
- ASTM C 231/C231M–Air Content of Freshly Mixed Concrete by the Pressure Method
- ASTM C 173/C173M–Air Content of Freshly Mixed Concrete by the Volumetric Method
- ASTM C 31/C31M–Making and Curing Concrete Test Specimens in the Field

This program requires demonstration of the knowledge and hands-on skills covered in the Job Task Analysis (JTA). ACI will grant certification only to those applicants who meet both of the following requirements:

- A passing grade on the ACI written examination
- Successful completion of the ACI performance examination

The one-hour written examination is 'closed book' and consists of 55 multiple choice and true–false questions. There are between five and ten questions on each of the ASTM test methods and practices. To pass the written examination, BOTH of the following conditions must be met:

- At least 60% correct for each of the required test methods and practices
- A minimum score of 70% overall

The performance examination is also 'closed book' and requires actual demonstration of six of the required test methods and practices, plus a verbal description of Practice C172/C172M (sampling). The information outlined here may change with time. Visit www.concrete.org for more (and updated) information.

LABORATORY TEST 1

RELATIVE DENSITY (SPECIFIC GRAVITY) AND ABSORPTION OF COARSE AGGREGATE

Designation: ASTM C 127

Scope

This method covers the determination of specific gravity and absorption of coarse aggregate. The specific gravity may be expressed as bulk specific gravity, bulk specific gravity (saturated surface-dry (SSD)), or apparent specific gravity. The bulk specific gravity (SSD) and absorption are based on aggregate after soaking in water.

Apparatus

- Weighing balance
- Container
- Oven
- Wire bucket

Test Specimen

- Reject all material passing a 4.75-mm C 127 2 (No. 4) sieve by dry sieving and thoroughly washing to remove dust or other coatings from the surface.
- Minimum mass of specimen to consider is listed in Table A.2:

TABLE A.2 Minimum Mass of Specimen Based on Size of Aggregate	
NMAS, in. (mm)	**Minimum Mass, lb (kg)**
1/2 (12.5) or less	4.4 (2)
3/4 (19.0)	6.6 (3)
1.0 (25.0)	8.8 (4)
1.5 (37.5)	11 (5)
2 (50)	18 (8)

Procedure

1) A sample of aggregate is immersed in water for 24 ± 4 hours to fill the pores.
2) Remove the aggregates from the water, dry the surface of the particles, and determine the mass. Report the mass as 'B.'
3) Place the specimen in a wire-bucket and take the submerged mass at $23 \pm 2.0°C$, as shown in Figure A.1. Report the mass as 'C.'
4) Finally, the sample is oven-dried, and the mass determined. Report the mass as 'A.'

a) Soaked Aggregates

b) Saturate-Surface Drying

c) Weighing the SSD Aggregate (B)

d) Weighing inside Water (C)

e) Oven-Drying

f) Weighing the Oven-Dried Aggregate (A)

FIGURE A.1 Determining specific gravity and absorption of coarse aggregate. *Photos taken at the Farmingdale State College of the State University of New York.*

Results

Present the results in Table A.3 and perform the calculations as directed.

TABLE A.3
Data Sheet for Relative Density (Specific Gravity) and Absorption

	Mass, lb (g)	
Item	Specimen 1	Specimen 2
Oven-dried mass, A (lb or g)		
Saturated surface-dried mass, B (lb or g)		
Submerged mass, C (lb or g)		
Bulk dry relative density (specific gravity) $= \dfrac{A}{B-C}$		
Bulk SSD relative density (specific gravity) $= \dfrac{B}{B-C}$		
Apparent dry relative density (specific gravity) $= \dfrac{A}{A-C}$		
% Absorption $= \dfrac{B-A}{A} \times 100$		
Average bulk dry relative density		
Average bulk SSD relative density		
Average apparent dry relative density		
Average % absorption		

Report

Report the specific gravity in three decimal digits and % absorption in two decimal digits.

LABORATORY TEST 2

BULK DENSITY (UNIT WEIGHT) AND VOIDS IN AGGREGATE

Designation: ASTM C 29, C 127, and C 128

Scope

This test method covers determining the bulk density (unit weight) and air voids in coarse aggregates, fine aggregates, or a mixture of coarse and fine aggregates. Density is the mass per unit volume, and unit weight is the weight per unit volume. Void is the empty space among a batch of aggregates.

Apparatus
- Weighing balance
- Mold
- Tamping rod
- Scoop
- Plate

TEST SPECIMEN

There must be enough aggregates to fill the mold. The aggregates are to be in dry condition.

Procedure
1) Measure the empty weight and interior volume of the mold as shown in Figure A.2.
2) Fill the mold with aggregate, about one-third full, and stroke with the rod 25 times. Cover the whole area while stroking.

Weighing Empty Mold	Filling One-Third	Rodding 25 times
Finishing Using Hand	Leveling the surface	Weighing Aggregate Full

FIGURE A.2 Different steps of determining unit weight of aggregates.

3) Fill the mold another one-third and stroke another 25 times.
4) Fill the mold to overflowing and stroke another 25 times.
5) Level the surface with a cutting plate or the rod. The slight projection of the coarse aggregate should balance the voids in the top surface.
6) Weigh the filled mold.
7) The difference in weight in Step 6 and Step 1 is the weight of aggregate filled.

Results

Present the results in Table A.4 and perform the calculations as directed.

TABLE A.4
Data Sheet for Density and Voids in Aggregates

Item	Specimen 1	Specimen 2
Mass of the empty mold, T (lb or kg)		
Volume of the mold, V (ft³ or m³)		
Mass of the aggregate plus the mold, Q (lb or kg)		
Bulk density (rodded), $\rho_{Rod} = \dfrac{Q-T}{V}$ (pcf or kg/m³)		
Bulk dry specific gravity, G_{sb} determined earlier following ASTM C 127 or Test Method C 128		
Density of water, ρ_w (pcf or kg/m³)	62.4 lb/ft³ or 1,000 kg/m³	
Percent void, $\% \text{Void} = \left[\dfrac{(G_{sb}\,\rho_w) - \rho_{Rod}}{G_{sb}\,\rho_w} \right] \times 100$		
Average bulk density, ρ_{Rod} (pcf or kg/m³)		
Average % void		

Report

Report the results for the bulk density to the nearest 1 lb/ft³ (10 kg/m³) and the void content to the nearest 1%.

LABORATORY TEST 3

Sieve Analysis of Fine and Coarse Aggregates

Designation: ASTM C 136

Scope

This test method covers the determination of the particle size distribution of fine and coarse aggregates by sieving. Accurate determination of material finer than the 75-μm (No. 200) sieve cannot be achieved by use of this method alone. Test Method ASTM C 117 for material finer than 75-μm sieve by washing should be employed.

Apparatus

- Sieves
- Sieve shaker
- Container
- Scoop
- Weighing balance

Test Specimen

Coarse Aggregate: The size of the coarse aggregate should conform to Table A.5.
Fine Aggregate: The amount of the dry fine aggregate is 0.66 lb (300 g) minimum.

TABLE A.5
Minimum Mass of Specimen Based on Size of Aggregate

NMAS, in. (mm)	Minimum Mass, lb (kg)
0.375 (9.5)	2.0 (1.0)
0.5 (12.5)	4.4 (2.0)
3⁄4 (19.0)	6.6 (5.0)
1.0 (25.0)	8.8 (10)
1.5 (37.5)	11 (15)
2.0 (50)	18 (20)

Procedure

1) Dry the sample to constant mass at a temperature of 230°F (110°C).
2) Weigh appropriate amount of dry sample, as mentioned above.
3) Stack the given set of standard sieves, place and clamp for the stacked set of sieves on the mechanical sieve shaker. Take at least one sieve which allows passing of all the materials.
4) Place the sample aggregate on top of the stacked set of sieves.
5) Shake the set of sieves for 10–15 minutes.
6) Weigh the amount of aggregate retained on each of the sieves.

Figure A.3 shows the sieve analysis procedure and the sieve analysis curve.

| Pouring Aggregates in Stacked Sieves | Shaking | Weighing |

FIGURE A.3 Different stages of sieve analysis of aggregates. *Photos taken at the Colorado State University–Pueblo.*

Results

- Calculate the % retained, cumulative % retained, and % finer for each sieve. Record in Table A.6.
- Draw the % finer vs. sieve size curve, with the sieve size in logarithmic scale.
- Explain the results with focus on gradation type and its effects.

TABLE A.6
Data Sheet for Sieve Analysis

Sieve No	Sieve Size (mm)	Mass Retained (lb or g)	% Retained	Cumulative % Retained	% Finer
	Total = (lb or g)				

Report

- Depending upon the form of the specifications for use of the material under test, the report should include one of the following:
  - Total percentage of material passing each sieve
  - Total percentage of material retained on each sieve
  - Percentage of material retained between consecutive sieves
- Report percentages to the nearest whole number, except if the percentage passing the 75-μm (No. 200) sieve is less than 10%, in which case it should be reported to the nearest 0.1%.
- Report the fineness modulus, when required, to the nearest 0.01.

LABORATORY TEST 4

TIME OF SETTING OF HYDRAULIC CEMENT BY VICAT NEEDLE

Designation: ASTM C 191

Scope

These test methods determine the time of setting of hydraulic cement by means of the Vicat needle. Two test methods are given: Method A, with a manually operated Vicat apparatus, and Method B, with an automatic Vicat machine.

Apparatus

- Vicat apparatus for Method A
- Reference masses and devices for determining mass
- Glass graduates
- Plane non-adsorptive plate
- Flat trowel
- Conical ring
- Mixer, bowl, and paddle
- Moist cabinet or room
- Automatic Vicat needle apparatus for Method B

Procedure

1) Quickly form the cement paste into a ball with gloved hands and toss six times, from one hand to the other, keeping the hands about 6 in. (150 mm) apart.
2) Press the ball resting in the palm of the hand into the larger end of the conical ring, held in the other hand, completely filling the ring with paste.
3) Place the ring on its larger end onto the non-absorptive plate, and slice off the excess paste at the smaller end of the top of the ring by a single oblique stroke of the trowel, held at a slight angle with the top of the ring.
4) Smooth the top of the specimen with one or two light touches of the pointed end of the trowel.
5) Allow the setting specimen to remain in the moist cabinet or moist room for 30 minutes after molding.
6) Determine the penetration of the 1-mm needle at this time and every 15 minutes thereafter (every 10 minutes for Type III cements) until a penetration of 1 in. (25 mm) or less is obtained, as follows:
   - Lower the needle of the rod until it rests on the surface of the cement paste.
   - Tighten the setscrew and set the indicator at the upper end of the scale, or take an initial reading.
   - Release the rod quickly by releasing the set screw, and allow the needle to settle for 30 sec.
   - Take the reading to determine the penetration.

Calculation

The Vicat initial time of setting is the time elapsed between the initial contact of cement and water and the time when the penetration is measured or calculated to be 1 in. (25 mm).

The Vicat final time of setting is the time elapsed between initial contact of cement and water and the time when the needle does not leave a complete circular impression in the paste surface.

Report

Report the time of setting in minutes and the method used.

Note: Another method available is ASTM C 266 Test Method for Time of Setting of Hydraulic-Cement Paste by Gillmore Needles

LABORATORY TEST 5

COMPRESSIVE STRENGTH OF HYDRAULIC CEMENT MORTARS (USING 2-IN. (50 MM) CUBE SPECIMENS)

Designation: ASTM C 109 and AASHTO T 106

Scope

This test method covers determination of the compressive strength of hydraulic cement mortars, using 2-in. (50-mm) cube specimens.

Apparatus

- Weights and weighing devices
- Glass graduates
- Specimen molds (Figure A.4)
- Mixer, bowl, and paddle
- Flow table and flow mold
- Tamper
- Trowel
- Moist cabinet or room
- Testing machine

Procedure

1) Apply a thin coating of release agent to the interior faces of the mold.
2) Seal the surfaces using grease where the halves of the mold join.
3) Prepare the paste with one part of cement to 2.75 parts of graded, standard sand by weight. Use a water–cement ratio of 0.485 for all Portland cements, and 0.460 for all air-entraining Portland cements.
4) Start molding the specimens within 2.5 minutes after mixing of the batch.
5) Place a layer of mortar about half the depth of the mold in all of the cube compartments.
6) Tamp the mortar in each cube compartment 32 times in 4 rounds, for about 10 sec. Each round is to be at right angles to the other, and consisting of eight adjoining strokes.

FIGURE A.4   Cubic mold.

7) Fill the compartments with the remaining mortar and tamp as specified for the first layer.

8) During tamping of the second layer, place the mortar, which is forced out, to the tops of the molds after each round of tamping.

9) Draw the flat side of the trowel lightly once along the length of the mold to level the mortar. Cut off the mortar to a plane surface, flush with the top of the mold, by drawing the straight edge of the trowel with a sawing motion over the length of the mold.

10) Place the test specimens in the moist cabinet or moist room from 20–72 hours, with their upper surfaces exposed to the moist air but protected from dripping water.

11) If the specimens are removed from the molds before 24 hours, keep them on the shelves of the moist cabinet or moist room until they are 24 hours old.

12) Immerse the specimens, except those for the 24-hour test, in saturated lime water in storage tanks.

13) Test the specimens immediately after their removal from the moist cabinet in the case of 24-hour specimens, and from storage water in the case of all other specimens.

14) Wipe each specimen to a surface-dry condition, and remove any loose sand grains from the faces that will be in contact with the bearing blocks of the testing machine.

15) Carefully place the specimen in the testing machine below the center of the upper bearing block.

16) Apply the load rate at a relative rate of movement between the upper and lower platens, corresponding to a loading on the specimen with the range of 200–400 lb/s (900–1,800 N/s).

17) Record the total maximum load indicated by the testing machine.

Calculation

Calculate the compressive strength of mortar using Table A.7.

Report

- Report the flow to the nearest 1%, and the water used to the nearest 0.1%.
- Average compressive strength of all specimens from the same sample shall be reported to the nearest 10 psi (0.1 MPa).

TABLE A.7 Data Sheet for Compressive Strength of Mortar	
Item	Value
A = Average gross-area (in.^{2})	
W = Maximum load (lb)	
Compressive strength $= \dfrac{\text{Max.Load(lb)}}{2'' \times 2''} = \dfrac{\text{Max.Load}}{4}$ (psi)	

LABORATORY TEST 6

TEMPERATURE OF FRESHLY MIXED HYDRAULIC-CEMENT CONCRETE

Designation: ASTM C 1064/C 1064M

Scope
This test method covers the determination of temperature of freshly mixed Portland cement concrete.

Apparatus
- Container
- Temperature measuring device

Test Specimen
a) Prior to sampling the freshly mixed concrete, dampen (with water) the sample container.
b) Place the freshly mixed concrete into the container.

Test Procedure
1) Place the temperature measuring device so that the end of the temperature sensing portion is submerged a minimum of 3 in. (75 mm) into the freshly mixed concrete.
2) Close the void left by the placement by gently pressing the concrete around the temperature measuring device at the surface of the concrete, so as to prevent ambient air temperature from affecting the reading, as shown in Figure A.5.
3) Leave the temperature measuring device in the freshly mixed concrete for a minimum period of 2 minutes, or until the temperature reading stabilizes, and then read and record the temperature.
4) Complete the temperature measurement of the freshly mixed concrete within 5 minutes after obtaining the sample.

Report
Record the measured temperature of the freshly mixed concrete to the nearest 1°F (0.5°C).

FIGURE A.5 Measuring temperature of freshly mixed hydraulic cement concrete. *Courtesy of Lehigh Hanson.*

Study Questions for the ACI Concrete Field Testing Technician—Grade I Exam

1. The sensor of the temperature measuring device shall have at least in. concrete cover in all directions.
2. Complete the temperature measurement of freshly mixed concrete within _____ minutes after obtaining the sample.
3. How long must the temperature measuring device remain in the freshly mixed concrete for an accurate reading?
4. The temperature of concrete may be measured in a wheelbarrow. True or False?
5. The temperature of concrete may be measured in a wall from. True or False?
6. A maximum temperature of freshly mixed concrete is specified in the ASTM C 1064. True or False?

Answers

1. 3
2. 5
3. 2 minutes, or until the temperature reading stabilizes
4. True
5. True
6. False (nothing such is specified)

Note that the ACI Concrete Field Testing Technician—Grade I Exam uses the multiple-choice and true/false type of questions only. The study questions presented here in fill-in-the-blank format is to save page space.

LABORATORY TEST 7

Sampling Freshly Mixed Concrete

Designation: ASTM C 172

Scope

This practice covers procedures for obtaining representative samples of fresh concrete as delivered from stationary, paving and truck mixers, and from agitating and non-agitating equipment, used to transport central-mixed concrete and from continuous mixing equipment.

Sampling Test Specifications

- The elapsed time shall not exceed 15 minutes between obtaining the first and final portions of the composite sample.
- Transport the individual samples to the place where fresh concrete tests are to be performed or where test specimens are to be molded. They shall be combined and remixed with a shovel the minimum amount necessary to ensure uniformity and compliance with the maximum time limits specified.
- Start tests for slump, temperature, and air content within 5 minutes after obtaining the final portion of the composite sample.
- Start molding specimens for strength tests within 15 minutes after fabricating the composite sample.

Procedure

Sampling from Stationary Mixers

- Take two or more portions at regularly spaced intervals during the discharge of the middle portion of the batch, and mix them up into one composite sample with a shovel.
- No samples should be taken before 10% or after 90% of the batch has been discharged.
- To sample the concrete, pass the container through the discharge stream, or completely move the discharge stream into the sample container.
- If the concrete is coming out too quickly to send the stream entirely into the container, discharge the stream into a large container which could hold the entire batch, and then sample from that container.
- Take care not to restrict the flow of concrete from the mixer, container, or transportation unit, so as to cause segregation.

Sampling from Paving Mixers

- Sample the concrete after the paving mixer has discharged, not during the discharge like other methods.
- Take samples from at least five different portions of the pile, and then mix them up into one composite sample.
- Try to avoid contaminating the concrete with the subgrade material, especially prolonged contact with subgrade material which can absorb a lot of moisture, like clay. A way to avoid this is to place containers on top of the subgrade and catch the concrete in them. In some cases, containers may need to be supported

so that they don't spill. They must be big enough to hold a representative sample of the concrete.

Sampling from Revolving Drum Truck Mixers
- Sample the concrete by collecting two or more portions of concrete at regular intervals, during the discharge of the middle part of the batch. Take these samples within a 15-minute period, and mix them up into one composite sample.
- Do not take samples until all water has been added to the mixer, and do not select samples from the first 10% and last 10% of the batch.
- Samples should be obtained by either diverting the entire stream into the container, or by passing the container through the stream at regular intervals.
- The rate of discharge of the batch should match the rate of revolution of the drum.

Sampling from Continuous Mixers
- Sample the concrete after the discharge of at least 140 L (5 ft³) of concrete, following all mixture proportioning adjustments.
- Sample the concrete at the frequency specified by collecting two or more portions, taken at regularly spaced intervals during discharge of the concrete.
- Take the portions so they are obtained within the time limit, and combine them into one composite sample for test purposes.
- Do not obtain samples from the very first or last portions of a mixer's continuous discharge. Sample by repeatedly passing a receptacle through the entire discharge stream or by completely diverting the discharge into a sample container.
- After obtaining the composite sample, wait a minimum of 2 minutes and a maximum of 5 minutes before beginning tests.

Sampling from Open Top Containers
For open-top containers, samples of the concrete may be taken by any of the previous three procedures. Choose the method that best fits the situation.

Procedure for Concrete with Large Maximum Size Aggregates
When the concrete being sampled has aggregate in it which is larger than the appropriate size for the molds or equipment being used, most of the sample will need to be run through a sieve to remove the extra-large aggregate. Afterward, perform a unit weight test on a small portion of the concrete with the large aggregate included.

Study Questions for the ACI Concrete Field Testing Technician—Grade I Exam

1. The elapsed time shall not exceed minutes between obtaining the first and final portions of the composite sample.
2. While sampling from stationary mixers, no samples should be taken before or after 90% of the batch has been discharged.
3. The tests for slump, temperature, and air content must be started within minutes after obtaining the final portion of the composite sample.
4. The molding of specimens for strength tests must be started within minutes after fabricating the composite sample.
5. While sampling from paving mixers, samples should be taken from at least_____ different portions of the pile.

ANSWERS:

1. 15
2. 10%
3. 5
4. 15
5. five

LABORATORY TEST 8

SLUMP OF HYDRAULIC-CEMENT CONCRETE

Designation: ASTM C 143

Scope

This test method is intended to provide the user with a procedure to determine slump of plastic hydraulic-cement concretes. It covers determination of slump of hydraulic-cement concrete, both in the laboratory and in the field.

Apparatus

- Conical slump cone: has the shape of the frustum of a cone and has an outer diameter of 4 in. (100 mm) at the top and 8 in. (200 mm) at the bottom, with a height of 12 in. (300 mm)
- Tamping rod: 2-ft-long (600 mm) with tamping or both ends rounded to hemispherical tip, measuring 5/8 in. (16 mm) in diameter.

Procedure

1) Dampen the interior of the mold.
2) Place the mold on a flat, moist, nonabsorbent (rigid) surface.
3) Fill the mold in three layers, each approximately one-third the volume of the mold.
   a) For the first layer, fill one-third the volume of the mold. Rod it uniformly over the cross-section with 25 strokes of the tamping rod, progressing spirally toward the center. Rod the layer throughout its depth.
   b) For the second layer, fill up to two-thirds of the volume of the mold. Rod it with 25 strokes so that the strokes just penetrate into the underlying (first) layer.
   c) For the third (last and top) layer, fill and heap concrete above the top of the mold. Rod it with 25 strokes so that the strokes just penetrate into the underlying (second) layer.

4) After the top layer has been rodded, strike off the surface of the concrete by means of a screeding and rolling motion of the tamping rod.
5) Continue to hold the mold down firmly and remove concrete from the area surrounding the base of the mold to preclude interference with the movement of slumping concrete.
6) Remove the mold immediately from the concrete by carefully raising it 12 in. (300 mm) in 5 ± 2 sec in a vertical motion, so as not to disturb the concrete cone.
7) Immediately measure the slump by determining the vertical difference between the top of the mold and the displaced original center of the top surface of the specimen.
8) Disregard the test if collapse or shearing off of concrete, from one side or a portion of the mass, occurs.

Note: Complete the slump test from the start of filling the mold to the removal of the mold in 2.5 minutes.

Figures A.6 and A.7 show the slump test procedure and slump test result interpretation, respectively.

Filling the Slump Cone   Tamping   Filled Cone   Removing Cone

FIGURE A.6   Slump test procedure. *Photos taken at the Farmingdale State College of the State University of New York.*

Collapse   Shear-Type Failure   True Slump

FIGURE A.7   Slump test result interpretation. *Photos taken at the Colorado State University–Pueblo.*

Report

Report the slump to the nearest 0.25 in. (5 mm).

Study Questions for the ACI Concrete Field Testing Technician – Grade I Exam

1. The mold for making the slump is in the shape of _____.
2. The height of the slump mold is _____.
3. The slump mold is filled in layers.
4. After the final rodding, the sides of the slump mold should be tapped lightly with the tamping rod. True or False?
5. How much time is allowed for lifting the slump mold?
6. How much time is allowed to conduct the slump test from beginning to completion?

Answers:

1. The frustum of a cone
2. 12 in. (300 mm)
3. 3
4. False
5. 5 ± 2 sec
6. 2.5 minutes

LABORATORY TEST 9

AIR CONTENT OF FRESHLY MIXED CONCRETE BY THE PRESSURE METHOD

Designation: ASTM C 231

Scope

This test method covers determination of the air content of freshly mixed concrete from observation of the change in volume of concrete with a change in pressure, exclusive of any air that may exist inside voids within aggregate particles. For this reason, it is applicable to concrete made with relatively dense aggregate particles.

Apparatus

- Air meters: Type A or Type B
- Spray tube
- Trowel
- Tamping rod: straight steel rod 5/8 in. (15.6 mm) in diameter and 16 in. (400 mm) long
- Mallet
- Strike-off bar
- Strike-off plate

Procedure

1) Obtain the sample of freshly mixed concrete in accordance with applicable procedures of Practice C 172.
2) Dampen the interior of the measuring bowl and place it on a flat, level, firm surface.
3) Place the concrete in the measuring bowl in three layers of approximately equal volume. Consolidate each layer of concrete by 25 strokes of the tamping rod, evenly distributed over the cross-section. After each of the three layers is rodded, briskly tap the sides of the measuring bowl 10–15 times with the mallet.
4) Instead of rodding, vibration can be used. For the case of vibration:
   a) Place the concrete in two layers of approximately equal volume.
   b) Consolidate each layer by three insertions of the vibrator, evenly distributed. Do not let the vibrator touch or rest on the measuring bowl.
5) Thoroughly clean the flanges or rims of the bowl and the cover assembly so that when the cover is clamped in place, a pressure-tight seal will be obtained.
6) Clamp the cover to the bowl, ensuring a pressure-tight seal.

Then, depending on the type of Meter (A or B) used, follow Step 7 to onward.

Using Type A Meter

7) Add water over the concrete by means of the tube until it rises to about the halfway mark of the standpipes.
8) Incline the apparatus assembly about 30 degrees from vertical and, using the bottom of the bowl as a pivot, outline several complete circles with the upper end of the column, simultaneously tapping the cover lightly to remove any entrapped air bubbles above the sample.
9) Return the apparatus assembly to a vertical position and fill the water column slightly above the zero mark while lightly tapping the sides of the bowl.

10) Bring the water level to the zero mark of the graduated tube before closing the vent at the top of the water column.

11) Apply more than the desired test pressure by means of the small hand pump, relieving local restraints by tapping the sides of the measuring bowl sharply.

12) When the pressure gage indicates the exact test pressure, read the water level and record to the nearest division or half-division on the gage of the standpipe.

13) Gradually release the air pressure through the vent at the top of the water column and tap the sides of the bowl lightly for about 1 minute.

14) Record the water level to the nearest division or half-division on the gage of the standpipe.

15) Calculate the final air content by subtracting the aggregate correction factor from the apparent air content and record the results.

Using Type B Meter

7) Close the air valve. Using a rubber syringe, inject water through one petcock until water emerges from the opposite petcocks.

8) Jar the meter gently until all air is expelled from this same petcock.

9) Close the air valve on the air chamber and pump air into the air chamber until the gage hand is on the initial pressure line.

10) Allow a few seconds for the compressed air to cool to normal temperature.

11) Stabilize the tapping the gage lightly by hand.

12) Open the air valve between the air chamber and the measuring bowl.

13) Tap the sides of the measuring bowl briskly with the mallet to relieve local restraints.

14) Lightly tap the pressure gage by hand to stabilize the gage hand.

15) Read the percentage of air on the dial of the pressure gage.

16) Close the air valve and then release the pressure in the bowl by opening both petcocks before removing the cover.

17) Calculate the final air content by subtracting the aggregate correction factor from the reading of the dial gage and record the results.

Figure A.8 shows different steps of measuring the air content of a concrete mix using Type B Meter.

Compacting   Leveling   Injecting Water   Pumping

FIGURE A.8   Measuring the air content of a concrete mix by the pressure method. *Photos taken at the Farmingdale State College of the State University of New York.*

Report

Report the percentage of air on the dial of the pressure gage. Apply an aggregate correction factor, if required.

Study Questions for the ACI Concrete Field Testing Technician—Grade I Exam

1. If the concrete sample is to be consolidated by rodding, the measuring bowl is filled in _____ layers.
2. If the concrete sample is to be consolidated by vibration, the measuring bowl is filled in _____ layers.
3. When rodding the concrete sample, what is the specified number of strokes required for each layer?
4. After rodding each layer, what should be done to the measuring bowl before adding another layer of concrete?
5. When consolidating the concrete by vibration, how many times should the vibrator be inserted into each layer?

Answers:

1. 3 equal
2. 2 equal
3. 25
4. Tap the sides of the measure 10–15 times with the mallet
5. 3

LABORATORY TEST 10

Air Content of Freshly Mixed Concrete by the Volumetric Method

Designation: ASTM C 173

Scope

This test method covers the determination of the air content of freshly mixed concrete containing any type of aggregate, whether it be dense, cellular, or lightweight.

Apparatus

- Air meter (Figure A.9)
- Bowl
- Funnel
- Tamping rod with both ends rounded to hemispherical tip
- Strike-off bar
- Calibrated cup
- Syringe
- Mallet
- Scoop
- Measuring vessel for isopropyl alcohol of 70% by volume
- Mallet

Procedure

1) Wet the inside of the bowl and dry it to a damp, but not shiny, appearance.
2) Using the scoop, fill half of the bowl with freshly mixed concrete.
3) Rod the half-filled bowl 25 times with the tamping rod.
4) Tap the sides of the bowl 10–15 times with the mallet to close any voids left by the tamping rod, and to release any large bubbles of air that may have been trapped.
5) Fill the rest of the half-filled bowl, rod it 25 times, penetrating the prior layer about 1 in. (25 mm), and tap the sides of the bowl 10–15 times with the mallet.

FIGURE A.9 Equipment for measuring the air content of a concrete mix using volumetric method. *Photos taken at the Colorado State University–Pueblo.*

6) Add a slight excess of concrete, 1/8 in. (3 mm) or less, above the rim and strike off the excess concrete with the strike-off bar.

7) Wipe the flange of the bowl clean.

8) Wet the inside of the top section of the meter, including the gasket.

9) Attach the top section to the bowl and insert the funnel. Add at least 1 pint (0.5 L) of water, followed by the selected amount of isopropyl alcohol. Record the amount of alcohol added.

10) Continue adding water until it appears in the graduated neck of the top section.

11) Remove the funnel. Adjust the liquid level until the bottom of the meniscus is level with the zero mark. A rubber syringe is useful for this purpose.

12) Attach and tighten the watertight cap.

13) Quickly invert the meter, shake the base, and return the meter to the upright position. Do not invert the meter for more than 5 sec at a time.

14) Repeat the inversion, shaking and upright process for a minimum of 45 sec and until the concrete is free from the bowl.

15) Using the hand on the flange to rotate the meter, vigorously roll the meter ¼ to ½ turn forward and back several times, quickly starting and stopping the roll.

16) Turn the base about 1/3 turn and repeat this rolling process.

17) Repeat Steps 15 and 16 for approximately 1 minute while listening for aggregates sliding the meter.

18) Set the meter upright, loosen the cap, and allow the liquid level to stabilize. The liquid level is stable when it does not change more than 0.25% within a 2-minute period.

19) If the liquid level does not stabilize within 6 minutes, discard the test and conduct a new test using additional alcohol.

20) If there is more foam present in the neck than that which is equivalent to two full air percent divisions, discard the test and conduct a new test using additional alcohol.

21) If the air content is greater than the 9% range, add calibrated cups of water to the meter to bring the liquid level into the graduated range of the meter. Record the number of cups of water added.

22) When the liquid level is stable, read the level to the bottom of the meniscus to the nearest 0.25%. This is the initial meter reading.

23) Re-tighten the cap and repeat Steps 15–22.

24) If the second reading of the liquid level does not change more than 0.25% from the initial meter reading, record the second reading as the final meter reading.

25) If the second reading differs from the first reading by more than 0.25%, record the second reading as the initial meter reading and repeat Steps 15–22.

26) If the third reading of the liquid level has not changed more than 0.25% from the initial meter reading, record the third reading as the final meter reading. Otherwise, discard the test and conduct a new test using additional alcohol.

27) Once the final meter reading has been obtained, disassemble the meter, dump out the contents of the base, and examine the base for portions of undisturbed, tightly packed concrete. If such material is present, the test is invalid.

28) The final air content reading is equal to the final meter reading minus the correction for large amounts of alcohol plus the number of calibrated cups of water added.

Results

- When less than 2.5 pints (1.2 L) of isopropyl alcohol is used, the final meter reading is the air content of the sample of concrete tested.
- When 2.5 pints (1.2 L) or more of isopropyl alcohol is used, subtract the correction from Table A.8 from the final meter reading to obtain the air content of the concrete sample.

TABLE A.8
Correction of Air Content for 2.5 pints (1.2 L) or More of Isopropyl Alcohol Used

Pints	Liters	Subtract (%)
2.5	0.9	0.15
3	1.4	0.3
4	1.9	0.6
5	2.4	0.9

Report

Report the air content to the nearest 0.25%.

Study Questions for the ACI Concrete Field Testing Technician – Grade I Exam

1. What type of aggregate is acceptable for this test? Dense, cellular, or lightweight?
2. Why is the isopropyl alcohol added to the meter?
3. How many times does the rolling operation occur?
4. Is the tip of the tamping rod flat, rounded, or hemispherical?
5. What type of alcohol and concentration must be used in this test?

Answers:

1. Any type of aggregate is acceptable, whether it be dense, cellular, or lightweight
2. To dispel any foam that may be on the surface of the water
3. A minimum of two and maximum of three times
4. Hemispherical
5. Isopropyl alcohol of 70% by volume

LABORATORY TEST 11

Density, Yield, Relative Yield, and Cement Content of Freshly Mixed Concrete

Designation: ASTM C 138

Scope

This test method covers determining the density and yield of the freshly produced concrete. Yield is defined as the volume of concrete produced from a mixture of known quantities of the component materials. Some other parameters, such as relative yield cement content and air void, can also be determined.

Relative yield (R_y) is the ratio of the actual volume of concrete obtained to the volume as designed for the batch calculated. A value for R_y greater than 1.00 indicates an excess of concrete being produced whereas a value less than this indicates that the batch is short of its designed volume. In practice, a ratio of yield in cubic feet per cubic yard of design concrete mixture is frequently used, for example, 27 ft³/yd³.

Apparatus

- Weighing balance
- Mold
- Tamping rod with both ends rounded to hemispherical tip
- Scoop
- Strike-off plate
- Mallet

Procedure (Figure A.10)

1) Measure the empty mass (M_m) and interior volume (V_m) of the mold.For consolidation, rodding is to be used for a slump greater than 3 in. (75 mm). Vibration is to be used for a slump smaller than 1 in. (25 mm). Either of these can be used for slumps between 1 and 3 in. (25–75 mm).

2) For rodding:

   a) Fill the mold with freshly mixed concrete, about one-third full, and stroke with the rod 25 times when nominal 0.5-ft³ (14-L) or smaller measures are used, 50

| Weighing Empty Mold | Filling One-Third | Rodding 25 Times |
| Tamping Outside | Leveling the Final Surface | Weighing the Mold-Full |

FIGURE A.10 Different steps of determining density of concrete. *Photo by James Lee.*

strokes when nominal 1-ft^3 (28-L) measures are used, and one stroke per 3 in.^{2} (20 cm^3) of surface for larger measures.

b) Tap the outsides of the mold 10–15 times with the mallet to close any voids left by the tamping rod, and to release any large bubbles of air that may have been trapped.

c) Fill the mold another one-third, rod another 25 times, penetrating the prior layer about 1 in. (25 mm), and tap the outsides another 10–15 times with the mallet.

d) Fill the mold so as to avoid overflowing, rod another 25 times, and tap another 10–15 times with the mallet.

3) For vibration:

a) Fill and vibrate the mold in two approximately equal layers. Place all of the concrete for each layer in the mold before starting vibration of that layer.

b) Insert the vibrator at three different points for each layer. When compacting the bottom layer, do not allow the vibrator to rest on or touch the bottom or sides of the mold. When compacting the final layer, the vibrator shall penetrate into the underlying layer approximately 1 in. (25 mm).

4) Carefully level the surface with a cutting plate to level the surface smoothly.

5) Measure the mass of the concrete-filled mold (M_c).

6) The difference in weight in Step 5 and Step 1 is the weight of concrete filled ($M_c - M_m$).

Results

Present the results in Table A.9 and perform the calculations as directed.

TABLE A.9 Data Sheet for Density, Yield, Relative Yield and Cement Content in Concrete	
Parameters	**Value**
Empty mass of the mold, M_m (lb or kg)	
Interior volume of the mold, V_m (ft^3 or m^3)	
Mass of the concrete-filled mold, M_c (lb or kg)	
Density, $D = \dfrac{M_c - M_m}{V_m}$ (lb/ft^3, kg/m^3)	
Total mass of all materials batched, M	
Yield, $Y = \dfrac{M}{27 \times D}$ (yd^3)	
Yield, $Y = \dfrac{M}{D}$ (ft^3)	
Yield, $Y = \dfrac{M}{D}$ (m^3)	
Volume of concrete for which the batch was designed, Y_d (yd^3 or m^3)	
Relative yield, $R_y = \dfrac{Y}{Y_d}$	
Mass of cement in the batch, C_b (lb or kg)	
Cement content, $C = \dfrac{C_b}{Y}$	

Air void (A) of the freshly mixed concrete can also be determined if the theoretical density of the concrete, computed on an air-free basis, (lb/ft³ or kg/m³) is known using the following equation:

$$\text{Air Content, } A = \frac{D_T - D}{D_T} \times 100$$

where D_T is the sum of the product of the bulk specific gravity and water density of each ingredient in the concrete mix, $\sum (G_{sb} \rho_w)$. For the aggregate components, the bulk specific gravity and mass should be based on the saturated, surface-dry condition. For cement, a value of 3.15 may be used.

Report

Report the following:

- Air content
- Cement content
- Concrete
- Relative yield
- Unit weight
- Yield

Study Questions for the ACI Concrete Field Testing Technician—Grade I Exam

1. For consolidation, which method is to be used for slumps greater than 3 in. (75 mm)?
2. For a measure of 0.5 ft³ (14 L) or smaller, what is the required number of strokes of the tamping rod for consolidation of each layer?
3. Given the volume of the measure be 0.504 ft³, mass of the empty measure be 19.6 lb, and the mass of the measure plus the concrete be 90.2 lb, the density of the concrete is:
4. The density of a freshly mixed concrete is 145 pcf. For a design batch of 10 yd³, the total mass of all materials batched was 39,000 lb. What is the yield per batch, in ft³?
5. For the above problem, what is the yield per cubic yard in ft³/yd³?

Answers:

1. Rodding
2. 25
3. Density, $D = \dfrac{M_c - M_m}{V_m} = \dfrac{90.2 - 19.6}{0.504} = 140$ pcf
4. Yield, Y (ft³) $M/D = 39{,}000$ lb/145 pcf $= 269$ ft³
5. Relative Yield, $R_y = Y/Y_d = 269$ ft³/10 yd³ $= 26.9$ ft³/yd³

LABORATORY TEST 12

MAKING AND CURING CONCRETE TEST SPECIMENS IN THE FIELD

Designation: ASTM C 31

Scope

This practice covers procedures for making and curing cylinder and beam specimens from representative samples of fresh concrete for a construction project.

Apparatus

- Cylinder molds
- Beam molds
- Tamping rod
- Vibrator
- Mallet
- Scoop, blunted trowel, or shovel
- Metal pan or bowl

Procedure

Making Cylindrical Specimens

1) Select the tamping rod and mode of consolidation (rodding or vibration). Vibration of concrete is required for slumps less than 1.0 in. (25 mm).
2) Place the concrete in the molds using a scoop, blunted trowel, or shovel. Make specimens in layers as indicated in Table A.10.
3) If consolidation is done by rodding, place the concrete in the mold, in the required number of layers of approximately equal volume. Rod each layer with the rounded end of the rod using the number of strokes and size of rod specified in Table A.10. For each upper layer, allow the rod to penetrate through the layer being rodded and into the layer below, about 1.0 in. (25 mm).
4) If consolidation is done by vibration, please see the standard for details.
5) After each layer is rodded, tap the outsides of the mold lightly 10–15 times with the mallet in order to close any holes left by rodding and to release any large air bubbles that may have been trapped.

Making Beam or Prism Specimens

1) Select the tamping rod and mode of consolidation (rodding or vibration). Vibration of concrete is required for slumps less than 1 in.
2) Place the concrete in the molds using a scoop, blunted trowel, or shovel. Make specimens in layers as indicated in Table A.11.

TABLE A.10
Rodding Requirements for Concrete Cylinder

Diameter, in. (mm)	Number of Layers of Equal Depth	Number of Roddings per Layer
4 (100)	2	25
6 (150)	3	25
9 (225)	4	50

TABLE A.11	
Layer Requirements for Concrete Beam	
Depth, in. (mm)	Number of Layers of Equal Depth
6–8 (150–200)	2
>8 (200)	3 or more

3) If consolidation is done by rodding, place the concrete in the mold, in the required number of layers of approximately equal volume. Rod each layer with the rounded end of the rod, one stroke per 2 in.^{2} (1,250 mm^2). For each upper layer, allow the rod to penetrate through the layer being rodded and into the layer below, about 1.0 in. (25 mm).
4) If consolidation is done by vibration, please see the standard for details.
5) After each layer is rodded, tap the outsides of the mold lightly 10–15 times with the mallet in order to close any holes left by rodding and to release any large air bubbles that may have been trapped.

Curing
1) Immediately after molding and finishing the cylinders, the specimens are to be stored for a period up to 48 hours, in a temperature range from 60 and 80°F (16 and 27°C), and in an environment preventing moisture loss from the specimens.
2) For concrete cylinders with a specified strength of 6,000 psi (40 MPa) or greater, the initial curing temperature shall be between 68 and 78°F (20 and 26°C).
3) Upon completion of this curing and within 30 minutes after removing the molds, cure specimens with free water maintained on their surfaces at all times at a temperature of 73 ± 3°F (23 ± 2°C) using water storage tanks or moist rooms, except when capping with sulfur mortar capping compound and immediately prior to testing.
4) For a period not to exceed 3 hours immediately prior to testing, standard curing temperature is not required, provided that free moisture is maintained on the cylinders and the ambient temperature is between 68 and 86°F (20 and 30°C).
5) Beams are to be cured the same as cylinders, except that they shall be stored in water saturated with calcium hydroxide at 73 ± 3°F (23 ± 2°C) at least 20 hours prior to testing.

Transportation of Specimens to Laboratory
- Specimens shall not be transported until at least 8 hours after final set.
- During cold weather, protect the specimens from freezing with suitable insulation.
- Prevent moisture loss during transportation by properly wrapping the specimens.
- Transportation time shall not exceed 4 hours.

Study Questions for the ACI Concrete Field Testing Technician – Grade I Exam
1. How much time must elapse until specimens can be transported after the final set?
2. What is the maximum time allowed for transportation of the specimen to the laboratory?

3. What types of specimens shall be stored in water saturated with calcium hydroxide at $73 \pm 3°F$ ($23 \pm 2°C$) at least 20 hours prior to testing?
4. While consolidating any upper layer, how deep is the rod is allowed to penetrate through to the layer below?
5. After rodding, the mold's exterior of which layer(s) is tapped lightly 10 to 15 times with the mallet?

Answers:

1. 8 hours
2. 4 hours
3. Beams
4. About 1.0 in. (25 mm)
5. All layers

LABORATORY TEST 13

MAKING AND CURING CONCRETE TEST SPECIMENS IN THE LABORATORY

Designation: ASTM C 192

Scope

This practice covers procedures for making and curing test specimens of concrete in the laboratory under accurate control of materials and test conditions, using concrete that can be consolidated by rodding as described herein.

Apparatus

- Metal pan or bowl
- Scoop, blunted trowel, or shovel
- Tamping rod
- Mallet

Procedure

If Hand-Mixing

1) Mix the batch in a watertight, clean, and damp metal pan or bowl, with a bricklayer's blunted trowel.
2) Mix the cement, powdered insoluble admixture, if used, and fine aggregate without the addition of water until they are thoroughly blended.
3) Add the coarse aggregate and mix the entire batch without water until the coarse aggregate is uniformly distributed throughout the batch.
4) Add water and the admixture solution, if used, and mix the mass until the concrete is homogeneous in appearance.

If Machine Mixing

1) Add the coarse aggregate, some of the mixing water, and the solution of admixture, when required.
2) Start the mixer, then add the fine aggregate, cement, and water with the mixer running.
3) Mix the concrete after all ingredients are in the mixer for 3 minutes, followed by a 3-minute rest, and then followed by a 2-minute final mixing time.
4) Cover the open end or top of the mixer to prevent evaporation during the rest period.
5) Take precautions to compensate for mortar retained by the mixer, so that the discharged batch, as used, will be correctly proportioned.
6) To eliminate segregation, deposit machine-mixed concrete in the clean, damp mixing pan and remix by a shovel or trowel until it appears to be uniform.

Making Cylindrical Specimens (Figure A.11)

1) Mold specimens, as near as practicable, to the place where they are to be stored during the first 24 hours.
2) Place the concrete in the molds using a scoop, blunted trowel, or shovel. Make specimens in layers as indicated in Table A.12.
3) If consolidation is done by rodding, place the concrete in the mold, in the required number of layers of approximately equal volume. Rod each layer with

the rounded end of the rod, using the number of strokes and size of rod specified in Table A.12.

4) If consolidation is done by vibration, please see the ASTM standard for details. Vibration of concrete is required for slumps less than 1 in.

5) After each layer is rodded, tap the outsides of the mold lightly 10–15 times with the mallet in order to close any holes left by rodding and to release any large air bubbles that may have been trapped.

Making Beam or Prism Specimens (Figure A.12)

1) Mold specimens, as near as practicable, to the place where they are to be stored during the first 24 hours.

2) Place the concrete in the molds using a scoop, blunted trowel, or shovel. Make specimens in layers as indicated in Table A.13.

3) If consolidation is done by rodding, place the concrete in the mold, in the required number of layers of approximately equal volume. Rod each layer with

Weighing   Mixing   Compacting   Finishing

FIGURE A.11   Preparing concrete cylinder. *Photos taken at the Farmingdale State College of the State University of New York.*

TABLE A.12
Rodding Requirements for Concrete Cylinder

Diameter, in. (mm)	Number of Layers of Equal Depth	Number of Roddings per Layer
3 or 4 (75–100 mm)	2	25
6 (150 mm)	3	25
9 (225 mm)	4	50

Mixing   Compacting   Finishing

FIGURE A.12   Preparing concrete beam specimen. *Photos taken at the Farmingdale State College of the State University of New York.*

TABLE A.13

Layer Requirements for Concrete Beam

Depth, in. (mm)	Number of Layers of Equal Depth
Up to 8 (200)	2
>8 (200)	3 or more

TABLE A.14

Rodding Requirements for Concrete Beam

Top Surface Area, in.^{2}	Diameter of Rod, in. (mm)	Number of Roddings per Layer
25 or less	3/8 (9.4 mm)	25
26–49	3/8 (9.4 mm)	One for 1.0 in.^{2}
50 or more	5/8 (15.6 mm)	One for 2.0 in.^{2}

the rounded end of the rod, using the number of strokes and size of rod speci-fied in Table A.14.

4) If consolidation is done by vibration, please see the standard for details.

5) After each layer is rodded, tap the outsides of the mold lightly 10–15 times with the mallet in order to close any holes left by rodding and to release any large air bubbles that may have been trapped.

Curing

1) To prevent evaporation of water from unhardened concrete, cover the speci-mens immediately after finishing.

2) Remove the specimens from the molds 24 ± 8 hours after casting.

3) Then, moist-cure the specimens at $73.5 \pm 3.5°F$ ($23.0 \pm 2.0°C$) from the time of molding until the moment of testing.

4) Flexural strength test specimens, in addition to the above curing, should be immersed in water saturated with calcium hydroxide at $73 \pm 3°F$ ($23 \pm 2°C$) for a minimum period of 20 hours, immediately prior to testing.

LABORATORY TEST 14

COMPRESSIVE STRENGTH OF CYLINDRICAL CONCRETE SPECIMENS

Designation: ASTM C 39

Scope

This test method covers determination of compressive strength of cylindrical concrete specimens, such as molded cylinders and drilled cores. It is limited to concrete having a unit weight greater than 50 lb/ft³ (800 kg/m³).

Apparatus

Loading machine

Procedure

1) Compression tests of moist-cured specimens should be made as soon as practicable after removal from moist storage in the moist condition.
2) Place the test specimen on the lower bearing block, carefully aligning the axis of the specimen with the center of the loading, as shown in Figure A.13.
3) Set the load indicator to zero.
4) Apply the load continuously and without shock, at a rate of approximately 20–50 psi/sec (0.15–0.35 MPa/sec).
5) Apply the load until the specimen fails, and record the maximum load carried by the specimen during the test.
6) Calculate the compressive strength of the specimen by dividing the maximum load by the average cross-sectional area, to the nearest 10 psi (0.1 MPa).
7) If the specimen length-to-diameter ratio (*L/D*) is 1.75 or less, correct the result obtained by the appropriate correction factor, listed in Table A.15.

Results

Use the data sheet presented in Table A.16 to record and calculate the test results.

Testing   Failed Sample

FIGURE A.13   Testing concrete cylinder under compression. *Photos taken at the Farmingdale State College of the State University of New York.*

TABLE A.15

Specimen Length (*L*) to Diameter (*D*) Correction Factor

L/D	1.75	1.50	1.25	1.0
Correction Factor	0.98	0.96	0.93	0.87

TABLE A.16

Data Sheet for Compressive Strength of Cylindrical Concrete Specimens

Item	Value
Diameter, *D* (in. or mm)	
Cross-sectional area, $A = \dfrac{\pi D^2}{4}$ (in.^{2} or mm^2)	
Maximum load, *P* (lb or N)	
Compressive strength, *P/A* (psi or MPa)	

Report

- Compressive strength, calculated to the nearest 10 psi (0.1 MPa).
- Type of fracture, if different from that of the usual cone.
- Age of specimen.

LABORATORY TEST 15

FLEXURAL STRENGTH OF CONCRETE (USING SIMPLE BEAM WITH CENTER-POINT LOADING)

Designation: ASTM C 293

Scope

This test method covers determination of the flexural strength of concrete specimens by the use of a simple beam with center-point loading.

Apparatus

Loading machine

Procedure

1) Flexural tests of moist-cured specimens should be made as soon as practical after removal from moist storage.
2) When using molded specimens, turn the test specimen on its side with respect to its position as molded and center it on the support blocks, as shown in Figure A.14.
3) Bring the load-applying blocks in contact with the surface of the specimen at the center-point and apply a load of between 3 and 6% of the estimated ultimate load.
4) Load the specimen continuously and without shock, at a rate that constantly increases the extreme fiber stress between 125 and 175 psi/min (0.9 and 1.2 MPa/min).
5) Record the peak load.

Results

Use the data sheet presented in Table A.17 to record and calculate the test results.

Report

• Average width to the nearest 0.05 in. (1 mm)

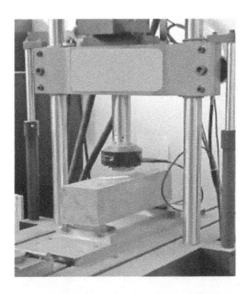

FIGURE A.14   Positioning and centering the beam on the support blocks

Item	Value
TABLE A.17 **Data Sheet for Flexural Strength of Concrete Using the Center-Point Loading**	
Span length, L (in. or mm)	
Average width of specimen at the fracture, b (in. or mm)	
Average depth of specimen at the fracture, d (in. or mm)	
Maximum applied load, P (lb or N)	
Modulus of rupture, $R = \dfrac{3}{2} \dfrac{PL}{bd^2}$ (psi or MPa)	

- Average depth to the nearest 0.05 in. (1 mm)
- Span length in in. (mm)
- Maximum applied load in pound-force
- Modulus of rupture calculated to the nearest 5 psi (0.05 MPa)
- Curing history
- Age of specimens

LABORATORY TEST 16

FLEXURAL STRENGTH OF CONCRETE (USING SIMPLE BEAM WITH THIRD-POINT LOADING)

Designation: ASTM C 78

Scope

This test method covers the determination of the flexural strength of concrete using a simple beam with third-point loading.

Apparatus

Loading machine

Procedure

1) Flexural tests of moist-cured specimens should be made as soon as practical after removal from moist storage.
2) When using molded specimens, turn the test specimen on its side with respect to its position as molded and center it on the support blocks, as shown in Figure A.15.
3) Bring the load-applying blocks in contact with the surface of the specimen at the third-points and apply a load of between 3% and 6% of the estimated ultimate load.
4) Load the specimen continuously and without shock, at a rate that constantly increases the extreme fiber stress between 125 and 175 psi/min (0.86 and 1.21 MPa/min).
5) Record the peak load.

Results

Use the data sheet presented in Table A.18 to record and calculate the test results.

REPORT

- Average width to the nearest 0.05 in. (1 mm)
- Average depth to the nearest 0.05 in. (1 mm)

a) Third-Point Loading   b) A Failed Beam

FIGURE A.15 Concrete modulus of rupture testing. *Photos taken at the Farmingdale State College of the State University of New York.*

TABLE A.18

Data Sheet for Flexural Strength of Concrete Using Third-Point Loading

Item	Value
Span length, L (in. or mm)	
Average width of specimen at the fracture, b (in. or mm)	
Average depth of specimen at the fracture, d (in. or mm)	
Maximum Applied Load, P (lb or N)	
Modulus of rupture if the fracture initiates in the tension surface within the middle-third of the span length, $R = \dfrac{3PL}{2bd^2}$ (psi or MPa)	
Modulus of rupture if the fracture occurs outside the middle-third by no more than 5% of the span length, $R = \dfrac{3Pa}{bd^2}$ (psi or MPa)	

Note: a = the average distance between the line of fracture and the nearest support, measured on the tension surface of the beam, in in. (mm). If the fracture occurs in the tension surface outside of the middle-third of the span length by more than 5% of the span length, discard the results of the test.

- Span length in in. (mm)
- Maximum applied load in pound-force
- Modulus of rupture calculated to the nearest 5 psi (0.05 MPa)
- Curing history
- Age of specimens

LABORATORY TEST 17

SPLITTING TEST OF CONCRETE

Designation: ASTM C 496

Scope
This test determines the indirect tensile strength of concrete cylinders.

Apparatus
- Loading machine
- Plywood strips

Procedure
1) Measure the dimensions (length and diameter) of the specimen.
2) Place the specimen horizontally on the loading frame such that the load can be applied along the diameter of the specimen, as shown in Figure A.16.
3) Bring the load cell in touch with the specimen so that the plywood strips barely touch the specimen.
4) Apply the load continuously until failure at a rate of about 100–200 psi/min (15–30 kPa/min) of splitting tensile stress occurs. In other words, the load should be applied at a rate that the total loading time lies between 3 and 6 minutes per test.
5) Record the maximum load applied.

Results
Use the data sheet presented in Table A.19 to record and calculate the test results.

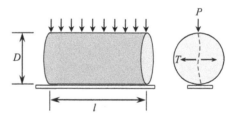

FIGURE A.16   Splitting test of concrete specimen.

TABLE A.19 Data Sheet for Splitting Test of Concrete	
Item	**Value**
Length of the sample, l (in. or mm)	
Diameter of the sample, D (in. or mm)	
Peak force needed to crack the sample diagonally, P (lb or N)	
Tensile strength, $T = \dfrac{2P}{\pi l D}$ (psi or MPa)	

Report

- Diameter and length in in. (mm)
- Maximum load in lb (N)
- Splitting tensile strength calculated to the nearest 5 psi (0.05 MPa)
- Estimated proportion of coarse aggregate fractured during test
- Age of specimen
- Curing history
- Type of fracture

LABORATORY TEST 18

REBOUND NUMBER OF HARDENED CONCRETE

Designation: ASTM C 805

Scope

This test method covers the determination of a rebound number of hardened concrete using a spring-driven steel hammer. This test method is applicable to assess the in-place uniformity of concrete, delineate regions in a structure of poor quality or deteriorated concrete, and estimate in-place strength development.

Apparatus

Rebound hammer

Procedure

1) Hold the instrument firmly so that the plunger is perpendicular to the test surface, as shown in Figure A.17.
2) Gradually push the instrument toward the test surface until the hammer impacts.
3) Maintain pressure on the instrument.
4) Read the rebound number on the scale to the nearest whole number and record the rebound number.
5) Take 10 readings from each test area.

Results

Discard readings differing by more than 6 units, and determine the average of the remaining readings. If more than two readings differ from the average by six units, discard the entire set of readings.

Report

- Identification of the concrete member such as location, curing condition, etc.
- Average rebound number
- Predicted compressive strength based on the manufacturer provided correlations

FIGURE A.17 Rebound number testing of concrete. *Courtesy of James Instruments, Inc., 3727 N. Kedzie, Chicago, IL. Used with permission.*

LABORATORY TEST 19

PENETRATION OF BITUMINOUS MATERIALS

Designation: ASTM D 5

Scope

This test method covers determination of the penetration of semi-solid and solid bituminous materials.

Apparatus

- Container
- Penetration machine
- Clean cloth

Test Specimen

1) If the sample is not sufficiently fluid when received, heat the sample with care, stirring when possible to prevent local overheating, until it has become sufficiently fluid to pour.
2) Pour the sample into the sample container to a depth such that, when cooled to the temperature of test, the depth of the sample is at least 120% of the depth to which the needle is expected to penetrate.
3) Allow the sample to cool in air at a temperature between 60 and 85°F (15 and 30°C).

Where the conditions of the test are not specifically mentioned, the temperature, load, and time are understood to be 77°F (25°C), 0.22 lb (100 g), and 5 sec, respectively. Other conditions may be used for special testing as listed in Table A.20:

In such cases, the specific conditions of the test should be reported.

Procedure

1) Clean a penetration needle with toluene or other suitable solvent, dry with a clean cloth, and insert into the penetrometer.
2) Keep the sample container completely covered with water in the bath.
3) Either note the reading of the penetrometer dial or bring the pointer to zero. Position the needle by slowly lowering it until its tip just contacts the surface of the sample, as shown in Figure A.18.
4) Quickly release the needle holder for the specified period and adjust the instrument to measure the distance penetrated in tenths of a millimeter. If the container moves, ignore the result.

TABLE A.20 Other Testing Conditions		
Temperature, °F (°C)	Load, lb (g)	Time, sec
32 (0)	0.44 (200)	60
39.2 (4)	0.44 (200)	60
113 (45)	0.10 (50)	5
115 (46.1)	0.10 (50)	5

5) Make at least three determinations at points on the surface of the sample not less than 0.5 in. (10 mm) from the side of the container and not less than 0.5 in. (10 mm) apart.

Penetration Grading

Penetration grades are listed as a range of penetration units (one penetration unit = 0.1 mm), such as 40–50 if the penetration ranges 4–5 mm. Some other requirements of penetration grading are listed in Table A.21.

Report

Report (to the nearest whole unit) the average of three penetrations whose values do not differ by more than the value listed in Table A.22.

Conditioning Specimen in Water   Penetrating an Asphalt Binder Specimen

FIGURE A.18 Penetration testing of asphalt binder. *Photos taken at the Farmingdale State College of the State University of New York.*

TABLE A.21
Penetration Grading for Asphalt Binder

Penetration Grade	Penetration at 77°F (25°C)	
	Minimum Penetration Unit	Maximum Penetration Unit
40–50	40	50
60–70	60	70
85–100	85	100
120–150	120	150
200–300	200	300

TABLE A.22
Maximum Difference between Highest and Lowest Penetration

Penetration, mm	0–49	50–149	150–249	250–500
Maximum difference between highest and lowest penetration, mm	2	4	12	20

LABORATORY TEST 20

Viscosity Determination of Asphalt at Elevated Temperatures Using a Rotational Viscometer

Designation: ASTM D 4402

Scope

This test method outlines a procedure for measuring the apparent viscosity of asphalt from 38–260°C (100–500°F) using a rotational viscometer and a temperature-controlled thermal chamber for maintaining the test temperature.

Apparatus

- Container
- Rotational viscometer (shown in Figure A.19)
- Oven
- Clean cloth

Procedure

1) Set the temperature controller to the desired test temperature.
2) Preheat the sample chamber and the selected apparatus-measuring geometry.
3) Add the volume of sample specified by the manufacturer for the apparatus-measuring geometry to be used in the sample chamber.

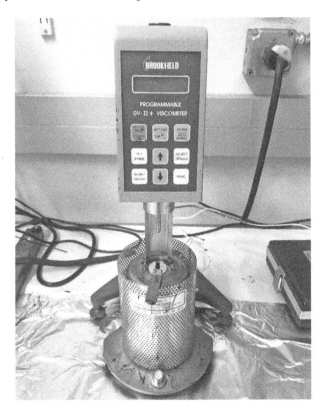

FIGURE A.19   Rotational viscometer. *Photo by Dr. Mehedi Hasan.*

4) Insert the selected preheated apparatus-measuring geometry into the liquid in the chamber, and couple it to the viscometer.
5) Bring the asphalt sample to the desired temperature within 30 minutes.
6) Start the motor rotation of the viscometer at a speed which will develop a resisting torque that is between 10% and 98% of the full-scale instrument capacity. Maintain this speed, and allow the sample to equilibrate for an additional 5 minutes.
7) Measure either the viscosity or the torque at 1-minute intervals for a total of 3 minutes.

Results

If the rotational viscometer has a digital output displaying viscosity in centipoise (cP), multiply it by 0.001 to obtain the viscosity in pascal-sec (Pa·s).

Report

Report the test temperature, apparatus-measuring geometry type and size, torque in mNm, and speed in sec-1 (r/min) with viscosity results in pascal sec (Pa·s), millipascal sec (mPa·s), or centipoise (cP).

LABORATORY TEST 21

VISCOSITY OF ASPHALTS BY VACUUM CAPILLARY VISCOMETER

Designation: ASTM D 2171

Scope

This test determines the absolute viscosity of asphalt by a Vacuum Capillary Viscometer at 60°C.

Apparatus

- Container
- Capillary viscometer (shown in Figure A.20)
- Oven
- Clean cloth

Procedure

1) Maintain a bath at a temperature of 140°F (60°C).
2) Preheat the clean viscometer at 275°F (135°C).
3) Charge the viscometer by pouring the asphalt sample.
4) Place the charged viscometer in an oven or bath, maintained at 275°F (135°C), for a period of 10 minutes.
5) Remove the viscometer and place it vertically in the bath so that the uppermost timing mark is at least 20 mm (0.75 in.) below the surface of the bath water.
6) Connect the vacuum system to the viscometer.

FIGURE A.20 Capillary viscometer. *Courtesy of Raysky Scientific Instruments, Labfreez Group, Guangdong, China. Used with permission.*

7) Wait 30 minutes for the temperature to reach equilibrium.

8) Start the flow of asphalt in the viscometer.

9) Measure the time elapsed, to 0.1 sec, for the leading edge of the meniscus to pass between successive pairs of timing marks.

10) Report the first flow time exceeding 60 sec between a pair of timing marks.

Results

Determine the calibration factor from the ASTM D 2171 corresponding to the pair of timing marks. Calculate the viscosity as the product of the calibration factor (Poises/sec) and the flow time (sec).

Report

Report the temperature and viscosity to three significant digits.

LABORATORY TEST 22

STANDARD METHOD OF TEST FOR THEORETICAL MAXIMUM SPECIFIC GRAVITY AND DENSITY OF HOT-MIX ASPHALT PAVING MIXTURES

Designation: AASHTO T 209

Scope

This test method covers the determination of the theoretical maximum specific gravity and density of loose hot-mix asphalt paving mixtures at 25°C (77°F).

Apparatus

- Pycnometer
- Vacuum and vibration machine
- Wire basket
- Weighing balance

Test Specimen

The size of the sample should conform to the requirements listed in Table A.23. Samples larger than the capacity of the container may be tested a portion at a time.

Procedure (Figure A.21)

1) Fill the pycnometer with 77°F water.

TABLE A.23 Specimen Size Requirements	
NMAS, in. (mm)	**Minimum Mass, lb (kg)**
2.0 (50)	13 (6.0)
1.5 (37.5)	8.8 (4.0)
1.0 (25)	5.5 (2.5)
0.75 (19.0)	4.4 (2.0)
0.50 (12.5)	3.3 (1.5)
0.375 (9.5)	2.2 (1.0)
No. 4 (4.75)	1.1 (0.5)

Immerse Loose Mixture   Tightening the Lid   Vacuum-Suction of the Immersed Mixture

FIGURE A.21 Different phases of theoretical maximum specific gravity testing. *Photos taken at the Farmingdale State College of the State University of New York.*

TABLE A.24

Data Sheet for Theoretical Maximum Specific Gravity of Hot-Mix Asphalt

Item	Value
Dry weight of the loose asphalt coated samples, A (g or lb)	
Mass of pycnometer + water, D (g or lb)	
Mass of pycnometer + water + asphalt, E (g or lb)	
Maximum specific gravity, $G_{mm} = \dfrac{A}{A-(E-D)} = \dfrac{A}{A+D-E}$	

2) Place the metal lid on the pycnometer; let stand a couple of minutes.
3) Dry the pycnometer thoroughly, and obtain the mass of the pycnometer filled with water. This is D.
4) Split the sample on a pan and obtain an appropriate amount of sample. This is A.
5) Fill the pycnometer containing the asphalt sample with water, about ¾ full.
6) De-air the sample using a vacuum and vibration. This process should take about 15 minutes.
7) After de-airing, remove the pycnometer from the vibrator and fill with water.
8) Place metal lid on the pycnometer.
9) Dry the pycnometer thoroughly, and obtain the mass of the pycnometer filled with water and the asphalt mix. This is E.

Results

Use the data sheet presented in Table A.24 to record and calculate the test results.

Report

Report the specific gravity value to three digits after the decimal point.

LABORATORY TEST 23

RESISTANCE TO PLASTIC FLOW OF BITUMINOUS MIXTURES USING MARSHALL APPARATUS

Designation: ASTM D 1559 and AASHTO T 245

Scope

This method covers the measurement of the resistance to plastic flow of cylindrical specimens of asphalt mixture, loaded on the lateral surface by means of the Marshall apparatus. This method is for use with mixtures containing asphalt binder or asphalt cutback, and aggregate up to a 1-in. (25-mm) maximum size.

APPARATUS

- Marshall mold, hammer, and stabilometer
- Scoop
- Spatula
- Weighing balance
- Oven
- Disc paper
- Wire basket
- Damp towel

Procedure

Specimen Preparation

1) Determine the mixing temperature at which the viscosity of binder is 170 centistokes (cSt).
2) Determine the compaction temperature at which the viscosity of binder is 280 cSt.
3) Weigh about 1.2 kg (2.6 lb) of aggregate, and place in the oven at the compaction temperature.
4) Put the asphalt binder, mold, and other required equipment in the oven at the compaction temperature.
5) After heating, mix the aggregate and desired amount of binder by using a mechanical mixer, or by doing so manually, as shown in Figure A.22.

Adding Binder   Mixing   Compacting   Compacted Specimen

FIGURE A.22 Marshall specimen preparation. *Photos taken at the Farmingdale State College of the State University of New York.*

6) Place a sheet of release paper inside the mold. Then, add the mixes in the mold and spade with a heated spatula 15 times around the perimeter and 10 times in the middle of the mold.

7) Place a sheet of release paper on the top and the collar. Then, put the entire system in the pedestal.

8) Clamp the mold and apply the desired number of blows as follows:
   - Sample size = 4 in. (100 mm) diameter cylinder, 2.5 in. (112.5 mm) in height
   - Tamper foot = Flat and circular, with a diameter of 3.875 in. (97 mm)
   - Compaction pressure = Specified as an 18-in. (450-mm) free fall drop distance of a hammer assembly, with 10-lb sliding weight.
   - Number of blows = Typically 35, 50, or 75 on each side, depending on traffic loading.
   - Simulation method = The tamper foot strikes the sample on the top and covers almost the entire sample's top area. After a specified number of blows, the sample is turned over and the procedure is repeated.

9) After cooling to room temperature, the specimen is extruded. Then, the release paper is removed. Finally, the bulk-specific gravity of the compacted specimen (G_{mb}) is determined by weighing it in air (A), in water (C), and in saturated surface-dry condition (B).

Specimen Testing for Stability and Flow

1) The test is conducted at 140°F (60°C). Therefore, the specimen needs to be warmed up to 140°F (60°C) by immersing in a water bath for 30–40 minutes, or by placing it in the oven for 2 hours.

2) Place the specimen in the loading frame (Figure A.23), and place the upper segment of the loading frame on the specimen. Prepare the dial gauges to read the load and deformation.

3) Apply the load on the specimen at a loading rate of 2 in./minute (50 mm/minute). Basically, the load is increased until it reaches a maximum, then, when the load just begins to decrease, the loading is stopped, and the maximum load is recorded.

Bulk Specific Gravity Testing   Stability and Flow Testing

FIGURE A.23 Testing of Marshall specimen. *Photos taken at the Farmingdale State College of the State University of New York.*

4) During the loading, an attached dial gauge measures the specimen's plastic flow due to the loading. The flow value is recorded in 0.01 in. (0.25 mm) increments at the same time the maximum load is recorded.

5) If the specimen height is other than that of 2.50 in. (63.5 mm), multiply the stability value by a correction factor.

Report

- Average Marshall stability of at least three replicate specimens
- Average Marshall flow of at least three replicate specimens

LABORATORY TEST 24

Bulk Specific Gravity and Density of Non-Absorptive Compacted Bituminous Mixtures

Designation: ASTM D 2726

Scope
This test method covers the determination of the bulk-specific gravity and density of specimens of compacted bituminous mixtures.

Apparatus
- Balance
- Oven
- Wire basket
- Damp towel

Procedure
1) Completely submerge the specimen in the water bath at $25 \pm 1°C$ ($77 \pm 1.8°F$) for 3–5 minutes, and then determine the mass by weighing in water, as shown in Figure A.24. Designate this mass as C. If the temperature of the specimen differs from the temperature of the water bath by more than $2°C$ ($3.6°F$), the specimen should be immersed in the water bath for 10–15 minutes, instead of 3–5 minutes.
2) Surface dry the specimen by blotting quickly with a damp cloth towel, and then determine the mass by weighing in air. Designate this mass as B.
3) After determining the mass in water and in a saturated surface-dry condition, thoroughly dry the specimen to a constant mass at $110 \pm 5°C$ ($230 \pm 9°F$). Allow the specimen to cool and weigh it in air. Designate this mass as A.

Results
Use the data sheet presented in Table A.25 to record and calculate the test results.

Report
Report the specific gravity value to three places after the decimal point.

Weighing in Water   SSD Drying   SSD Weighing   Oven-Drying   Weighing in Air

FIGURE A.24   Bulk specific gravity testing. *Photos taken at the Farmingdale State College of the State University of New York.*

TABLE A.25

Data Sheet for Bulk Specific Gravity of Hot-Mix Asphalt

Item	Value
Dry mass of the compacted samples, A (g or lb)	
Surface-dry mass, B (g or lb)	
Mass of specimen underwater, C (g or lb)	
Bulk specific gravity, $G_{mb} = \dfrac{A}{B-C}$	

LABORATORY TEST 25

Density of Hot Mix Asphalt (HMA) Specimens by Means of the SuperPave Gyratory Compactor

Designation: ASTM D 6925 and AASHTO T 312

Scope

This procedure covers preparing specimens, using samples of plant-produced HMA, for determining the volumetric properties of HMA in accordance with the AASHTO T 312.

Apparatus

- Superpave mold and compactor
- Scoop
- Spatula
- Oven
- Disc paper

Procedure (Figure A.25)

1) Combine the appropriate aggregate fractions to the desired specimen weight. Generally, 4.6–4.8 kg (10.1–10.6 lb) of aggregate is required. The weight may need to be adjusted to result in a compacted specimen of 150 mm (6.0 in.) in diameter and 115 mm (4.5 in.) in height, at the number of gyrations, N_{des}.

2) Place the aggregate, asphalt binder, container, and all necessary mixing implements in an oven and bring to the required mixing temperature. Specimens are to be mixed at the temperature at which the viscosity of binder is 0.170 ± 0.02 Pa·s.

3) Mix the aggregate and asphalt binder quickly and thoroughly.

4) After mixing, subject the loose mix to short-term conditioning for 2 hours ± 5 minutes at the compaction temperature ± 3°C (5.4°F). Specimens are to be compacted at the temperature at which the viscosity of binder is 0.28 ± 0.03 Pa·s.

5) Place the compaction mold assembly in an oven at the required compaction temperature ± 5°C for a minimum of 45 minutes prior to the compaction of the first mixture specimen.

Heating

Compacting

Ejecting

Finished Specimen

FIGURE A.25 Superpave gyratory specimen preparation. *Photos taken at the Farmingdale State College of the State University of New York.*

TABLE A.26 Number of Gyrations for Different Traffic Levels			
Design ESALs (millions)	$N_{initial}$	N_{design}	N_{max}
Less than 0.3	6	50	75
0.3 to less than 3.0	7	75	115
3.0 to less than 30	8	100	160
30 or more	9	125	205

6) Remove the heated mold and plate(s) from the oven.

7) Place the base plate and a paper disc in the bottom of the mold.

8) Pour the mix into the mold in a single lift.

9) Level the mix in the mold.

10) Place a paper disc and the heated upper plate on top of the leveled sample.

11) Load the mold into the compactor; make sure to check the settings, particularly:

    a.   The number of gyrations, N_{des} as listed in Table A.26.

    b.   The pressure must be 600 ± 18 kPa (87 ± 2.6 psi).

    c.   The angle must be$1.16 \pm 0.02°$.

12) Start the compaction process.

13) The machine will stop after the desired number of gyrations. Record the number of gyrations and specimen height. Discard the specimen if its height is not 115 ± 5 mm (4.53 ± 0.2 in.) at N_{des}.

14) Open the door and extrude the specimen from the mold; a brief cooling period may be necessary before fully extruding some specimens, to ensure the specimens are not damaged.

15) Carefully remove the paper discs.

16) Cool the compacted specimen to room temperature.

17) Identify the specimen with chalk or a marker.

18) Determine the bulk-density of the compacted specimen (G_{mb}) using the ASTM D 2726 test standard.

Report

- Mixtures parameters, such as aggregate size, binder type, proportions, etc.
- Number of gyrations
- Specimen height
- G_{mb}

LABORATORY TEST 26

NOTCHED BAR IMPACT TESTING OF METALLIC MATERIALS

Designation: ASTM E 23

Scope

These test methods describe notched-bar impact testing of metallic materials by the Charpy (simple-beam) test and the Izod (cantilever-beam) test. They give the requirements for test specimens, test procedures, test reports, test machines, etc.

Apparatus

Pendulum type testing machine

Test Specimen

Charpy (Simple-Beam) Impact Test Specimens: 2.2 in. (55 mm) long, 0.42 in. (10 mm) by 0.42 in. (10 mm) cross-section. The notch is made at the middle of the specimen. The notch can be done in three different fashions (Type A, B, or C) shown in the standard.

Izod (Cantilever-Beam) Impact Test Specimens: 3.0 in. (75 mm) long, 0.42 in. (10 mm) by 0.42 in. (10 mm) cross-section. The notch is made as shown in the standard.

Specimen Conditioning

Izod testing is not recommended at temperatures other than room temperature.

Procedure

1) Raise the pendulum to the latched position and lock it.
2) Move the pointer to the maximum capacity of the range being used.
3) Position the specimen simple-beam style for Charpy and cantilever-beam style for Izod.
4) Stand a safe distance away and release the pendulum (Figure A.26).
5) Read the indicated value – this is the toughness of the specimen.

Report

- If a specimen does not separate into two pieces, report it as unbroken. Unbroken specimens with absorbed energies of less than 80% of the machine capacity may be averaged with values from broken specimens.

Test Machine   Clamping Specimen   Failed Specimen

FIGURE A.26   Charpy (simple-beam) impact test. *Photos taken at the Farmingdale State College of the State University of New York.*

- If the absorbed energy exceeds 80% of the machine capacity, report the value as approximate – do not average it with other values. Absorbed energy values above 80% of the scale range are inaccurate and should be reported as approximate.
- If an unbroken specimen does not pass between the machine anvils, (for example, it stops the pendulum), the result should be reported as exceeding the machine capacity.

LABORATORY TEST 27

TENSION TESTING OF METALLIC MATERIALS

Designation: ASTM E 8

Scope

This test determines the modulus of elasticity, yield stress, tensile strength, etc. of a metal such as steel, aluminum, etc.

Apparatus

- Loading machine
- Extensometer or dial gage
- Slide calipers

Procedure

1) Measure the dimensions (gage length and initial diameter) of the dog-bone type specimen.
2) Clamp the specimen in the loading frame.
3) Attach the extensometer or dial gage to the specimen, as shown in Figure A.27.
4) Apply the tension load at a rate less than 100 ksi/min (690 kPa/min). A slower load provides a greater opportunity to record a good amount of data.
5) Record load-deformation data (preferably, as much as possible in order to have a smooth stress–strain curve).
6) Continue applying the load until failure occurs.

Results

Use the data sheet presented in Table A.27 to record and calculate the test results.

Report

- Stress versus strain graph
- Modulus of elasticity
- Yield stress
- Ultimate stress

Side View   Attaching Dial Gage   Failed Sample

FIGURE A.27 Tension testing of a metallic material. *Photos taken at the Colorado State University–Pueblo.*

TABLE A.27

Data Sheet for Tension Testing of a Metallic Material

Load, P (lb or N)	Deformation, ΔL (in. or mm)	Stress, $\sigma = \dfrac{P}{A_o}$ (psi or MPa)	Strain, $\varepsilon = \dfrac{\Delta L}{L_o} m$

Initial Diameter, $D_o =$

Initial Cross-Sectional Area, $A_o = \dfrac{\pi D_o^2}{4} =$

Initial Gage Length, $L_o =$
Final Diameter, $D_f =$

Final Cross-Sectional Area, $A_f = \dfrac{\pi D_f^2}{4} =$

Total Elongation, $\Delta L \ (Total) =$
Initial Slope of Stress–Strain Diagram, $E =$

Percent Elongation $= \left(\dfrac{\Delta L (Total)}{L_o} \right) \times 100 \ =$

Percent Reduction in Area (RA), $\% \ RA = \left(\dfrac{A_i - A_f}{A_i} \right) \times 100 \ =$

- Failure stress
- Percent elongation at failure
- Percent reduction in area at failure
- The failure shape of the sample

LABORATORY TEST 28

TORSION TESTING

Designation: ASTM E 143

Scope
This test method covers the determination of the shear modulus of structural materials.

Apparatus
- Torsion machine
- Slide calipers

Procedure
1) Measure the diameter and the gage length of the specimen. The wall thickness is required for a hollow shaft.
2) Clamp the specimen in the loading frame, as shown in Figure A.28.
3) Apply torque and measure the angle of twist versus the applied torque continuously.

Results
Use the data sheet presented in Table A.28 to record and calculate the test results.

Report
- Shear strain versus shear stress graph
- Shear modulus

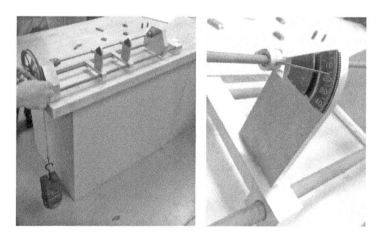

FIGURE A.28   Torsion test procedure. *Photos taken at the Colorado State University–Pueblo.*

TABLE A.28
Data Sheet for Tension Testing of a Metallic Material

Torque, T (lb·ft or N·m)	Angle of Twist, ϕ (radians)	Shear Stress, $\tau = \dfrac{Tr}{J}$ (psi or Pa)	Shear Strain, $\gamma = \dfrac{\phi r}{L}$

Radius of the sample, $r =$

Gage length, $L =$

Polar moment of inertia, $J = \dfrac{\pi r^4}{2} =$

Initial Slope of Shear Stress–Strain Diagram, $G =$

LABORATORY TEST 29

DETERMINATION OF THE FLEXURAL PROPERTIES OF STRUCTURAL WOOD BEAMS

Designation: ASTM D 198

Scope

This test is intended for rectangular cross-sections, but is also applicable to beams of irregular shapes. This test method covers the determination of the many properties of structural beams made of solid or laminated wood, or of composite. This section discusses the Flexural test. Others are:

- Compression (short column)
- Compression (long member)
- Tension
- Torsion
- Shear modulus

Apparatus
- Loading machine
- Extensometer or dial gage
- Slide calipers or tape measure
- Bearing plates

Test Specimen

This test method is intended primarily for beams of rectangular cross-section, but is also applicable to beams of other cross-sections. The size of the beams can be up to 12 in. (300 mm) deep by 6 in. (150 mm) wide.

Procedure
1) Measure the dimensions of the specimen.
2) Support the beam at its ends by metal bearing plates, to provide roller-type supports as shown in Figure A.29. The size of the bearing plates may vary with the size and shape of the beam. A bearing plate of 0.5 in. (13 mm) thick by 6 in. (150 mm) long

FIGURE A.29   Bending test on wooden beam. *Courtesy of TestResources, Shakopee, MN, USA, TestResources.net. Used with permission.*

TABLE A.29

Data Sheet for Flexural Properties of Structural Wood Beam

Item	Value
Span length, L (in. or mm)	
Distance from reaction to nearest load point, a (in. or mm)	
Depth of the beam, d (in. or mm)	
Width of beam, b (in. or mm)	
Moment of inertia, $I = bh^3/12$ (in.^{4} or mm^4)	
Load at proportional limit, P_p (lb or N)	
Deflection of beam at the proportional limit, Δ (in. or mm)	
Modulus of elasticity, $E = \dfrac{P_p a}{48\Delta I}\left(3L^2 - 4a^2\right)$ (psi or MPa)	
Maximum load at failure, P (lb or N)	
Flexural strength, $S_R = \dfrac{3Pa}{bd^2}$ (psi or MPa)	

which extends the entire width of the beam is good for a rectangular beam of 12 in. (300 mm) deep by 6 in. (150 mm) wide.

3) Lateral support is required to prevent lateral buckling if the depth-to-width ratio of the beam is three or greater.
4) The beam is subjected to a bending moment until complete failure occurs.
5) Conduct the test at a constant rate to achieve the maximum load in about 10 minutes, but no less than 6 minutes nor more than 20 minutes.
6) Obtain load-deflection data until the failure occurs.

Results

Use the data sheet presented in Table A.29 to record and calculate the test results.

Report

- Depth of the section
- Width of the section
- Beam span
- Flexural strength
- The modulus of elasticity
- Distance from reaction to nearest load point

LABORATORY TEST 30

COMPRESSION PARALLEL TO GRAIN (SHORT COLUMN, NO LATERAL SUPPORT) OF STRUCTURAL WOOD

Designation: ASTM D 198

Scope

This test method covers the determination of the compressive properties of elements taken from structural members made of solid or laminated wood, or of composite constructions when such an element has a slenderness ratio (length to least radius of gyration) of less than 17.

Apparatus

- Loading machine
- Extensometer or dial gage
- Slide calipers or tape measure
- Bearing plates

Test Specimen

- Commonly used in structural applications, that is, in sizes greater than nominal 2-in. by 2-in. (50-mm by 50-mm) cross-section.
- The specimen should be a short column having a maximum length, l, less than 17 times the least radius of gyration, r, of the cross-section of the specimen.

Procedure

The structural member is subjected to a force uniformly distributed on the contact surface of the specimen, in a direction generally parallel to the longitudinal axis of the wood fibers. The force generally is uniformly distributed throughout the specimen during loading to failure, without flexure along its length. The following steps can be followed:

1) Measure the dimensions of the specimen.
2) Place the specimen in the load mechanism, taking care to have the long axis of the specimen and the grips coincide.
3) Apply the load at a constant rate of head motion, so that the fiber strain is 0.001 in./in. per minute. For measuring only compressive strength, the test may be conducted at a constant rate to achieve the maximum load in about 10 minutes, but not less than 5 nor more than 20 minutes.
4) Record load-deformation data as accurately as possible.
5) Continue applying the load until failure occurs.
6) Record the maximum load.

Results

Use the data sheet presented in Table A.30 to record and calculate the test results.

TABLE A.30

Data Sheet for Compression Test on Wood Beam

Item	Value
Initial width of the section, b (in. or mm)	
Initial thickness of the section, h (in. or mm)	
Initial cross-sectional area of the specimen $A = bh$ (in.^{2} or mm^2)	
Initial length of the sample, L (in. or mm)	
Applied load at the proportional limit, P_p (lb or N)	
Deformation of the specimen, ΔL (in. or mm)	
Strain at the proportional limit, $\varepsilon = \Delta L/L$	
Applied peak load, P (lb or N)	
Compressive strength $= P/A$ (psi or MPa)	
Modulus of elasticity $= P_p/A\varepsilon$ (psi or MPa)	

Report

- Width of the section
- Thickness of the section
- Beam length
- Compressive strength
- The modulus of elasticity

LABORATORY TEST 31

Tension Parallel to Grain of Structural Wood

Designation: ASTM D 198

Scope

This test method covers the determination of the tensile properties of structural elements, made primarily of lumber equal to and greater than nominal 1 in. (19 mm) thickness.

Apparatus

- Loading machine
- Extensometer or dial gage
- Slide calipers or tape measure

Procedure

The structural member is clamped at the extremities of its length and subjected to a tensile load such that in sections between clamps, the tensile forces should be axial and generally uniformly distributed throughout the cross-sections, without flexure along its length. The following steps can be followed:

1) Measure the dimensions of the specimen.
2) Place the specimen in the grips of the load mechanism, taking care to have the long axis of the specimen and the grips coincide.
3) The load may be applied at a constant rate of grip motion so that maximum load is achieved in about 10 minutes, but not less than 5 nor more than 20 minutes.
4) Record load-deformation data as accurately as possible.
5) Continue applying the load until failure occurs.
6) Record the maximum load.

Results

Use the data sheet presented in Table A.31 to record and calculate the test results.

TABLE A.31 Data Sheet for Tension Test on Wood Beam	
Item	**Value**
Initial width of the section, b (in. or mm)	
Initial thickness of the section, h (in. or mm)	
Initial cross-sectional area of the specimen $A = bh$ (in.^{2} or mm^2)	
Initial gage length, L (in. or mm)	
Applied load at the proportional limit, P_p (lb or N)	
Deformation of the specimen, ΔL (in. or mm)	
Strain at the proportional limit, $\varepsilon = \Delta L/L$	
Applied peak load, P (lb or N)	
Tensile strength $= P/A$ (psi or MPa)	
Modulus of elasticity $= P_p/A\varepsilon$ (psi or MPa)	

Report

- Width of the section
- Thickness of the section
- Beam length
- Tensile strength
- The modulus of elasticity

Comments

The shear modulus test and torsion test are not discussed because they are used less often in practice. Read ASTM D 198 for more information.

LABORATORY TEST 32

SMALL CLEAR SPECIMENS OF TIMBER: STATIC BENDING

Designation: ASTM D 143

Scope

These methods cover the determination of various strength and related properties of wood by testing small clear specimens, such as static bending. Others include:

- Compression parallel to grain
- Impact bending
- Toughness
- Compression perpendicular to grain
- Hardness
- Shear parallel to grain
- Cleavage
- Tension parallel to grain
- Tension perpendicular to grain
- Nail withdrawal
- Specific gravity and shrinkage in volume
- Radial and tangential shrinkage
- Moisture determination

Apparatus

- Loading machine
- Slide calipers or tape measure

Test Specimen

The static bending tests should be made on $2 \times 2 \times 30$-in. ($50 \times 50 \times 750$-mm) primary method specimens, or $1 \times 1 \times 16$-in. ($25 \times 25 \times 400$-mm) secondary method specimens.

Procedure

1) Use center loading and a span length of 28 in. (700 mm) for the primary method, and 14 in. (360 mm) for the secondary method (Figure A.30).
2) Both supporting knife edges should be provided with bearing plates and rollers of a thickness such that the distance from the point of support to the central plane is not greater than the depth of the specimen.
3) A center point load is applied, and the resulting deformation is recorded.
4) The load is continued until a 6-in. (150-mm) deflection occurs, or until the specimen fails to support a load of 200 lb (890 N) for primary method specimens and to a 3-in. (75-mm) deflection or until the specimen fails to support a load of 50 lb (220 N) for secondary method specimens.
5) Document load-deformation data until the failure is recorded.

Results

Use the data sheet presented in Table A.32 to record and calculate the test results.

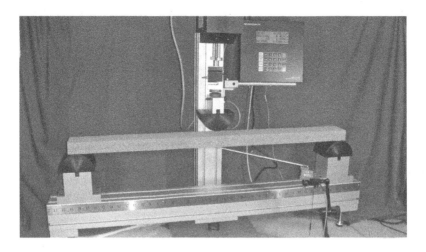

FIGURE A.30 Static bending test on small-clear wood specimen. *Courtesy of TestResources, Shakopee, MN, USA, TestResources.net. Used with permission.*

TABLE A.32	
Data Sheet for Flexural Strength Using the Center-Point Loading	
Item	Value
Span length, L (in. or mm)	
Average width of specimen at the fracture, b (in. or mm)	
Average depth of specimen at the fracture, d (in. or mm)	
Maximum applied load, P (lb or N)	
Modulus of rupture, $R=3/2$ Pl/ $R = \dfrac{3}{2}\dfrac{PL}{bd^2}$ (psi or MPa)	

Report

- Width of the section
- Thickness of the section
- Beam length
- Flexural strength

LABORATORY TEST 33

SMALL CLEAR SPECIMENS OF TIMBER: COMPRESSION PARALLEL TO GRAIN

Designation: ASTM D 143

Scope

This test method covers determination of the compression strength of wood parallel to grain for small clear specimens of timber.

Apparatus

- Loading machine
- Slide calipers or tape measure

Test Specimen

The compression parallel-to-grain tests should be made on 2 by 2 by 8-in. (50 by 50 by 200-mm) primary method specimens, or 1 by 1 by 4-in. (25 by 25 by 100-mm) secondary method specimens.

Procedure

1) Specimen is placed uniaxially along the loading frame with the load cell parallel to the grain.
2) The load should be applied continuously throughout the test at a rate of motion of the movable crosshead of 0.003 in./in. (mm/mm).
3) Load-compression curves should be taken over a central gage length not exceeding 6 in (150 mm) for primary method specimens, and 2 in. (50 mm) for secondary method specimens.
4) Load-compression readings should be continued until the proportional limit is well passed.
5) The maximum load value is recorded to determine the compression capacity.

Results

Use the data sheet presented in Table A.33 to record and calculate the test results.

TABLE A.33 Data Sheet for Compression Test Parallel-to-Grain on Wood Beam	
Item	**Value**
Initial width of the section, b (in. or mm)	
Initial thickness of the section, h (in. or mm)	
Initial cross-sectional area of the specimen, $A = bh$ (in.^{2} or mm^2)	
Initial length of the sample, L (in. or mm)	
Applied load at the proportional limit, P_p (lb or N)	
Deformation of the specimen, ΔL (in. or mm)	
Strain at the proportional limit, $\varepsilon = \Delta L/L$	
Applied peak load, P (lb or N)	
Compressive strength $= P/A$ (psi or MPa)	
Modulus of elasticity $= P_p/A\varepsilon$ (psi or MPa)	

Report

- Width of the section
- Thickness of the section
- Beam length
- Compressive strength
- The modulus of elasticity

LABORATORY TEST 34

SMALL CLEAR SPECIMENS OF TIMBER: COMPRESSION PERPENDICULAR TO GRAIN

Designation: ASTM D 143

Scope

This test method covers determination of the compression strength of wood perpendicular to grain for small clear specimens of timber.

Apparatus

- Loading machine
- Slide calipers or tape measure

Test Specimen

The test should be made on $2 \times 2 \times 6$-in. ($50 \times 50 \times 150$-mm) specimens.

Procedure

1) The specimens should be placed so that the load will be applied through the bearing plate to a radial surface.
2) Apply load through a metal bearing plate 2 in. (50 mm) in width, placed across the upper surface of the specimen at equal distances from the ends and at right angles to the length.
3) Apply load continuously throughout the test at a rate of motion of the movable crosshead of 0.012 in. per minute (0.305 mm per minute).
4) Load-compression curves should be taken for all specimens up to 0.1-in. (2.5-mm) compression, after which the test is discontinued.

Results

Use the data sheet presented in Table A.34 to record and calculate the test results.

Report

Width, thickness, and compressive strength of the section

TABLE A.34 Data Sheet for Compression Test Perpendicular-to-Grain on Wood Beam	
Item	**Value**
Initial cross-sectional area of loading plate, A (in.^{2} or mm^2)	
Applied peak load, P (lb or N)	
Compressive strength $= P/A$ (psi or MPa)	

LABORATORY TEST 35

SMALL CLEAR SPECIMENS OF TIMBER: TENSION PARALLEL TO GRAIN

Designation: ASTM D 143

Scope

This test method covers determination of the tensile strength of wood parallel to grain for small clear specimens of timber.

Apparatus

- Loading machine
- Extensometer or dial gage
- Slide calipers or tape measure

Test Specimen

An 18-in (450 mm) dog-bone type specimen is prepared. The central 2.5 in. (63 mm) is the gage length with thickness of 0.375 in. (9 mm).

Procedure

1) Place the annual rings at the critical section on the ends of the specimens, perpendicular to the greater cross-sectional dimension.
2) Fasten the specimen in the special grips.
3) Measure the deformation over a 2-in. (50-mm) central gage length on all specimens.
4) Apply load continuously throughout the test at a rate of motion of the movable crosshead of 0.05 in. (1 mm) per minute.
5) Load-deformation data is recorded until the failure occurs.

Results

Use the data sheet presented in Table A.35 to record and calculate the test results.

TABLE A.35
Data Sheet for Tension Test on Wood Beam

Item	Value
Initial width of the section, b (in. or mm)	
Initial thickness of the section, h (in. or mm)	
Initial cross-sectional area of the specimen, $A = bh$ (in.^{2} or mm^2)	
Initial gage length, L (in. or mm)	
Applied load at the proportional limit, P_p (lb or N)	
Deformation of the specimen, ΔL (in. or mm)	
Strain at the proportional limit, $\varepsilon = \Delta L/L$	
Applied peak load, P (lb or N)	
Tensile Strength $= P/A$ (psi or MPa)	
Modulus of elasticity $= P_p/A\varepsilon$ (psi or MPa)	

Report

- Width of the section
- Thickness of the section
- Beam length
- Tensile strength
- The modulus of elasticity

LABORATORY TEST 36

Small Clear Specimens of Timber: Tension Perpendicular to Grain

Designation: ASTM D 143

Scope
This test method covers determination of the tensile strength of wood perpendicular to grain for small clear specimens of timber.

Apparatus
- Loading machine
- Slide calipers or tape measure

Test Specimen
A 2-in. cube is shaped. The diameter of the circular groove is 1.0 in.

Procedure
1) Fasten the specimens to be tested in the grips.
2) Apply load continuously throughout the test at a rate of motion of the movable crosshead of 0.10 in. (2.5 mm) per minute.
3) Record the maximum load only.

Results
Use the data sheet presented in Table A.36 to record and calculate the test results.

Report
- Width, thickness, beam length, and tensile strength of the section

TABLE A.36 Data Sheet for Tension Test on Wood Beam	
Item	**Value**
Initial width of the section, b (in. or mm)	
Initial thickness of the section, h (in. or mm)	
Initial cross-sectional area of the specimen, $A = bh$ (in.^{2} or mm^2)	
Applied peak load, P (lb or N)	
Tensile strength $= P/A$ (psi or MPa)	

LABORATORY TEST 37

MODULUS OF RUPTURE OF BRICK

Designation: ASTM C 67

Scope

This test method covers the determination of the modulus of rupture of brick and structural clay tile.

Apparatus

- Loading machine
- Slide calipers or tape measure

Procedure

1) Measure the average dimensions.
2) Support the specimen flatwise using roller-type support; the load is to be applied on the wider face. The span is to be 1 in. (25 mm) less than the specimen length.
3) Apply the load at the center, at 0.05 in. (1.27 mm) per minute, until failure occurs using the steel bearing plate.
4) Record the maximum load applied.

Results

Use the data sheet presented in Table A.37 to record and calculate the test results.

REPORT

- Dimensions and the modulus of rupture of the specimen

TABLE A.37 Data Sheet for Modulus of Rupture of Brick	
Item	Value
Span of specimen, L (in. or mm)	
Average width, b (in. or mm)	
Average depth, d (in. or mm)	
Maximum load, W (lb or N)	
Modulus of rupture, $MOR = \dfrac{3}{2}\dfrac{WL}{bd^2}$ (psi or MPa)	

LABORATORY TEST 38

Compressive Strength of Brick

Designation: ASTM C 67

Scope

This test method covers the determination of the compressive strength of brick and structural clay tile.

Apparatus

- Loading machine
- Slide calipers or tape measure

Procedure

1) Measure the average dimensions.
2) Apply a sulfur cap with a thickness of about 0.25 in. (6 mm)
3) Apply a compressive load on the wider face until failure occurs. The failure should take place within 2–3 minutes.
4) Record the maximum load applied.

Results

Use the data sheet presented in Table A.38 to record and calculate the test results.

Report

- Dimensions and the compressive strength of the specimen

TABLE A.38
Data Sheet for Compressive Strength of Brick

Item	Value
Average gross-area, A (in.^{2} or mm^2)	
Maximum load, W (lb or N)	
Compressive strength = W/A, (psi or MPa)	

LABORATORY TEST 39

24-Hour Absorption of Brick

Designation: ASTM C 67

Scope

This test covers the 24-hour absorption of brick and structural clay tile.

Apparatus

- Oven
- Damp cloth
- Weighing balance

Procedure

1) Measure the dry weight of the specimen, W_d.
2) Submerge the specimen in water for 24 hours.
3) Remove the specimen and wipe off the surface using a damp cloth.
4) Immediately weigh the specimen. Designate this weight as W_s.

Results

Use the data sheet presented in Table A.39 to record and calculate the test results.

Report

The 24-hour absorption of bricks.

TABLE A.39
Data Sheet for 24-Hour Absorption of Brick

Item	Value
Dry weight of the specimen, W_d (lb or kg)	
Surface-dry weight, W_s (lb or kg)	
Absorption $(\%) = \dfrac{W_S - W_d}{W_d} \times 100$	

LABORATORY TEST 40

Initial Rate of Absorption (Suction) of Bricks

Designation: ASTM C 67

Scope

This test method covers determination of the initial rate of absorption (suction) of brick and structural clay tile.

Apparatus

- Tape measure
- Damp cloth
- Metal bars
- Tray

Procedure

1) Measure the dimensions of the specimen. Let the length be L in. and width be B in.
2) Measure the dry weight of the specimen, W_d in g.
3) Set two metal bar-supports in the tray.
4) Add water so that the depth of the water is 1/8 in. (3 mm) above the bottom of the brick.
5) Set the brick in place, flatwise on metal support, without splashing.
6) Remove the brick after 1 minute has elapsed.
7) Wipe off the surface water with a damp cloth and weigh, W_s.

Results

Use the data sheet presented in Table A.40 to record and calculate the test results.

Report

The average initial rate of absorption.

TABLE A.40
Data Sheet for Initial Rate of Absorption (Suction) of Bricks

Item	Value
Length of brick, L (in.)	
Width of brick, B (in.)	
Dry weight of the specimen, W_d (g)	
Surface-dry weight after soaking 1 minute, W_s (lb or g)	
Initial rate of absorption, $= \dfrac{30(W_s - W_d)}{BL}$ (lb/min or g/min)	

Bibliography

GENERAL

Bhavikatti, S. 2010. *Basic Civil Engineering*. New Age International Publishers, India.

Domone, P. and Illston, J. 2010. *Construction Materials Their Nature and Behavior*, 4th Edition. Spon Press, London.

Zhang, H. 2011. *Building Materials in Civil Engineering*. Woodhead Publishing Limited, Sawston, Cambridge, UK.

CEMENT, CONCRETE AND CONCRETE STRUCTURES

De Schutter, G., Bartos, P., Domone, P. and Gibbs, J. 2008. *Self-Compacting Concrete*. Whittles Publishing, Caithness, Scotland.

Forde, M. 2009. *ICE Manual of Construction Materials*. Thomas Telford, London.

Mehta, P. K. and Monteiro, P. J. 2014. *Concrete: Microstructure, Properties, and Materials*, 4th Edition. McGraw-Hill Education, New York.

Mindess, S., Young, J. and Darwin, D. 2003. *Concrete*, 2nd Edition. Pearson, Upper Saddle River, NJ.

Neville, A. M. 1995. *Properties of Concrete*, 4th Edition. Pearson Education, Harlow, UK.

Price, W. F. 2001. *The Use of High-Performance Concrete*. E&FN Spon, London.

ASPHALT MATERIALS

The Asphalt Handbook, 7th Edition. 2007. Asphalt Institute, Lexington, KY.

Construction of Hot Mix Asphalt Pavements, 2nd Edition. 2001. Asphalt Institute, Lexington, KY.

Hot-Mix Asphalt Paving Handbook, 2nd Edition. 2016. Builders Book, Inc.

Huang, Y. 2003. *Pavement Analysis and Design*, 2nd Edition. Pearson, Upper Saddle River, NJ.

Mallick, R. B. and El-Korchi, T. 2013. *Pavement Engineering: Principles and Practice*, 2nd Edition. CRC Press, Boca Raton, FL.

O'Flaherty, C. A. 2000. *Highways: The Location, Design, Construction and Maintenance of Road Pavements*. Elsevier/Butterworth Heinemann.

Thom, N. H. 2008. *Principles of Pavement Engineering*. Thomas Telford, London.

STEELS

Callister, W. D. 2007. *Materials Science and Engineering, An Introduction*, 7th Edition. Wiley, New York.

Ledesma-Carrión, D. 2017. *Optimization of the Electric Arc Furnace for the Production of Steel: Steelmaking Process*. LAP LAMBERT Academic Publishing, Germany.

Williams, A. 2016. *Steel Structures Design for Lateral and Vertical Forces*, 2nd Edition. McGraw Hill.

WOODS AND WOOD STRUCTURES

APA. 2010. *A Guide to Engineered Wood Products, C800*. APA–The Engineered Wood Association, Tacoma, WA.

APA. 2015. *Standard Specification for Structural Glued Laminated Timber of Softwood Species*. ANSI 117-2015, APA–The Engineered Wood Association, Tacoma, WA.

APA. 2016. *Engineered Wood Construction Guide*. Form E30, 2016, APA–The Engineered Wood Association, Tacoma, WA.

APA. 2017. *Standard for Wood Products - Structural Glued Laminated Timber*. ANSI A190.1-2017, APA–The Engineered Wood Association, Tacoma, WA.

APA. 2018. *Standard for Performance-Rated Cross-Laminated Timber*. ANSI/APA PRG 320-2018, APA–The Engineered Wood Association, Tacoma, WA.

Barnett, J. R. and Jeronimidis, G. 2003. *Wood Quality and Its Biological Basis*. Blackwell Publishing, Oxford, pp. 30–52.

Kettunen, P. O. 2006. *Wood Structure and Properties*. Trans Tech Publications Ltd, Switzerland, p. 401.

MASONRY

ACI. 2011. *Building Code Requirements and Specification for Masonry Structures*. ACI 530-11, American Concrete Institute, Farmington Hills, MI.

Ekwueme, C. and Brandow, G. E. 2009. *Design of Reinforced Masonry Structures*, 6th Edition. The Masonry Society, Longmont, CO.

Kreh, D. 1998. *Building with Masonry*. Taunton Press, Newtown, CT.

Index

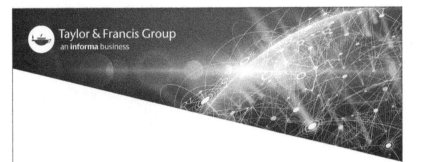